# Kinetic Modelling in Systems Biology

# CHAPMAN & HALL/CRC
Mathematical and Computational Biology Series

**Aims and scope:**

This series aims to capture new developments and summarize what is known over the whole spectrum of mathematical and computational biology and medicine. It seeks to encourage the integration of mathematical, statistical and computational methods into biology by publishing a broad range of textbooks, reference works and handbooks. The titles included in the series are meant to appeal to students, researchers and professionals in the mathematical, statistical and computational sciences, fundamental biology and bioengineering, as well as interdisciplinary researchers involved in the field. The inclusion of concrete examples and applications, and programming techniques and examples, is highly encouraged.

**Series Editors**

Alison M. Etheridge
*Department of Statistics*
*University of Oxford*

Louis J. Gross
*Department of Ecology and Evolutionary Biology*
*University of Tennessee*

Suzanne Lenhart
*Department of Mathematics*
*University of Tennessee*

Philip K. Maini
*Mathematical Institute*
*University of Oxford*

Shoba Ranganathan
*Research Institute of Biotechnology*
*Macquarie University*

Hershel M. Safer
*Weizmann Institute of Science*
*Bioinformatics & Bio Computing*

Eberhard O. Voit
*The Wallace H. Couter Department of Biomedical Engineering*
*Georgia Tech and Emory University*

Proposals for the series should be submitted to one of the series editors above or directly to:
**CRC Press, Taylor & Francis Group**
4th, Floor, Albert House
1-4 Singer Street
London EC2A 4BQ
UK

# Published Titles

**Bioinformatics: A Practical Approach**
*Shui Qing Ye*

**Cancer Modelling and Simulation**
*Luigi Preziosi*

**Computational Biology: A Statistical Mechanics Perspective**
*Ralf Blossey*

**Computational Neuroscience: A Comprehensive Approach**
*Jianfeng Feng*

**Data Analysis Tools for DNA Microarrays**
*Sorin Draghici*

**Differential Equations and Mathematical Biology**
*D.S. Jones and B.D. Sleeman*

**Exactly Solvable Models of Biological Invasion**
*Sergei V. Petrovskii and Bai-Lian Li*

**Handbook of Hidden Markov Models in Bioinformatics**
*Martin Gollery*

**Introduction to Bioinformatics**
*Anna Tramontano*

**An Introduction to Systems Biology: Design Principles of Biological Circuits**
*Uri Alon*

**Kinetic Modelling in Systems Biology**
*Oleg Demin and Igor Goryanin*

**Knowledge Discovery in Proteomics**
*Igor Jurisica and Dennis Wigle*

**Modeling and Simulation of Capsules and Biological Cells**
*C. Pozrikidis*

**Niche Modeling: Predictions from Statistical Distributions**
*David Stockwell*

**Normal Mode Analysis: Theory and Applications to Biological and Chemical Systems**
*Qiang Cui and Ivet Bahar*

**Pattern Discovery in Bioinformatics: Theory & Algorithms**
*Laxmi Parida*

**Spatiotemporal Patterns in Ecology and Epidemiology: Theory, Models, and Simulation**
*Horst Malchow, Sergei V. Petrovskii, and Ezio Venturino*

**Stochastic Modelling for Systems Biology**
*Darren J. Wilkinson*

**Structural Bioinformatics: An Algorithmic Approach**
*Forbes J. Burkowski*

**The Ten Most Wanted Solutions in Protein Bioinformatics**
*Anna Tramontano*

Chapman & Hall/CRC Mathematical and Computational Biology Series

# Kinetic Modelling in Systems Biology

Oleg Demin
Igor Goryanin

CRC Press
Taylor & Francis Group
Boca Raton  London  New York

CRC Press is an imprint of the
Taylor & Francis Group, an **informa** business

A CHAPMAN & HALL BOOK

CRC Press
Taylor & Francis Group
6000 Broken Sound Parkway NW, Suite 300
Boca Raton, FL 33487-2742

First issued in paperback 2019

© 2009 by Taylor & Francis Group, LLC
CRC Press is an imprint of Taylor & Francis Group, an Informa business

No claim to original U.S. Government works

ISBN-13: 978-0-8493-3816-8 (hbk)
ISBN-13: 978-0-8493-3816-8 (pbk)

### Library of Congress Cataloging-in-Publication Data

Demin, Oleg.
    Kinetic modelling in systems biology / Oleg Demin, Igor Goryanin.
       p. cm. -- (Mathematical and computational biology series)
    Includes bibliographical references and index.
    ISBN 978-1-58488-667-9 (alk. paper)
    1. Biological systems--Computer simulation. 2. Biological systems--Mathematical models. I. Goryanin, Igor. II. Title. III. Series.

QH323.5.D456 2009
570.1'13--dc22
                                          2008039738

**Visit the Taylor & Francis Web site at**
**http://www.taylorandfrancis.com**

**and the CRC Press Web site at**
**http://www.crcpress.com**

*To Our Families*

# Table of Contents

# About the Authors

**Oleg Demin** is a Russian biophysicist who leads the Group of Kinetic Modeling of Complex Biochemical Systems at the Moscow State University. Oleg is also CSO of Institute for Systems Biology SPb.

Oleg graduated (M.Sc.) in 1992 as a biophysicist from the Biophysical Department, Faculty of Biology, Moscow State University, and as an applied mathematician from the Faculty of Applied Mathematics and Cybernetics, Moscow State University, where he was developing approaches to quantitative description of biological systems. He obtained his Ph.D. in 1995 at Moscow State University. During this time he constructed kinetic model of mitochondria and developed an approach to describe regulatory properties of oscillatory biochemical systems.

In 1995–1999 Oleg worked as a visiting scientist in the Department of Microbial Physiology, Free University of Amsterdam, The Netherlands.

He joined the A.N. Belozersky Institute of Physico-Chemical Biology, Moscow State University, in 1999, working firstly as research scientist (1999–2002), a senior research scientist (2002–2004) and later as a leader of the Group of Kinetic Modeling of Complex Biochemical Systems. Since 2004 Oleg has also been the CSO of the Institute for Systems Biology SPb, Russia. During this time, Oleg was working on development of various approaches of modelling techniques and their application to the optimization of biotechnological processes, pharmaceutical research and drug development.

**Igor Goryanin** is a Russian biophysicist who holds a Henrik Kascer Chair in computational systems biology at the University of Edinburgh and leads the Computational Systems Biology Group, School of Informatics. Igor is also director for the Edinburgh Centre for Bioinformatics and codirector for the Centre for Systems Biology at Edinburgh. Igor graduated (M.Sc.) in 1985 as an applied mathematician from the Computer

Science Department, Moscow Engineering Physical Institute (MEPHI), where he was developing numerical methods and algorithms for analysis of stiff ordinary differential equations. Igor Goryanin spent more than twelve years working in the Institute of Biophysics, Russian Academy of Science, and obtained his Ph.D. in 1995 at the same institute. During this time he developed DBSolve, a software for mathematical stimulation and analysis of the cellular metabolism and regulation (Igor is an author for DBSolve). From 1989 to 1995 he was also CEO and cofounder of Biobank Inc., Russia.

In 1995–1997 Igor worked as a visiting computer scientist at the Mathematics & Computer Science Division, Argonne National Laboratories. He joined GlaxoSmithKline (formerly known as GlaxoWellcome) in 1997, working firstly as senior bioinformatics analyst/scientist (1997–1999), a senior research bioinformatics scientist, group leader, project manager (1999–2001) and later as a head of cell simulations and pathway modelling (2001–2005). During his time at GlaxoSmithKline, Igor was working on application of modelling and informatics techniques for the pharmaceutical research and development and drugs manufacturing industry. The whole-cell modelling of organisms approach developed by Igor has been successfully used to improve drug R&D and manufacturing process in production plants; i.e., designing antimicrobial assays and antimicrobial drug targets identification, rational organism design, rational biomarker design and target prioritisation, and reconstructing cellular networks for cancers, metabolic and lipid disorders.

In 2005, Goryanin moved to Edinburgh to take the position of a Henrik Kascer Chair in computational systems biology. In 2006 Igor developed one of the first master's courses in computational systems biology in the U.K., which is currently taught at the University of Edinburgh.

# Introduction

At the time of publication of this book, systems biology is in its adolescence. More than ten books on this subject have recently been published covering a wide spectrum of scientific endeavour. Until recently, even the definition of systems biology was vague, but this has now converged to the science that discovers how, at all levels of biological hierarchy, functionality emerges from the interactions between components of biological systems. Now, the big challenge is to develop a multi-year vision and strategy for the future. We consider kinetic modelling as one of the major pillars for the future development of systems biology.

This book, organized into nine chapters, provides an overview of the method of kinetic modelling in systems biology, including practical applications.

Chapter 1, an introduction, gives a historical overview of knowledge management and mathematical modelling in biology. It emphasizes the importance of computational techniques in reducing and guiding experimental design and data analysis.

Chapter 2 introduces the basic biological cellular network concepts (DNA, genes, proteins, regulatory mechanisms) in the context of cellular functioning. Different types of visual categories and graphical standards are presented. It describes the process of pathway reconstruction in detail (static modelling).

Chapter 3 presents the Edinburgh Pathway Editor (EPE) software package with the main concepts explained in detail. The EPE is extensively used throughout the book for pathway visualization and illustration.

In chapter 4 we present the process of construction and verification of kinetic models. We discuss the basic principles of kinetic model construction, development of a system of ordinary differential equations describing the dynamics, and derivation of the rate law for biochemical reactions.

We go on to describe the verification of kinetic model using *in vitro* and *in vivo* experimental data.

Chapter 5 is an introduction to DBSolve. The main features and user interface are described and examples are provided, from simple pathway to complex behaviour oscillation, chaos and bifurcations.

In chapter 6 ('Kinetic Modelling of Enzymes and Transporters') we discuss the basic principles of modeling of individual enzymes and transporters. To demonstrate how different types of experimental data can be incorporated into a kinetic model, we present kinetic models of adenine nucleotide translocator from mitochondria and of the following *Escherichia coli* enzymes: histidinol dehydrogenase, imidazologlycerol-phosphate synthetase, isocitrate dehydrogenase, isocitrate dehydrogenase kinase/phosphatase, phosphofructokinase-1 and galactosidase. Development of kinetic models of these enzymes and transporters illustrates the basic principles of kinetic description of enzymatic reactions; namely, how kinetic data measured under different conditions (pH, temperature and others) can be combined to construct a quantitative description predicting the kinetic behaviour of the enzyme and transporter under any set of conditions.

In chapter 7 ('Pathways Kinetic Modelling') we give a detailed explanation of how to construct kinetic models of intracellular systems (metabolic, signalling or gene regulatory pathways) on the basis of models of individual enzymes. We present kinetic models of the mitochondrial Krebs cycle and the *Escherichia coli* branched-chain amino acid biosynthesis to illustrate this approach.

Chapter 8 ('Kinetic Models of Organelles') illustrates how the principles of kinetic modelling described in detail in previous chapters can be applied to collect all the available information on energy metabolism of whole organelles such as mitochondria, to construct a kinetic model and predict response of the organelle to changes in external conditions. We present here a kinetic model describing the functioning of oxidative phosphorylation in mitochondria respiring on succinate. In the framework of the model we have simulated not only biochemical reactions but reactions coupled to production and consumption of electric potential difference as well as transport processes.

Over many years, the KM approach has been successfully applied for different problems in biotechnology and biomedicine. In chapter 9 ('Applications of Kinetic Modelling'), we present several example applications, ranging from drug safety mechanisms, in which we analyse the hepatotoxic effect of salicylate, to multiple target identification analysis for

*Mycobacterium tuberculosis* and optimization of *E. coli* amino acid biosynthesis for isoleucine and valine production.

On the companion CD, the Edinburgh Pathway Editor v 1.0 installation can be found, with pathway diagrams from this book. Pathway diagrams are available in several graphical formats, including *pwz* files for reuse. The full DBsolve 7 installation with examples is also included. All models from this book are available on CD in DBsolve SLV and SBML formats.* So, the reader can repeat all *in silico* simulations presented in the book as a learning exercise and use the models as templates for further modelling in research and practical applications.

Turning to acknowledgements, firstly, we thank Brendan Hamill and Luna de Ferrari for extensive help in the preparation of this book. Special credit is due to Alexander Mazein for his pathway diagrams.

We thank members of our groups and, especially, acknowledge the contributions of Anatoly Sorokin, Hongwu Ma, Galina Lebedeva, Ekaterina Mogilevskaya, Nail Gizzatkulov, Evgeniy Metelkin, Kirill Peskov, Aleksey Goltsov and Tatiana Plyusnina.

We thank Evgeny Selkov, Sr., Alex Selkov and Gene Selkov, Jr., for their multi-year contribution to the subject and many fruitful discussions, which also included colleagues from GlaxoSmithKline: Charlie Hodgman, Jana Wolf, Frank Tobin, Serge Dronov, Hugh Spence, Chris Preston and Malcolm Skingle.

We thank our colleagues worldwide for stimulating discussions: Hans Westerhoff, Boris Kholodenko, Ian Humphery-Smio, Nikolay Markevich, Evgeny Nikolaev, Ross Overbeek, Marta Cascante, Mathias Reuss, Hiroaki Kitano, Mariko Hatakayama, Yoshi Sakaki, Bernhard Palsson, Douglas Kell, colleagues from the University of Edinburgh, and Grahame Bulfield, for his support of Systems Biology research.

We hope that this book will help the reader to understand the kinetic modelling approach, apply it to solve real-life problems, and give the opportunity to think about future challenges.

---

* These Models are accompanied with files of experimental data (in DAT format) and files with dynamic visualisation of modeling results (in XML and PLT formats).

## CHAPTER 1

# Systems Biology, Biological Knowledge and Kinetic Modelling

Nowadays, systems biology is extending to cover almost all biological sciences, from cellular molecular biology (Westerhoff 2003), to whole organ and organism-level biomedical studies (Kitano 2007). An excellent review (Westerhoff and Palsson 2004, has been published on convergence between molecular biology and formal understanding (mathematical modelling) of biological systems as a whole. The authors claimed that the process started in the middle of the twentieth century when methods of molecular biology and theoretical generalization emerged. In our opinion, the history of systems biology (or knowledge-based interdisciplinary systems thinking) started much earlier.

The current view on biology as knowledge acquisition about complex biological systems goes back several centuries to the time when Antony van Leeuwenhoek (1632–1723) (figure 1.1) was able to observe behaviour of living organisms and cells.

More than three hundred years ago, van Leeuwenhoek made many discoveries using his new tool, the microscope. He discovered infusoria, then bacteria, and, two years later, spermatozoa.

In his letter of June 12 1716 to the Royal Society (van Leeuwenhoek 1979), he wrote '[M]y work, which I've done for a long time, was not pursued in order to gain the praise I now enjoy, but chiefly from a craving

FIGURE 1.1    Antony van Leeuwenhoek.

after knowledge, which I notice resides in me more than in most other men. And wherewithal, whenever I found out anything remarkable, I have thought it my duty to put down my discovery on paper, so that all ingenious people might be informed thereof.'

Unfortunately, there was no online publishing, nor were there computers or even scientific journals at that time (figure 1.2).

Antony could only publish his research achievements and inventions in his letters; they were later translated into English by the editor of the

FIGURE 1.2    First issue of *Nature*.

*Philosophical Transactions of the Royal Society of London* (van Leewenhoek, 1979), Robert Hooke, who was also the author of *Micrographia* and who introduced the term 'cell'.

It is significant that, in 2008, the same journal is available through on-line publishing. The new worldwide scientific information facilitates exchanges. Free on-line publishing is expanding. To share biological knowledge is not only to have access to the text (letters, manuscripts, ideas)—there are images as well. Three hundred years ago these had to be drawn by hand, but today, thanks to digital photography, images and movies can be directly stored on the computer and shared worldwide (e.g., YouTube).

The problem is that biological knowledge is not only text or images. There is a large amount of quantitative data available from modern high-throughput 'omics' technologies. These data should be shared, analyzed and knowledge extracted and managed to provide new insights in biology.

The collection of quantitative biological data in systems biology started almost a century ago when Michaelis and Menten (figure 1.3) published their paper on enzyme kinetic mechanisms (Michaelis and Menten 1913). They showed that the rate of the enzyme reaction is hyperbolic and dependent on the concentration of substrate (figure 1.4 and figure 1.5).

## DEPENDENCE OF ENZYME REACTION RATE ON THE SUBSTRATE CONCENTRATION

At that time it was very difficult to analyze and understand nonlinear behavior (it is something of a challenge even today), so the idea was to plot the data on a double reciprocal plot (figure 1.6). In this case the dependence becomes linear, and the intersection (by using a ruler) allows calculation of the $V_{max}$, and $K_m$ values. This technique is easy to understand. The data can be compared, stored and published. Linear regression procedures or other methods of elementary statistical analysis could be applied to calculate deviations and standard errors.

Afterwards, biologists and medical researchers introduced many more constants such as $K_d$, $K_i$, $K_a$, $IC_{50}$ and similar constants, but all of them essentially have the same nature: some biological parameters that can be measured in reproducible experiments and then compared and published. At approximately the same time, we see the development of journals in which scientists started to publish their reports and articles.

FIGURE 1.3 Leonor Michaelis and Maud Menten.

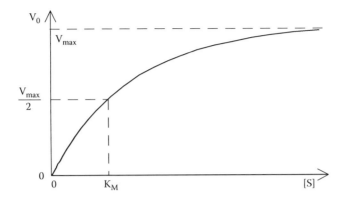

FIGURE 1.4   Michaelis–Menten curve.

Now, we could say it is our duty to convert all biological knowledge to a computer-readable form for analysis, comparison and hypothesis generation for new biological experiments.

In modern science, all enzymes, pathways and whole-cell models are created, analysed and simulated using computers. These analyses gave rise to enzyme kinetics, metabolic control analysis and pathway modelling and are now part of the field of systems biology. Simultaneous experimental (*in vitro* measurements), theoretical (ODE, regression analysis) and informatic (peer review journals, libraries) achievements were all required for scientific progress.

Another type of biological data emerged in the 1950s from the famous work of Watson and Crick (figure 1.7) on the elucidation of the structure of DNA (Watson and Crick 1953). The subsequent development of automated DNA sequencers and the generation of huge amount of sequence data would not have been possible without computers and databases.

Simultaneous new experimental (ABI), theoretical (DNA, sequence alignment algorithms) and informatics (databases) techniques were required to enable this progress. These gave rise to the new sciences of bioinformatics and computational biology and became the foundations for systems biology studies.

When the human genome was first sequenced, it was clear that publications in traditional paper journals were not enough to share the new biological knowledge on sequences. To cope with

$$v = \frac{V_{max}[S]}{[S] + K_m}$$

FIGURE 1.5   Michaelis-Menten equation.

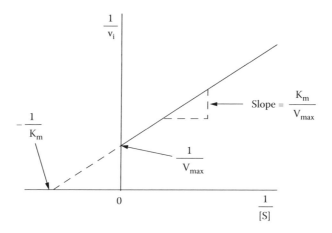

FIGURE 1.6    Double reciprocal Michaelis–Menten graph.

the problem, a number of sequence databases appeared, initially on CD, later with remote access via the Web. We are now experiencing the next phase of the distribution and sharing of biological data and knowledge, with thousands of sequenced organisms stored on the World Wide Web for users to access, search, retrieve and compare. Traditional paper publications no longer serve the purpose of biological data exchange or knowledge transfer. They can now be considered as an advertisement or an award for the teams who generate new biological data and knowledge. It is obvious now that the knowledge generation could not have been achieved without the World Wide Web (figure 1.8). Other new technologies are further expanding the social interaction of the World Wide Web for science. Examples of these are wikis

FIGURE 1.7    James D. Watson and Francis Crick.

FIGURE 1.8    Tim Berners-Lee, father of the World Wide Web.

(wiki means 'quick' in Hawaiian; Wiki Wiki Web) that allow users to easily create, edit and link Web pages, just by typing. Wikis are one of the main staples of the Web 2.0 or 'editable' web. The prime example is Wikipedia (Wikipedia: The free encyclopedia 2008), an online encyclopaedia written collaboratively that has made the wiki software familiar to 2.5 million contributors. Wikis provide an easy way to coordinate multi-partner projects in biology and to share information between wet and dry lab scientists in systems biology. Wikis save time in administration and in science, fostering a positive relation with data and documentation, which feels owned and hence curated by all stakeholders.

It is also clear that contemporary biological knowledge requires integration of different data sets: from 'omics' experiments to imaging, from computational models to simulated data, from theoretical mechanistic understanding to informatics tools. For this new type of knowledge integration about living systems we believe the modern computer science approaches should be married with mechanistic understanding of the functioning of biological systems (kinetic modelling).

Indeed, modern biology is closely associated with computer science (as are other scientific disciplines like physics and chemistry). Although biology is now growing out of its infancy, modern experimental equipment still does not allow direct observation of all biological samples, and so inferred models become an essential part of biological sciences. As we have mentioned earlier, the process of conversion of biological knowledge (figure 1.9 and figure 1.10) to models started from the Watson and Crick (1953) double-helix model

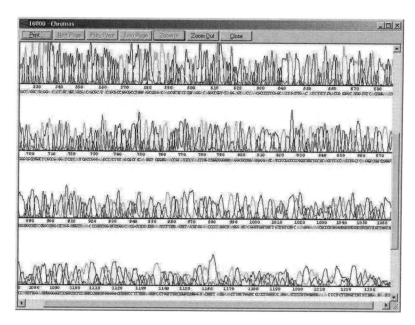

FIGURE 1.9    Output from first generation of DNA sequencing machines.

of DNA and then continued when bioinformaticians developed 3D models of proteins (Meyer 1997, figure 1.11) and later models of biological pathways and networks (Reich and Selkov 1981, figure 1.12). Modern systems of integrative biology depend heavily on the use of models of biological processes.

## WHAT ARE THE MODEL LIMITATIONS? OR, IN OTHER WORDS, WHAT CAN BE MODELLED?

It is possible to model (and make predictions) only for biological systems that we can observe with or without the help of experimental techniques. So models are always following this path in the development of new experimental techniques: visualisation, then understanding, and finally modelling. Models can be of several types, granularity and scope, but all of them should have predictive power. We can now observe a diverse range of biological objects: organs, tissues, cells, organelles, proteins, lipids, carbohydrates and small molecules. Modern imaging devices have a wide range: from MRI, CT and ultrasound scanning ($10^{-3}$ m), to MicroCT ($10^{-5}$ m), to confocal microscopy ($10^{-6}$ m), to electron microscopy ($10^{-9}$ m) to X-ray crystallography ($10^{-10}$ m). Image data provide a good foundation to model different space granularity. Biological processes have different time spans: from microseconds for signal transduction processes, to minutes for

FIGURE 1.10   The cover of *Nature* magazine, in which the first draft of the human genome was published.

metabolic reactions, to hours for gene regulations and circadian rhythms, to days for development and years for life span and disease. Diverse data could serve as a foundation for models different in time resolution.

In the past, to make a discovery, sometimes it was sufficient to provide images with explanation; now it is not enough. To understand biological systems we need quantitative measurements in time. It is important to measure concentration, location and states of other biological objects in time and in space. In this book we will try to describe the current state-of-the-art in understanding of biological networks and how all these diverse experimental data could be integrated to understand biology better by kinetic modelling.

First, we need to think about a catalogue (list) of entities involved in biological processes. What types of objects are known in biology? Traditionally, intracellular components (like cellular organelles) are visible

FIGURE 1.11 The Expasy logo. One of the first web-based sequence databases.

using microscopes and involved in the structural organisation of the cell. Organelles serve different biological purposes, but from the kinetic modelling perspective we could consider them simply as different compartments. In some cases we need to include organelle geometry to achieve spatial modelling.

We treat entities as biological when they are not only involved in biological processes but are directly encoded by the genome (i.e., DNA, RNA, small RNAs, peptides derived from proteins by cleavage, proteins, protein complexes, cellular organelles).

To make the picture complete we include nonbiological entities that are involved in biological processes in living organisms. We call these molecular entities 'small molecules'. The term 'small molecule' refers to any constitutionally or isotopically distinct atom, molecule, ion, ion pair, radical, radical ion, complex, conformer, etc., identifiable as a separately distinguishable chemical entity. The small molecules in question are either

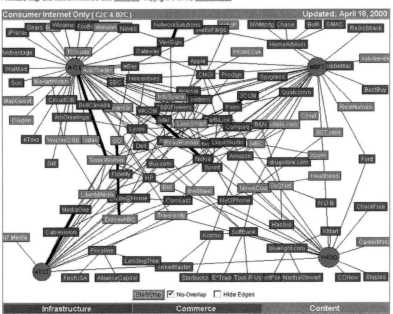

FIGURE 1.12 Networks of computers in the Internet have common properties with cellular networks.

products of nature or synthetic products (xenobiotics) used to intervene in the processes of living organisms (like drugs). For further details, see Chapter 2 and the References for links to biological ontologies, controlled vocabularies and chemical databases.

Having defined biological entities, we can now try to classify biological processes for kinetic modelling purposes. The essential distinguishing feature of all biological living (or alive) organisms is permanent change.

All changes have particular nonlinear characteristics that provide sustainability of life, cycles of reproduction, development and proliferation (the 'cycles of life'). In cell or organism death there are changes like constant degradation and decay. We cannot classify these changes as biological processes because there is no intrinsic control (compared with processes like apoptosis). So, all entities are always changing, even if the resulting changes could become zero. These changes are caused by dynamic interactions between entities.

We consider any two entities as connected if, for any length of time, there is a physical or chemical relationship connecting them.

All these temporal genotypic and phenotypic changes can be accounted for by kinetic models. In the following chapters we describe how the kinetic modelling approach could be applied to diverse activities such as biological information integration, model creation, generation of new hypotheses and, most importantly, for real life applications.

## CHAPTER 2

# Cellular Networks Reconstruction and Static Modelling

## PATHWAY RECONSTRUCTION

The first step in kinetic modelling is to develop a static model of the biological system. This is called 'pathway reconstruction'; i.e., to find out information about all players: cellular proteins, enzymes, small molecules, transcription factors, and all known interactions between them. Nonenzymatic spontaneous processes are usually included. The resulting network (i.e., a directed bond graph) should include all interactions connecting all known entities. A proper cellular network should only contain interconnected entities. Every biological entity should have a source and sink or at least participate in one of the reactions. Disconnected fragments, resulting from incomplete knowledge, could, optionally, be considered as part of the one whole cellular network. Cellular network reconstruction can be performed in two different ways: by annotation and integration of knowledge on biochemical reactions from the literature and databases ('literature-derived network') or by annotating genomes ('genome-derived network'). The best reconstruction is obtained by comparing and combining the two approaches. Before the genome sequencing era, the process of pathway reconstruction, or static model building, was very time consuming and laborious. One had to read all papers related to a particular pathway; identify the biological entities involved in the pathway, substrates and products; and then write them into the pathway map, or static model. To preserve this work, in 1980, Selkov and his team in Pushchino, Moscow, Russia, commenced the first project to collect and integrate into a single database (Enzymes and Metabolic Pathways database, EMP/MPW; Selkov 1996; shown in figure 2.1) all available information about enzymes and metabolic pathways for future reuse in kinetic modelling and pathway reconstruction. Since then, many bioinformatics databases have emerged and can be used for pathway reconstruction. Cellular networks may be reconstructed at different levels of granularity and details.

In the next section we describe the process of reconstructing a high-quality cellular network.

## THE HIGH-QUALITY NETWORK RECONSTRUCTION: DESCRIPTION OF THE PROCESS

Figure 2.2 shows the process of a whole organism network reconstruction (Ma et al. 2007). The network reconstruction process starts from organism gene annotation or "genome-based" network. The annotation process could be automated, and a network could be regularly updated from online databases. Unfortunately, the content of these databases usually does not match. For example, the number of human genes in the NCBI Entrez-Gene database (Maglott 2007) and in the HGNC (HUGO Gene Nomenclature Committee) database (Wu 2006) is different "It is therefore very important to integrate information from different databases to obtain a more complete enzyme gene list for the network reconstruction." (Ma et al. 2007, p. 135).

Special attention should be paid to genes with partial EC (Enzyme Commission) numbers such as 1.-.-.-. Only a chemically well characterised enzyme is assigned a full EC number by IUBMB (International Union of Biochemistry and Molecular Biology; IUBMB nomenclature 2008). In the post-genome era, this process is lagging behind the functional annotation of genes which is mainly based on the DNA sequence. For example, in the UniProt database there are more than 800 proteins annotated with partial EC numbers (Ashburner et al. 2000). For these genes we cannot obtain the corresponding reactions through the EC numbers. The reactions have to be added directly from functional annotation in Uniprot and many genes need to be manually annotated by reading the relevant literature.

Another problem in the reconstruction of the genome-based network is the often ambiguous relationship between EC numbers and reactions in the reaction databases. Two proteins with the same EC number in human

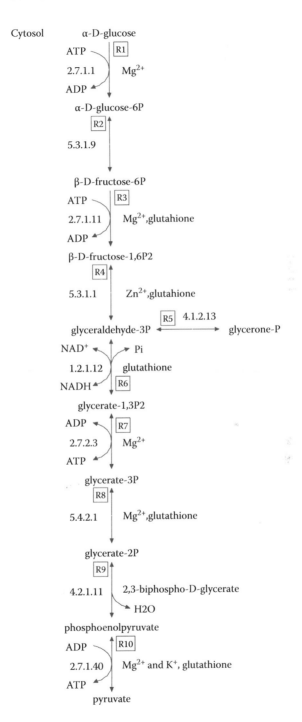

FIGURE 2.1 EMP glycolytic pathway.

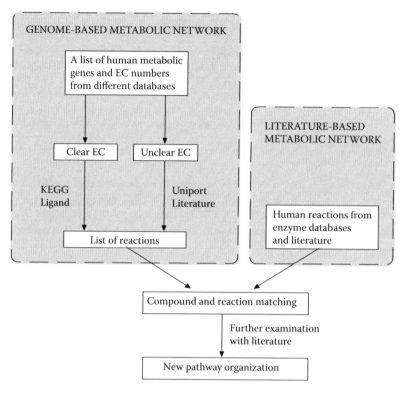

FIGURE 2.2   Reconstruction of a high-quality human metabolic network.

and another organism may not catalyze the same reaction. For example, the GBA3 gene in human, codes for the cytosolic beta-glucosidase which has the EC number 3.2.1.21, whereas in other organisms the proteins with this EC number also catalyze the degradation of cellulose. This degradation reaction does not occur in humans. Unfortunately, there is still no automatic way to obtain the human-specific EC number-reaction relationships. The KEGG ligand database is one of the most complete metabolic reaction databases, including more than 7000 reactions (Goto et al. 2002). Accordingly, KEGG (Kyoto Encyclopedia of Genes and Genomes; KEGG 2008) can be used to generate the 'first-pass' reaction list, then the reactions can be checked manually to include only organism-specific reactions.

The second step of the reconstruction is to refine the genome-based network, based on information from literature. Some databases such as the EMP database already have literature-based, organism-specific metabolic networks available. In such cases, the annotator can compare and integrate the two networks together to obtain a more complete and higher quality organism-specific network. However, this integration process is very time consuming, mainly due to the different compound and reaction nomenclature systems used in the data sets. The reactions and compounds in the genome-based network come mainly from the KEGG ligand database, while other databases such as EMP use their own nomenclature for reactions and compounds. In order to check whether two reactions from the two networks are the same, one needs to check whether all the compounds in the reaction equations are the same. A straightforward way to match the compounds in different databases is to compare them by name. However, even though most databases have a synonym list for each compound, the total number of matching compounds falls well short of 100 percent. Different methods to obtain more compound matches include using synonyms from other compound databases such as PubChem (Wheeler et al. 2006) and ChEBI (Brooksbank et al. 2005). This allows matching compounds by structure and 'fuzzy matching' between a generic compound and more specific compounds (for example, d-glucose and alpha d-glucose). Based on the matched compounds, one can find matching reactions between the genome-based network and the literature-based network by checking whether two reactions have the same reactants. These matching reactions allow us to compare the two networks at a higher level; that is, at pathway level. If a pathway reconstructed from literature contains one or more matching reactions with a genome-based pathway, then these two pathways will be functionally linked. We can then compare the two pathways to see whether they have some previously unidentified matching reactions or whether a reaction in one pathway complements a gap in another pathway. This pathway consolidation process can only be done by manual examination of the reactions in the pathways and visual inspection of the pathway maps. However, it is an important step to improve and maintain the quality of the reconstructed network and to better understand the biological function of the large-scale network. The Edinburgh Human Network reconstruction (Ma et al. 2007) is one example of this approach.

## VISUAL NOTATIONS: THREE CATEGORIES

As previously mentioned, the rapid development of high-throughput technologies (HTP) in recent years has resulted in the accumulation of huge amounts of data on living systems. At the same time, data from HTP are very shallow and usually lack the mechanistic details that are provided by a hypothesis-driven 'reductionist' approach. System biology 'is at the crossroads' (Latterich 2005), and we should select to know more about less or less about more. The flood of data, lack of mechanistic knowledge and the complexity of living systems mean that mathematical modelling and informatics approaches are essential components of modern biology. However, the wide diversity of experimental and modelling techniques, skills and questions of interest are recognised as huge obstacles for smooth iterations from biological hypothesis to mathematical model and experimental design. Visualization techniques that circumvent this obstacle and take into account the different focus of mathematical and biological communities are of paramount importance in systems biology.

The scientific community has recognised the importance of visualisation in modern biology. A recent survey on systems biology by Klipp and colleagues (Klipp et al. 2007) demonstrates that knowledge organisation, visualisation and sharing are among the areas in which standardization is most needed. It also revealed that 'about 40% of the survey respondents draw graphical representations (for example wiring schemes or network graphs) according to a defined nomenclature, and it is possible that this fraction would have been even larger if more experimentalists had participated in the survey'. Standardization is needed for model annotation, description, exchange and transfer between mathematical frameworks.

There are several reasons why scientists believe that visualisation standards can support further progress in systems biology. The questionnaire responders expressed a need for visual notation systems, which represent the main concepts of the biological domain as networks (metabolic, signalling, interactions, regulatory networks) and map these concepts to ordinary differential equations (ODE) or stochastic equations/schemes. We proposed (Mazein et al. 2008) a novel view on the visualisation of mathematical models and the corresponding biological networks. A set of three interconnected complementary categories of visual notation was introduced.

The first phenomenological level (figure 2.3) includes coarse-grained cellular maps which are the highest level representations of biological networks. At this level, most details are still unknown or undefined. It can be used as a high-level outline for the whole collection of pathways and biological models. Moreover, it is helpful as an easily understood and human-readable diagram of the biological network and as an outline for navigation. Accordingly, it often includes large clusters which were previously identified in networks manually or automatically, such as the TCA (citric acid cycle) or the purine metabolism pathway.

The next level of detail is that in which all known biological entities can be represented in the diagrams. This includes the so-called missing links,

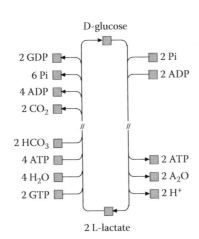

GLUCONEOGENESIS          GLYCOLYSIS

FIGURE 2.3    Phenomenological level: glycolysis.

in which information about the participation of a given entity is obtained not from experimental data but by indirect inference from modelling or by comparison with data from other species. This second level is the biological level (figure 2.4), in which we can represent detailed information about biological processes, provide experimental evidence and include links to bioinformatics databases.

Finally, the mathematical or kinetic level (figure 2.5) is constructed on the basis of the biological model and includes all known regulatory processes and states of proteins, down to the level of amino acid residues or DNA-binding elements, as well as detailed kinetic schemes. Usually the kinetic models have underlying mathematical descriptions based on rigorous mathematical formalisms (ODE, stochastic master equation or similar). See chapter 4 for more details on mathematical formalisms.

Using these three layers we are able to employ the formalism that is most appropriate to a given level of detail, using both entity relationship diagrams and process diagrams. So we could describe a process in detail on the kinetic level and, at the same time, to avoid overcomplexity, we can describe the same process at a more general level in the phenomenological diagram. In the mathematical sense, 'kinetic model' refers to a system of mechanistic ordinary differential equations that determine the temporal state of the corresponding system of biochemical reactions.

We will now analyze the notations in more detail. To be useful in knowledge management (gathering, representation and dissemination) any

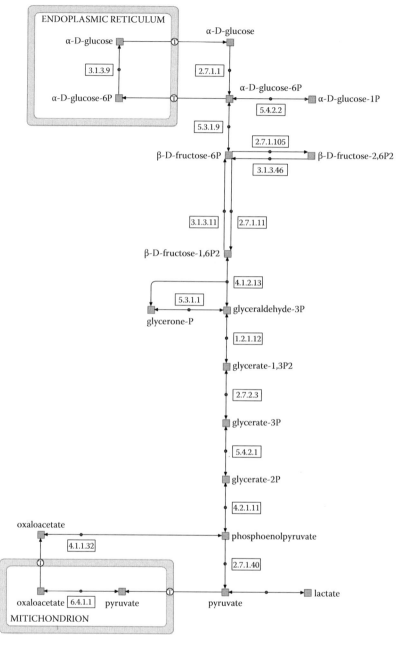

FIGURE 2.4 Biological level: glycolysis.

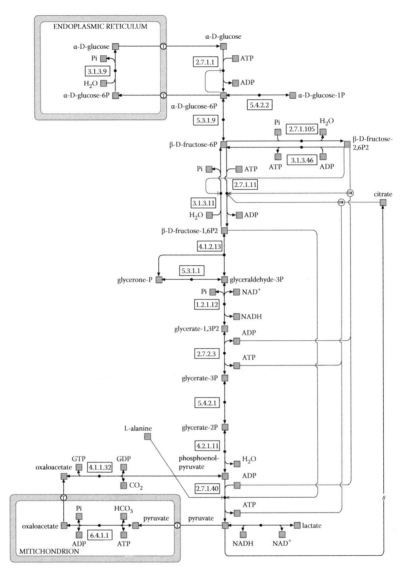

FIGURE 2.5   Kinetic level: glycolysis.

notational system '…must: (i) allow representation of diverse biological objects and interactions; (ii) be semantically and visually unambiguous; (iii) be able to incorporate notations, and (iv) allow software tools to convert a graphically represented model into mathematical formulae for analysis and simulation…' (Kitano et al. 2005, p.961). One of the main problems of all current notations in systems biology is their inability to resolve internal contradictions between requirements (i), (ii) and (iv).

The representation of the whole diversity of partially known information on biological objects and interactions on the one hand and, on the other, the ability to be exported to mathematical formulas requires different types of disambiguation. Is it possible to solve this problem if we consider the whole system of several different cross-linked notations rather than specific types of notational diagram? Below we consider the requirements for each of the proposed three categories of notation and demonstrate their ability to represent all knowledge required for a systems biology project.

Despite their importance, 'at a glance' diagrams are by necessity very ambiguous and sparse, so the outline or schematic view belongs to the phenomenological category. This type of diagram requires textual description or explanation attached to it and should not be treated as an independent source of knowledge. Phenomenological diagrams can be disambiguated by connections to diagrams in two other categories. All pathway databases, for example, KEGG, and manually created diagrams, used at step two come under the biological category. Usually, biological diagrams do not contain enough evidence and references to experimental data for hypothesis formulation, and they should be upgraded by human annotators by providing references to original experiments and/or establishing casual relationships by intelligent guesswork or assumptions. Ambiguity caused by missing knowledge or lack of detailed information is resolved by providing the ability to visually highlight such 'grey areas'. In this case, strong evidence and corroborating data should be assigned to each element on the diagram. Grey areas could be considered as targets for modelling or experimental analysis. In contrast to biological category diagrams, mathematical category diagrams replace all unknowns with assumptions and hypotheses to allow conversion to mathematical models for analyses and simulations. The establishment of mutual connections between mathematical and biological category diagrams supports information flow during the systems biology project life cycle.

As detailed in the following chapter, the proposed categories have been implemented as three contexts and plug-ins for Edinburgh Pathway Editor (EPE; Sorokin 2006).

## Communication between Diagrams

To achieve a synergistic relationship between the three categories of diagram and gain more benefit from complex visual notation categories than from separate ones, rules of communication between different categories of diagram may be defined. The procedure of data, information and knowledge transfer via these communications must also be defined. In the definition of connections between diagrams, a well-known Internet hyperlink paradigm is adopted. Each connection has two ends: referrer and target. The referrer element of each such link requests information from the target element to which that link points. The target element stores information and returns it on request. The referrer element is entirely responsible for proper usage of the information obtained from the target element, especially in the case where the referrer element has more than one link to target elements of the same type. The target element does not need to know anything about further information processing.

There are at least two purposes in establishing communication between diagrams within or between the three categories: semantic annotation and data transmission. They are named according to the type of information requested from the target element. There is little difference in implementation or behaviour of such links but their purposes and meaning are different, so we treat them separately.

The annotation link provides semantic integrity for the collection of diagrams by connecting concepts between different domains and assigning biological meaning to mathematical expressions and mathematical descriptions to biological contexts. Requirements for annotation links have been identified, enabling the provision of information required by MIRIAM (minimum information requested in the annotation of biochemical models; Le Novère et al. 2005) to complement a mathematical diagram with biological information. A less obvious example of an annotation connection between diagrams is an assignment of the mathematical description of mechanism to the biological process.

It is possible to maintain several links not only to elements of the same diagram but also to different models; for example, to compare their outcomes and analyse differences in behaviour caused by proposition of different mechanisms. In this case, the biological diagram serves as a registry of biological elements involved in the modelling and of constraints applied to the model by known relations between elements. The annotation link provides mathematical diagrams with biological identity references and biological diagrams with mathematical descriptions of mechanism. The annotation link need not always connect elements of similar type; for example, protein or protein state could be connected to a reaction element emphasizing that this protein is the carrier of that reaction.

The question of comparison of behaviour of different models leads us to the second purpose of linking different diagrams: providing data. The data link is designed to support data integrity within the notation system. Just as the annotation link enforces separation of biological knowledge from mathematical description and supports division of responsibilities between mathematicians and biologists, the data link enforces separation of experimental data from estimated or fitted values. As mentioned before, the biological information should be stored together with biological diagrams or maps, while results of mathematical analysis should be stored with mathematical diagrams. Clear examples of such separation could be values of kinetic constants. Values obtained from experiments represent known data and should be associated with the biological diagram, while values from fitting or model simulation strictly depend upon the structure and assumptions of the model and should be stored with the mathematical diagram.

The visualisation of links between different diagrams in a static environment such as a printed scientific publication is a difficult task and may possibly require some textual information. In a dynamic environment like the World Wide Web, or in visual editors such as EPE, it is possible to follow links by opening a target element or by visualising its properties. In EPE implementation of this system, the linked diagram is opened and the target element is highlighted by clicking on the referrer element of the link.

Implementation of three interconnected visual categories in EPE consists of a set of contexts and plug-ins and can be downloaded with the latest version of EPE from its Web site. Hyperlinks between diagrams have been a distinctive feature of EPE from the beginning, and this facilitated the implementation of the new system.

## Tools and Methods for Static Modelling

Pathway analysis starts with analysis of an underlying graph of interactions, using graph theoretical methods (Almaas et al. 2004) and the properties of networks (Wagner 2001). Klamt and colleagues have analysed fragility and identified targets of whole-cell models by means of minimal cut sets, by identifying a minimal (irreducible) set of reactions whose inactivation provokes network failure (Klamt 2004).

Linear algebra with convex analysis (Famili 2003) and further linear optimization methods have proved to be useful for the analysis of whole-cell metabolic networks. Stoichiometric structural analysis (Nikolaev et al. 2004) and flux coupling analysis (Burgard et al. 2004) have been employed to obtain a better understanding of large pathways and genome-scale models, mostly metabolic.

Current network convex-based pathway analysis has been implemented in two versions: those of elementary modes (Schuster et al. 1999) and extreme pathways (Schilling et al. 2000). Both of these approaches use linear algebra to characterise all of the steady-state flux distributions of a cellular network. Each version of pathway analysis has its advantages and disadvantages and both can be used to study the properties of biochemical networks. Calculation of the elementary modes and extreme pathways represents an NP-hard computational problem and is thus difficult to practise on a whole-organism scale. Under such conditions, linear optimization (Chvatal 1983) can be used to identify solutions on the basis of an objective function. In addition, subdividing the reconstructed biochemical network into conceptually defined modules has been used to address the computational problem of calculating large sets of network-based pathways (Schuster et al. 2003). The relationship between these two versions of pathway analysis has been illustrated by comparing and contrasting the elementary modes and extreme pathways (Papin et al. 2004).

A recent development (Kalir and Alon 2004) builds on efforts to reconstruct transcriptional regulatory networks on a qualitative level (static models); i.e., obtaining connectivity diagrams through high-throughput E-technologies such as expression profiling and location analysis. The experimental methods described in the paper are widely used and can be readily extended to most model organisms. In principle this makes it possible to build detailed models of any transcriptional regulatory network whose connectivity is known. An outstanding problem is that of integrating pathway models with whole-cell models in order to correctly understand cellular behaviour. To complement the small-scale *in silico* models, there is also a need to build models of entire cells or at least subsystems such as metabolism or transcriptional regulation as a whole (Covert et al. 2004).

## Databases, Ontology and Standards for Pathway Reconstruction

Traditionally, the process of pathway reconstruction (static model building) started from fully or partially annotated genome sequences for the organism or from existing pathway databases.

The EMP (EMPProject 2008) covers all aspects of enzymology and metabolism and represents the whole factual content of original journal publications. The database format has about 300 subject fields to encode the following categories: bibliographic description, biological source, host, biochemical genetics, cell cultivation conditions, metabolism, enzyme and reaction, enzyme assay and purification, enzyme kinetics, enzyme regulation, enzyme modification, enzyme structure, equilibrium and thermodynamics, physical chemistry and spectral properties and immunochemistry.

The format allows different types of tables and stoichiometric matrices to unambiguously encode metabolic pathways, reaction mechanisms, rate laws and a very wide spectrum of numeric data. Each EMP record is a translation of the whole factual content of an original journal publication into a structured, indexed, and easily searchable form. The metabolic part of EMP constitutes a separate database, EMP Pathways (formerly known as MPW). The information stored in EMP is indispensable for analysis and mathematical simulation of metabolic networks, metabolic design, drug development and bioengineering technology.

The KEGG (KEGG 2008) database should be mentioned as a major source of pathway data (mostly metabolic). The advantages are a limited number of pathway maps, convenient layout, and links to genome information. The drawback is that KEGG pathways are not organism or tissue specific. Other databases worthy of mention are SEED (SEED 2008), EcoCyc, and MetaCyc (MetaCyc 2008)

For metabolic reconstruction, it is important to correctly define chemical compounds involved in the enzyme reactions, so Chemical Entities of Biological Interest (ChEBI) is a useful freely available dictionary of molecular entities focused on 'small' chemical compounds. The term 'molecular entity' refers to any constitutionally or isotopically distinct atom, molecule, ion, ion pair, radical, radical ion, complex, conformer, etc., identifiable as a separately distinguishable entity. The molecular entities in question are either natural products or synthetic products used to intervene in the processes of living organisms. ChEBI encompasses an ontological classification, whereby the relationships between molecular entities or classes of entities and their parents and/or children are specified. ChEBI uses nomenclature, symbolism and terminology endorsed by international scientific bodies; the enzyme classification comes from the International Union of Pure and Applied Chemistry (IUPAC). Molecules directly encoded by the genome (e.g., nucleic acids, proteins and peptides derived from proteins by cleavage) are not as a rule included in ChEBI.

All data in the database is nonproprietary or is derived from a non-proprietary source. It is thus freely accessible and available to anyone. In addition, each data item is fully traceable and explicitly referenced to the original source. Another recently released standard is the Biological Pathway Data Exchange format (BioPAX, 2008) to simplify creation of static models. BioPAX is a collaborative effort to create a data-exchange format for biological pathway data. BioPAX Level 2 covers metabolic pathways, molecular interactions and protein posttranslational modifications and is backwards-compatible with Level 1. Future levels will expand support for signalling pathways, gene regulatory networks and genetic interactions. The goal of the BioPAX collaboration is to develop a common exchange format for biological pathway data.

## SBML

The Systems Biology Mark-up Language (SBML; Hucka et al. 2003) is a computer-readable format for representing models of biochemical reaction networks. SBML is applicable to metabolic networks, cell-signalling pathways, regulatory networks and many others. It is internationally supported and widely used. SBML has been evolving since mid-2000 through the efforts of an international group of software developers and users. Today, SBML is supported by over 100 software systems. Advances in biotechnology are leading to larger, more complex quantitative models. The systems biology community needs information standards if models are to be shared, evaluated and developed cooperatively. SBML's widespread adoption offers many benefits, including: (1) enabling the use of multiple tools without rewriting models for each tool, (2) enabling models to be shared and published in a form which other researchers can use even in a different software environment, and (3) ensuring the survival of models (and the intellectual effort put into them) beyond the lifetime of the software used to create them.

Systems biology has become an essential part of twenty-first-century biological research. It is an approach that has been accepted internationally everywhere from academic research departments to industrial pharmaceutical companies. A cornerstone of this movement is the use of computational modelling, a crucial tool for helping us cope with the vast size and complexity of natural organisms. The use of quantitative computational models by experimental scientists promises to pave the way for more rigorous analyses of biological functions and ultimately will lead to new and better treatments for disease. However, in order for this to happen, the computational models themselves must reach a wider audience.

An important first step is to reach agreement on how to communicate models. For models at the level of cellular biochemical reaction networks, the Systems Biology Mark-up Language has become a *de facto* machine-readable format for exchanging models between software tools. SBML is a critical enabler of research in computational systems biology, but it works at the software level; it is not intended for humans to read and write. Humans find visual representations of models more appealing. However, there is currently no standard visual notation for computational models of biological systems, despite the increasing number of software tools offering visual diagrammatic interfaces. Experience in other fields such as electrical engineering has demonstrated the essential need for standardizing the visual notation for diagrams of models.

Our team has been developing a new visual notation aimed at addressing exactly this need.

## SBGN: A Visual Notation for Network Diagrams

Similar to the SBML collaboration, a Systems Biology Graphical Notation (SBGN) project was recently initiated. The goal of the SBGN effort is to help standardise a graphical notation for computational models in systems biology. Such a standard notation will have broad impact. For example, it will add rigor and consistency to the usually *ad hoc* diagrams that often accompany research articles in publications. It will also help bring consistency to the user interfaces of different software tools and databases. SBGN is a natural complement to SBML.

The SBGN project produces a new standard leading to novel industrial developments in software tools for systems biology. The real payoff for SBGN will come when more people and software adopt such a common visual notation and it becomes as familiar to them as circuit schematics are to computer engineers. When researchers are saved the time and effort required to familiarise themselves with different notations, they can spend more time thinking about the underlying networks being depicted. SBGN will be linked to ontologies.

## CHAPTER 3

# Edinburgh Pathway Editor

B iological networks are systems of biochemical processes inside a cell or an organism that involve cellular constituents such as DNA, RNA, proteins and various small molecules. Pathway maps are often used to represent the structure of such networks with associated biological information. Several pathway editors have recently been developed, and they vary according to specific domains of knowledge. This chapter presents an introduction to the Edinburgh Pathway Editor (EPE; Sorokin et al. 2006). EPE was designed for the annotation, visualization, and presentation of a wide variety of biological networks that include metabolic, genetic and signal transduction pathways. EPE is based on a metadata-driven architecture. The editor supports the presentation and annotation of maps, in addition to the storage and retrieval of reaction kinetics information in relational databases that are either local or remote. EPE also has facilities for linking graphical objects to external databases and Web resources and is capable of reproducing most existing graphical notations and visual representations of pathway maps. In summary, EPE provides a highly flexible tool for combining visualization, editing, and database manipulation of information relating to biological networks. EPE is open-source software, distributed under the Eclipse open-source application platform license. EPE software with tutorial and examples can be found on the accompanying CD.

## INTRODUCTION

Given the ever-increasing complexity of biological networks, such as those represented by metabolic, signal transduction and gene transcription regulation pathways, visualisation techniques are critically important for further development in integrative systems biology and our understanding of cellular behaviour (Cook, Farley, and Tapscott 2001; Demir et al. 2002 Kitano et al. 2005; Kohn 1999, 2001; Wu et al. 2002). While significant progress has been made in this area, a number of limitations still exist in current pathway editors.

Biological networks are used in several domains of modern biology and medicine. Several of these domains have a long history, and researchers in these areas have their own established conventions with respect to data visualisation. Additionally, the importance of visual representations alone is sometimes overestimated. Attempts to visualise all information in a single large picture often result in a confusing diagram that is difficult to interpret. Moreover, at the pictorial level, it is not possible to show all available information in a single diagram.

While large overview maps of biological interactions are useful for identifying missing information and understanding global systems behaviour, it is very difficult to develop, annotate, and verify data using these maps. Hierarchical organisation of information and active (hyperlinked) content can address these difficulties.

A common problem with biological network editors occurs because an application is often tightly linked to the proprietary database or to the notation developed by its authors. This makes it difficult to extend the information stored within a model and integrate it with information from other proprietary software.

In the Edinburgh Pathway Editor (EPE), we have tried to address all of these important concerns in creating a tool that can be used by biologists and can also provide the required level of mathematical abstraction and sophistication to create mathematical models of biological networks.

The field of systems biology is developing rapidly, and new intracellular biomolecules, such as micro-RNAs that control gene expression, have recently been discovered. These emerging entities are providing a paradigm shift in our understanding of biological systems, and EPE is ready to accommodate new types of networks associated with such discoveries.

The attention of the reader is drawn to the existence of other pathway editors. For the most part, these are SBML compatible, at least with respect to export features, since SBML is a *de facto* standard in the field of biological modelling. These packages include CellDesigner (The Systems Biology Institute, 2008), TERANODE (Teranode Corporation, 2008), Bio Sketch Pad (Biocomputation Group, University of Pennsylvania, 2008), Systems Biology Workbench (SBW) JDesigner (Keck Graduate Institute, 2008), BioUML (Biosoft, 2008), BioTapestry (Institute for Systems Biology, 2008), Pathway Builder (Protein Lounge, 2008), NetBuilder (NetBuilder Development Team, 2008), PathwayLab (InNetics, 2008), VitaPad (Holford et al. 2005), PATIKA (Demir et al. 2002), PathwayStudio (Ariadne Genomics, 2008), and EPE (Sorokin et al. 2006).

## FEATURE SUMMARY OF EPE

We have sought to overcome the aforementioned limitations by developing EPE as an Eclipse standalone application. Eclipse (Eclipse 2008) is an opensource community whose projects provide a vendor-neutral open development platform and application frameworks for building software. An Eclipse standalone application is software that uses a limited part of the Eclipse platform as a basis for the user interface and for the implementation of low-level system processes. EPE uses the Eclipse Graphical Editing Framework (GEF) as the basis for its drawing functionality. GEF allows developers to create a rich graphical editor from an existing application model.

Generally speaking, Eclipse allows developers to maintain a convenient balance between software extensibility and maintainability through the use of extension points that allow researchers to develop specialised plug-in software for scientific computing. One metaphor for describing extension points and extensions is a headphone jack and headphone cable. The headphone jack is the extension point. The headphone with its cable that plugs into the jack is the extension. All types of headphones with their cables can plug into the headphone jack if the jack and cable are designed to fit together properly. When a software application must accommodate other plug-ins in order to extend its functionality, the application will declare an extension point. The extension point declares a contract (for example, a Java interface) to which extensions must conform. This allows software plug-ins built by different research groups to interact in an easy fashion. Further information on Eclipse extensions and extension points is available from the book *Official Eclipse 3.0 FAQs* (Arthorne and Laffra 2004).

EPE uses a small number of basic objects (table 3.1 and table 3.2); that is, discrete items that can usually be represented in a visual fashion and

TABLE 3.1   EPE Visual Notation for Biological Objects

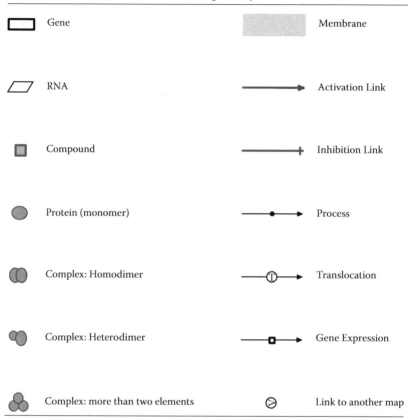

| | |
|---|---|
| Gene | Membrane |
| RNA | Activation Link |
| Compound | Inhibition Link |
| Protein (monomer) | Process |
| Complex: Homodimer | Translocation |
| Complex: Heterodimer | Gene Expression |
| Complex: more than two elements | Link to another map |

that have a number of properties such as a name for each object. These objects illustrate the main concepts of a biological network. Let us now consider the primary, high-level concepts of EPE, which are italicised in the following sentences. *Shapes* are biological items or subsystems, which are generally treated as "black boxes" with a number of *ports* for interfaces (The term 'black box' implies that a researcher does not need to know the details of the system visualised by the *shape* object in order to make use of them at this level of abstraction). Processes are concerned with the visualisation of a sequence of events; for example, events in a biochemical reaction or protein interaction. *Links* are used to represent any pair-wise relation between elements, such as *shapes*, that are represented in a diagram. *Links* include 'identity' or 'act on' relationships, when two objects represent the same entity or when one object regulates or modifies the activity of the other, respectively. *Labels* are items that represent textual

TABLE 3.2    EPE Visual Notation for Subcellular Locations

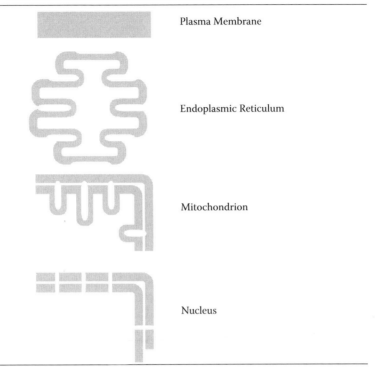

Plasma Membrane

Endoplasmic Reticulum

Mitochondrion

Nucleus

information and incorporate links to other maps and resources. The EPE concept of *context* separates metadata and standards for visualisation from pathway maps and pathway data and allows one to tune the 'drawing palette' for the selected type of map. A drawing palette includes a set of drawing objects and tools that are analogous to the erasers, paint brushes or pencils that are typically available in standard drawing packages. Given the notion of context, a researcher can conveniently create a new object with special customised properties. The concept of *context* refers to a collection of objects, their properties and their default values. It helps users create new objects on the basis of existing ones. The *context* property editor provides great flexibility in how information is stored and visualised.

Other types of information are not shown on pictorial pathway maps. For instance, EPE captures the provenance of relationships that include literature annotations and links to databases that corroborate the relationships depicted on the map. This information is normally stored in a database or as annotation comments that are supplementary to the map. EPE allows users to customise the list of object properties and to store these data within the object. The data are visualised by means of the graphical representation of linked pages or by pop-up windows. The general concept of properties allows the system to refer to values of other properties that are already defined, or even to properties of other objects, with the help of the simple language and conventions that we provide, and this greatly reduces the effort needed to update the information and to maintain its consistency. All objects can be marked as searchable, and it is possible to search for the information stored in the properties of those objects. Researchers can find objects on the map that have specific properties; they can also find objects that are associated with specific data. EPE provides two types of search facilities. The first is a simple mode that uses substring search of all objects in the map, folder (containing a collection of maps) or specified data source (all maps in a database). An advanced mode allows users to restrict searching according to the specific property of the specific object type. As an example, a researcher might use this search facility to obtain a list of all pathways in a database that involve the coenzyme NADPH or to graphically highlight all coenzymes in a single particular pathway. Moreover, with the advanced search function, it is possible to find all pathways that involve NADPH, open other maps that contain diagrams for these pathways, and then visually locate all NADPH species. It is also possible to change the visual characteristics of these objects in order to highlight them.

With simple search it is possible to locate all objects for which the searchable properties contain a specified string, so not only will NADP be located, but also NADPH and all enzymes containing NADP in the description of their function.

The information for model annotation comes from external databases, such as SwissProt or GenBank; scientific papers, which are generally referenced by a PubMed ID; and an annotator's or researcher's own knowledge and expertise. One common referencing example or parameter is the PubMed PMID number (U.S. National Institutes of Health, 2008). PubMed is the National Library of Medicine search service, which provides access to more than 16 million citations. PMID is an acronym for PubMed Identifier, which is a unique number that is assigned to each PubMed citation of life sciences and biomedical scientific journal articles. Other referencing examples or parameters include GenBank accession numbers (Benson et al. 2002), Protein Data Bank (PDB) record IDs (Research Collaboratory for Structural Bioinformatics, 2008), Swiss-Prot IDs (Wu et al. 2002), and Enzyme EC numbers (Nomenclature Committee of the International Union of Biochemistry and Molecular Biology, 2008). EPE provides a default implementation of a link to external databases and publishes corresponding Eclipse extension points. EPE allows data providers to enhance default behaviour with vendor-specific features such as complex search facilities or automatic field filling.

EPE permits users to create hyperlinks between pathway maps and provides a means for organizing information as a hierarchy of maps. This simplifies the process of analysis and verification by allowing researchers to focus on a small subset of data that may exist in a very large model.

EPE supports the addition of reaction kinetic information that is associated with biological processes. Additionally, EPE stores information about maps in a relational format. Databases such as Apache Derby (Apache Software Foundation, 2008), MySQL (MySQL AB, 2008) and Oracle are now supported for internal persistent storage. Apache Derby is treated as local storage, and MySQL and Oracle may be used for enterprise sharing. Any other type of persistent storage may be implemented as a plug-in through the published extension points.

EPE supports data sharing and distribution through Java Database Connectivity (JDBC) and also through an open Extensible Mark-up Language (XML) export format. The XML files can be used for archiving and backup. Pathway diagrams that are created by researchers can be saved as a model in SBML format or exported to common image formats including JPEG, PNG, WMF, and SVG. They may also be converted to a fully functional and hyperlinked HTML tree that is useful for sharing the diagrams via the World Wide Web and providing viewing capabilities for users who may not have the full editing EPE software installed. If necessary, it is also possible to generate the full list of reactions present on the diagram, in simple text form, or, conversely, to automatically import a reaction from an ASCII text string to the map.

A Java-based architecture powered by Eclipse makes it possible to run EPE on different platforms, from Mac to UNIX workstations. XML-based export and support for Oracle-based RDBMS (relational database management system) storage systems allow team development of large-scale models.

## A FLEXIBLE VISUAL REPRESENTATION

One of the main advantages of EPE is the ability to represent information using virtually any visual notation. The key object that facilitates this ability is the *context* of the map. The *context* can serve as a main repository for storing map metadata. It defines a list of objects that are used for drawing a map, the visual and data properties of these objects, links to the external databases and default values.

The Kyoto Encyclopedia of Genes and Genomes (KEGG; Ogata et al. 1999) provides well-known visualisations of the metabolic pathways in its collection of pathways. KEGG is a relatively simple format comprising three children of the *shape* object and two types of *labels*. The compound object extends the concept of *shape* and visually represents chemical compounds with small circles. It stores the name of the molecule and what we call a KEGGCID (KEGG Compound ID) property to reference the KEGG molecule ID. The special *label* is used to show the name of the compound and allows the system to open KEGG compound descriptions on the basis of the KEGGCID property. Biochemical reactions are visualised by an ordinary *process* object with additional properties to store EC (Enzyme Commission) numbers used to reference KEGG process descriptions. A special type of *shape* is used to indicate literature references, quality control and level of confidence of the pathway in the defined organism or tissue. In addition, literature annotation can be stored in the properties.

Only the main compounds are shown on a KEGG map; see, for example, the map in figure 3.1. This map was created in EPE and then exported to WMF (Windows Metafile Format) and converted to EPS (encapsulated postscript). Note that no $H_2O$, ATP, or NADPH molecules are shown, even though such molecules are involved in those enzyme processes. This lack of detail sometimes misleads scientists, especially when they are analyzing cellular data without paying much attention to the biology of the specific domain. A detailed analysis of the consequences of such a shallow approach is presented in Ma and Zeng (2003).

As mentioned earlier, EPE has the ability to organise knowledge in a hierarchical way as outline maps (figure 3.2). Cellular subsystems are denoted as boxes with defined inputs and outputs. The detailed map for a pathway is available by clicking on the text of the name in a box.

EPE can be used to visualise signal transduction and genetic regulation maps. Figure 3.3 demonstrates one of the simplest representations that researchers can use to show cellular signalling networks. Hyperlinks provide links to databases such as Swiss-Prot. All protein–protein interactions, modifications, and other events in this notation are visualised by the previously mentioned *process* object. In this case, *process* may have an additional type of object linked to it (a regulator) that does not exist in metabolic maps. A regulator object links with a *process* by an arrow pointing to the *process*. The properties of this object include type of regulation, literature details and experimental proof for the regulation event.

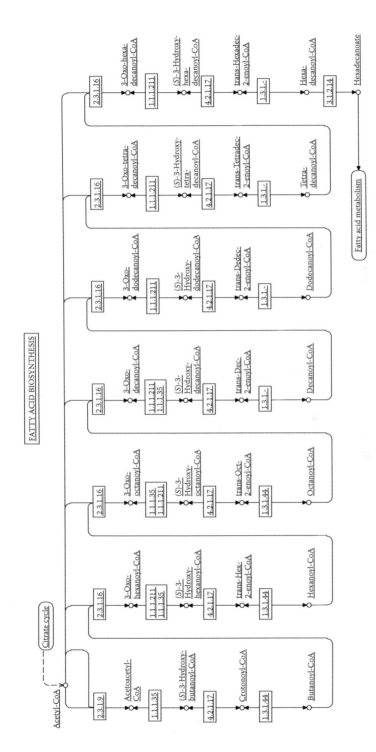

FIGURE 3.1 KEGG-like pathway map.

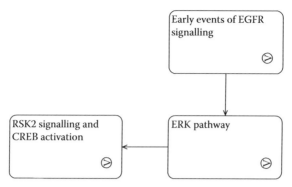

FIGURE 3.2 Phenomenological level. Outline for figure 3.4. Version 1. Modules represent linked subpathways (biological-level maps).

It is possible to create artistic forms of the diagrams which are similar to those of Pathway Builder (Institute for Systems Biology, 2008) or BioCarta (BioCarta, 2008). The EPE pathway diagram for the epidermal growth factor (EGF) (figure 3.4) can easily be created from a standard

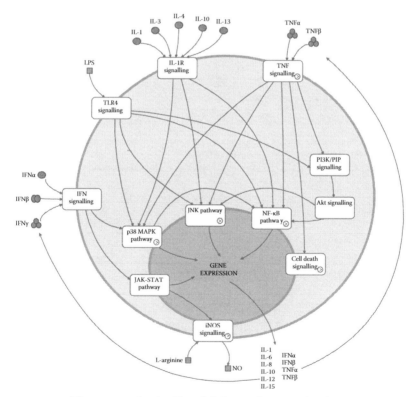

FIGURE 3.3 Phenomenological level. Macrophage activation.

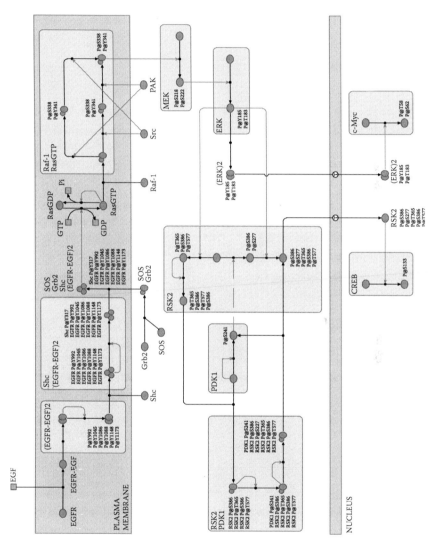

FIGURE 3.4  Biological level. EGFR signalling as a single comparatively large map.

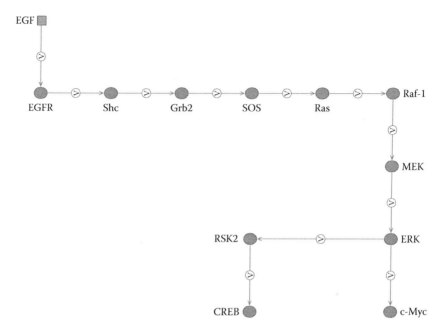

FIGURE 3.5 Phenomenological level. Outline for figure 3.4. Version 2. Protein–protein interactions.

pathway diagram by replacing shapes with graphical images or icons. The mathematical model behind the diagram remains unchanged after the conversion.

More examples of pathway diagrams are shown in figure 3.5 to figure 3.12. Clearly, we have not given a complete list of all properties needed to describe a genetic regulatory network. However, as mentioned

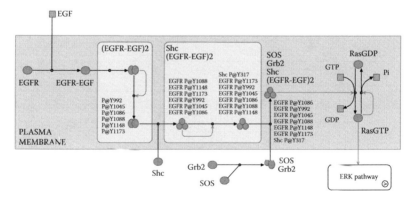

FIGURE 3.6 Biological level. EGFR signalling. Three comparatively small biological maps. First of three. Early events of EGFR signalling.

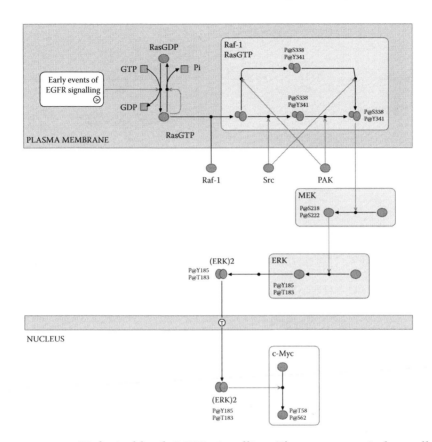

FIGURE 3.7   Biological level. EGFR signalling. Three comparatively small biological maps. Second of three. ERK pathway.

earlier, new types of data and properties can easily be added to the corresponding objects in the context. As a final visualisation, Figure 3.13 (human) shows a part of the complex compartmental genetic network. The very small squares represent input and output ports. The term 'compartment' refers to different regions in the cell, such as the cytosol, endosome lumen and endosome membrane. In this representation, a great number of different types of interactions between objects have been introduced. The context of this EPE map contains multiple objects and properties.

*Genes*, *proteins*, and *compartments* are children of the *shape* object in the sense that these objects were created by inheritance from *shape*. The *compartment* object is used to represent different parts of the cell. It is very important for the model to clearly define the cellular location of other objects on the map. The translocation of a protein from one compartment to another is usually accompanied by a protein modification event.

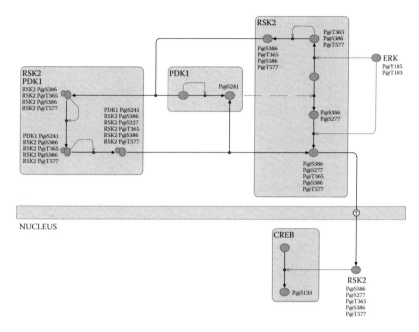

FIGURE 3.8 Biological level. EGFR signalling. Three comparatively small biological maps. Third of three. RSK2 signalling and CREB activation.

Each object on the interferon map has a *location* property which in turn refers to a *compartment* object.

In EPE, *protein* and *gene* objects contain references to corresponding databases and literature annotations. All objects that reference the same paper can easily be found by search procedures that are provided. This feature allows researchers to simplify the verification of the model and the literature curation process. Pair-wise relationships, such as those associated with protein translocation or process inhibition, are listed as children of a *link* object.

In addition, the *protein interaction* and *protein modification* objects are also represented as children of the *process* object. The *protein interaction* function is denoted by a small circle with two kinds of *links*. Ordinary *links* point to the *substrate* and the *product* objects that are involved in the interaction. *Regulatory links* point away from an object, such as an activator or inhibitor, which could change the rate of a process. An *activation link* ends with an arrowhead, and an *inhibition link* ends with a

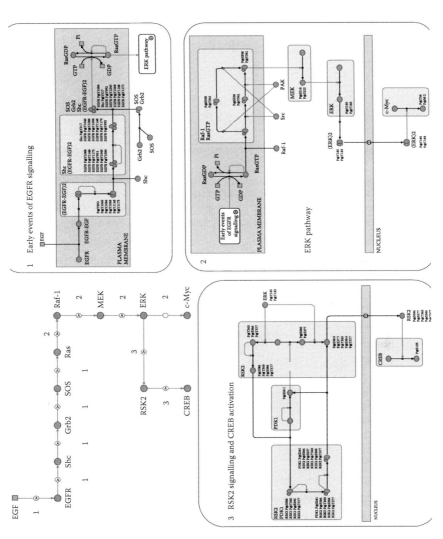

FIGURE 3.9 Links from phenomenological level to biological one.

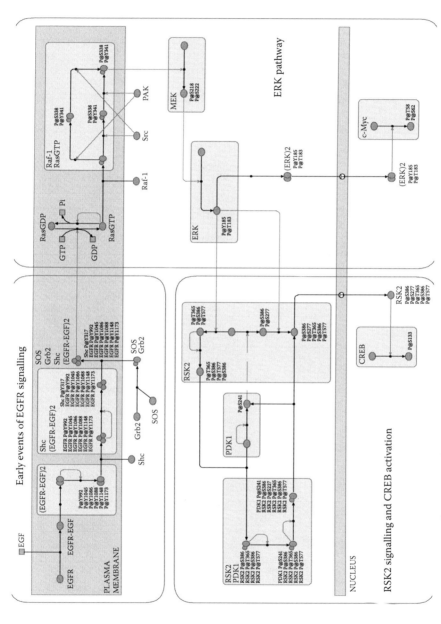

FIGURE 3.10   Phenomenological level. Overlapping between maps on biological level.

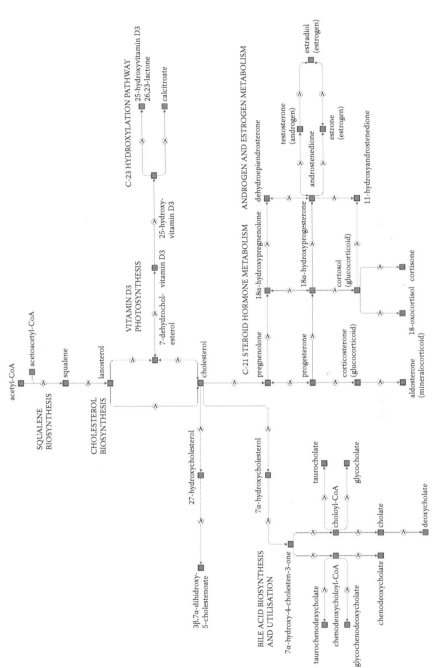

FIGURE. 3.11  Phenomenological level. Isoprenoid, steroid and bile acid metabolism.

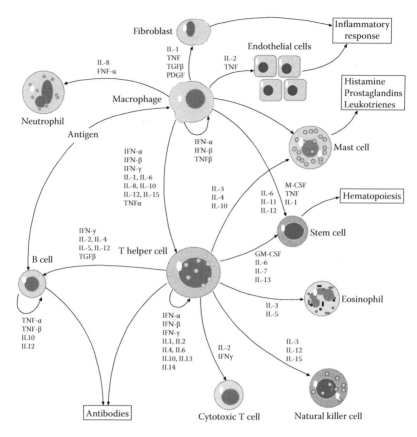

FIGURE 3.12  Phenomenological level. Cytokines.

bar head. Different modification processes are represented as open circles labelled with the letters P for phosphorylation, M for methylation, and A for acetylation.

EPE provides a special type of object known as a *gate* object to represent the inheritance hierarchy. The *gate* object is a child of the *process* object. It is a high-level logical object, so it cannot be added to the map by itself. The *gate* object is a parent for other types of *gates*, which are the *and gate*, *or gate*, and *not gate*. These gates help to reduce the complexity of the map through the addition of logical functions to regulation links. For example, instead of drawing two inhibition links to the protein binding process from two different proteins, it is easier to add an inhibition link from an *or gate* to the protein binding process, and

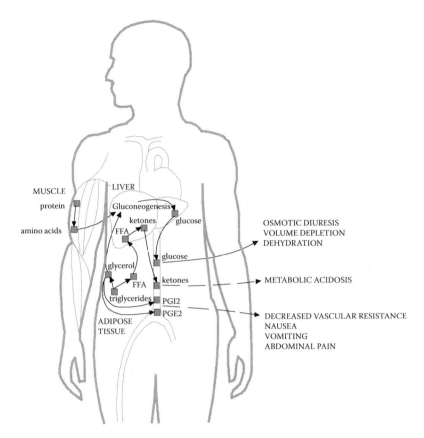

FIGURE 3.13   Phenomenological level. Pathogenesis of diabetic ketoacidosis.

then to add links from all known inhibitors to this *or gate.* Thus, the *gate* object is an example of the kind of a complex organization of information that is supported by EPE.

## CONCLUSION

EPE belongs to a new generation of software tools that will provide the scientific community with the necessary computational aids to manage biological data and perform comparative analyses between such data. The current progress in 'omics' technologies is leading to vast quantities of information that researchers will need to analyze in order to obtain a better understanding of biological phenomena. New data are also being generated at different biological levels—for example, at the phenotype, intramolecular and intercellular levels. Pathway maps provide a means to organise multidimensional views of a wealth of information and a means to assemble known and novel network characteristics. Pathway generation and analysis has the potential to lead to novel pathway biomarkers, predict possible drug adverse effects and ultimately reduce the time required for drug development.

With these goals in mind, we have developed the computer-assisted design tool described in this chapter. The tool is highly flexible and supports an extendable application for manual generation, annotation, and visualisation of biological networks of unlimited size and from different biological domains. The metadata-based architecture makes it easy for users to develop their own data structures to support data validation and knowledge extraction procedures. The flexible visualisation strategy allows researchers to tune the visual representation of a map for a variety of desired graphical notations. Object-linking techniques support the hierarchical presentation of information. Most significantly, EPE makes pathway maps and models more understandable and maintainable. We hope that EPE will help to address the growing need for systems biology tools suitable for biologists (Cassman 2005). EPE (Sorokin et al. 2006) is a first step towards a new generation of systems biology software that will provide a seamless, transparent front-end interface for theoreticians and experimentalists alike. The software, tutorials and demonstrations are available at the EPE Web site (Edinburgh Pathway Editor 2008) and on the CD accompanying this book.

## CHAPTER 4

# Construction and Verification of Kinetic Models

## INTRODUCTION

The science of biology has drastically changed over the last few decades in the sense that it is becoming an increasingly quantitative and progressively less descriptive science. In fact, the appearance of new instrumental techniques, such as HPLC, MS, capillary electrophoresis, DNA microarrays (Buchholz, Takors, and Wandrey 2001; Buchholz et al. 2002; Schaefer et al. 1999) and others, makes it possible to detect biologically significant compounds in small quantities with a high degree of accuracy. Assays based on these techniques have enabled the measurement of the level of expression of various genes and the concentration of a range of proteins and intermediates in intracellular metabolic and signalling pathways. In consequence, the number of publications devoted to quantitative characterization of intracellular processes has increased significantly. This avalanche-like increase in biological information has posed a new problem for biologists: how to analyze and to interpret the ever growing body of experimental data. Indeed, if the biological system under study consists of tens or hundreds of components and its behaviour is characterised by a set of hundreds or thousands of experimentally measured dependencies, then a mathematical model is needed for study and analysis of such a system. Like any other idealization, the mathematical model of any biological system is a formalised representation of our knowledge about the components of this system and the laws governing their functions and interactions. In this chapter, we describe an approach to construction of mathematical models, their study and application for reconstructing the dynamic and regulatory behaviour of biochemical systems using multilevel experimental data.

Models constructed using the kinetic modelling approach (Bakker et al. 1997; Chassagnole et al. 2001a; Chassagnole et al. 2001b; Demin et al. 2004; Kholodenko et al. 1999; Lebedeva et al. 2000; Markevich et al. 2004; Moehren et al. 2002; Noble et al. 2006; Rais et al. 2001; Riznichenko et al. 2000; Teusink, Bakker, and Westerhoff 1996; Westerhoff and Van Dam 1987) usually take into consideration all currently reported regulatory and dynamic features of the biological systems under study. The kinetic model considers cellular pathways as the whole set of processes catalyzed by intracellular enzymes and is described by a system of nonlinear differential equations. Hence, each enzyme is represented by a rate equation, which determines the rate of enzymatic reaction as a function of concentrations not only of substrate and product but also of the intermediates involved in considered pathways as inhibitors and/or activators of the enzyme. This description allows one to take into account all currently reported regulatory properties of the studied biological system. Since all parameters of rate equations have unambiguous physical and biological meanings, this makes it possible to model any mutations affecting the dynamic and regulatory features of the enzyme. Moreover, the kinetic model can be used to describe the responses of a biochemical system to any modifications including those required for constructing strains capable of overproduction of target products. The merits of kinetic models mentioned above arise from detailed description of all processes occurring in the biological system under study. The high level of detail gives rise to a large number of parameters, which requires a great body of experimental data for evaluation of these parameters. The high data capacity of kinetic models may be regarded as a disadvantage in comparison with other models. However, these are the models that look very promising for application to analysis and interpretation of a large amount of multilevel experimental data.

## BASIC PRINCIPLES OF KINETIC MODEL CONSTRUCTION

The term 'kinetic model' refers to a system of mechanistic ordinary differential equations that determine the temporal state of the corresponding system of biochemical reactions. In these equations, there is mass conservation between the production and consumption of each species:

$$dX/dt = V_{\text{production}} - V_{\text{consumption}}$$

where $V_{\text{production}}$ and $V_{\text{consumption}}$ are the respective rates of production and consumption of species $X$. $X$ designates any compound involved in both a metabolic pathway and in transcription or translation processes. This model describes dynamics of intracellular processes occurring in the cells of some cellular assemblage. Such an assemblage could represent, for example, a cellular suspension consisting of $10^4$ to $10^8$ cells. The concentration of species $X$ is defined as the total number of molecules of this species in all cells of the assemblage divided by the total volume of these cells.

When modelling intracellular processes in bacteria, one should take into account the effects of dilution due to the increasing numbers of cells in bacterial culture with time. In a culture in which the cells are growing exponentially, total intracellular volume $V$ will change in the following way:

$$V = V_0 \cdot e^{\mu \cdot t}$$

where $V_0$ is the total intracellular volume at time point $t = 0$ and $\mu$ is an experimentally measured growth rate. Hence, the concentration of any intracellular species, biosynthesis and consumption of which are stopped (i.e., the rate of these processes goes to zero), will decrease according to the equation:

$$X = X_0 \cdot e^{-\mu \cdot t}$$

where $X_0$ is the concentration of a given specie at time point $t = 0$. Therefore, to take account of the dilution effect caused by exponential growth of the cell culture, it is necessary to modify all equations of the kinetic model in the following way:

$$dX/dt = V_{\text{production}} - V_{\text{consumption}}, -\mu \cdot X$$

The construction of kinetic models for cellular systems can be accomplished in several stages. We will describe in detail each stage of the construction and illustrate it by several examples.

## Development of a System of Ordinary Differential Equations Describing the Dynamics of a Metabolic System

The first stage of model construction is elucidation of a static model of the system; i.e., identification of all cellular players (intermediates, enzymes, small molecules, and cofactors) and all nonenzymatic processes occurring in the cellular network (see chapter 2 for more details). The result of the elucidation is a network (i.e., a directed bond graph) of all interactions connecting all the species. For the network to be correct, each of the species must

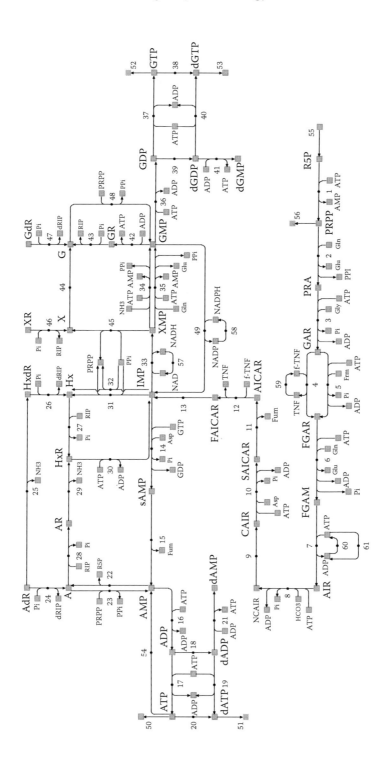

FIGURE 4.1   Purine biosynthesis pathway in *E. coli*.

participate in at least one reaction or serve as a cofactor. We will illustrate all details of static model development by the example of the purine biosynthesis pathway in *Escherichia coli*. This pathway is depicted in figure 4.1. Lists of reactions and metabolites of the pathway are shown in table 4.1 and table 4.2, respectively. Purine biosynthesis is a complex biochemical system comprising more than sixty reactions catalyzed by more than thirty enzymes. The following groups of reactions are usually distinguished in the purine biosynthesis pathway:

*De novo* biosynthesis of purine nucleotides (reactions 1–13 in figure 4.1);

Salvage pathways (reactions 22–32 and 42–48);

Interconversions of nucleotides (reactions 14–21, 33–41 and 49);

In our approach, the system of purine biosynthesis was supplemented with the following groups of reactions, which link this system to other metabolic systems:

Cell energy metabolism (reactions 54, 60 and 61);

Effluxes of nucleotides to DNA and RNA biosynthesis (reactions 50–53);

Influx of R5P from the pentose-phosphate pathway (reaction 55);

Efflux of PRPP to biosynthesis of pyrimidine nucleotides (reaction 56);

Reduction of cosubstrates (reactions 57–59).

Taking into account this information, we create a stoichiometric matrix, the columns of which correspond to the pathway reactions and whose rows correspond to the metabolites involved. The construction of a stoichiometric matrix for a given biochemical system is equivalent to development of a static model of this system. Using this model (i.e., the stoichiometric matrix), we can write down a system of differential equations, which describe the dynamics of the pathway under study:

$$\frac{d\mathbf{x}}{dt} = \mathbf{N} \cdot \mathbf{v}, \quad \mathbf{x}(0) = \mathbf{x}_0 \qquad (4.1)$$

Here, $\mathbf{x} = [x_1, \ldots, x_m]^T$ is a vector of intermediate concentrations; $\mathbf{x}_0 = [x_{10}, \ldots, x_{m0}]^T$ is a vector of initial concentrations of intermediates;

TABLE 4.1    Reactions of Purine Biosynthesis Pathway in *Escherichia coli*

| Reaction Number | Enzyme/Process Name | Gene |
|---|---|---|
| 1 | ribose-phosphate diphosphokinase | *prs* |
| 2 | amidophosphoribosyl transferase | purF |
| 3 | phosphoribosylamine-glycine ligase | *purD* |
| 4 | GAR transformylase 2 | purT |
| 5 | phosphoribosylglycinamide formyltransferase | *purN* |
| 6 | phosphoribosylformylglycinamide synthase | *purL* |
| 7 | phosphoribosylformylglycinamide cyclo-ligase | *purM* |
| 8 | N5-carboxyaminoimidazole ribonucleotide synthase | *purK* |
| 9 | N5-carboxyaminoimidazole ribonucleotide mutase | *purE* |
| 10 | phosphoribosylaminoimidazole-succinocarboxamide synthase | *purC* |
| 11, 15 | adenylosuccinate lyase | *purB* |
| 12, 13 | AICAR transformylase | *purH* |
| 14 | adenylosuccinate synthase | *purA* |
| 16,21 | adenylate kinase | *adk* |
| 17, 19, 37, 40 | nucleoside diphosphate kinase | *ndk* |
| 18, 20, 38, 39 | ribonucleoside-diphosphate reductase | *nrd* |
| 22 | AMP nucleosidase | *amn* |
| 23 | adenine phosphoribosyltransferase | *apt* |
| 24, 26, 27, 28, 43, 47 | purine nucleoside phosphorylase | *deoD* |
| 25, 29 | deoxyadenosine deaminase / adenosine deaminase | *add* |
| 30, 42 | guanosine kinase / inosine kinase | *gsk* |
| 31 | guanine phosphoribosyltransferase / hypoxanthine phosphoribosyltransferase | *hpt* |
| 32, 45, 48 | xanthine phosphoribosyltransferase / guanine phosphoribosyltransferase / hypoxanthine phosphoribosyltransferase | *gpt* |
| 33 | IMP dehydrogenase | *guaB* |
| 34, 35 | GMP synthase | *guaA* |
| 36, 41 | deoxyguanylate kinase / guanylate kinase | *gmk* |
| 44 | guanine deaminase | *ygfP* |
| 46 | xanthosine phosphorylase | *xapA* |
| 49 | GMP reductase | *guaC* |
| 50 | ATP consumption in DNA synthesis | |
| 51 | dATP consumption in RNA synthesis | |
| 52 | GTP consumption in DNA synthesis | |
| 53 | dGTP consumption in RNA synthesis | |
| 54 | ATP consumption in reactions of pyrophosphate transfer | |
| 55 | R5P biosynthesis in pentose phosphate pathway | |
| 56 | PRPP consumption in other biosynthetic processes | |
| 57 | NADH oxidation | |
| 58 | NADPH oxidation | |
| 59 | f-THF biosynthesis | |
| 60 | ADP phosphorylation | |
| 61 | ATP consumption by ATPases | |

TABLE 4.2    Metabolites of Purine Biosynthesis Pathway in *Escherichia coli*

| Metabolite Designation | Metabolite Name |
| --- | --- |
| ATP | adenosine triphosphate |
| ADP | adenosine diphosphate |
| AMP | adenosine monophosphate |
| dATP | deoxyadenosine triphosphate |
| dADP | deoxyadenosine diphosphate |
| dAMP | deoxyadenosine monophosphate |
| PRPP | 5-phosphoribosyl-1-pyrophosphate |
| PRA | 5-phosphoribosylamine |
| R5P | ribose-5-phosphate |
| R1P | ribose-1-phosphate |
| IMP | 5'-inosine monophosphate |
| NH3 | ammonia |
| NAD | nicotinamide adenine dinucleotide |
| NADP | nicotinamide adenine dinucleotide phosphate |
| NADPH | dihydronicotinamide adenine dinucleotide phosphate reduced |
| A | adenine |
| AR | adenosine |
| HxR | inosine |
| Hx | hypoxanthine |
| Pi | phosphate |
| P | phosphate |
| GMP | guanosine-5'-monophosphate |
| GDP | guanosine-5'-diphosphate |
| GTP | guanosine-5'-triphosphate |
| GR | guanosine |
| G | guanine |
| dGTP | deoxyguanosine-5'-triphosphate |
| dGDP | deoxyguanosine-diphosphate |
| dADP | deoxyadenosine-diphosphate |
| sAMP | adenylo-succinate |
| XMP | xanthosine-5-phosphate |
| XTP | xanthosine-5-triphosphate |
| Gln | glutamine |
| Glu | glutamate |
| PPi | pyrophosphate |
| GAR | 5-phospho-ribosyl-glycineamide |
| THF | tetrahydrofolate |
| fTHF | formyl-THF |
| FGAR | 5'-phosphoribosyl-N-formylglycineamide |
| FGAM | 5-phosphoribosyl-n-formylglycineamidine |

(*Continued*)

TABLE 4.2    Metabolites of Purine Biosynthesis Pathway in *Escherichia coli* (Continued)

| Metabolite Designation | Metabolite Name |
| --- | --- |
| AIR | 5-aminoimidazole ribonucleotide |
| NCAIR | N5-carboxyaminoimidazole ribonucleotide |
| CAIR | phosphoribosyl-carboxy-aminoimidazole |
| SAICAR | 5'-phosphoribosyl-4-(N-succinocarboxamide)-5-aminoimidazole |
| AICAR | aminoimidazole carboxamide ribonucleotide |
| FAICAR | phosphoribosyl-formamido-carboxamide |
| X | xanthine |
| XR | xanthosine |
| AdR | Deoxyadenosine |
| HxdR | deoxyinosine |
| GdR | Deoxyguanosine |
| CDP | cytidine-5'-diphosphate |
| dCDP | 2'-deoxycytidine-5'-diphosphate |
| dUDP | 2'-deoxyuridine-5'-diphosphate |
| dTTP | thymidine 5'-triphosphate |
| ASP | L-aspartate |
| dR1P | 2-deoxy-D-ribose-1-phosphate |
| Fum | fumarate |
| Gly | glycine |

$\mathbf{v} = [\mathbf{v}_1, \ldots, \mathbf{v}_n]^T$ is a vector of reaction rates; $\mathbf{N}$ is a stoichiometric matrix with $n$ columns and $m$ rows. In the case of the purine biosynthesis pathway, $m$ is equal to 43 and $n$ is equal to 61 and the vectors of intermediate concentrations and initial conditions are:

$\mathbf{x} =$ [R5P, PRPP, PRA, GAR, FGAR, FGAM, AIR, NCAIR, CAIR, SAICAR, AICAR,FAICAR, IMP, sAMP, AMP, ADP, ATP, dADP, dATP, dAMP, A, AdR, HxdR, Hx, HxR, XMP, GMP, GDP, GTP, dGDP, dGTP, dGMP, XR, GdR, X, G, GR, THF, f-THF, NADP, NADPH, NAD, NADH]$^T$

$\mathbf{x}_0 =$ [R5P$_0$, PRPP$_0$, PRA$_0$, GAR$_0$, FGAR$_0$, FGAM$_0$, AIR$_0$, NCAIR$_0$, CAIR$_0$, SAICAR$_0$, AICAR$_0$, FAICAR$_0$, IMP$_0$, sAMP$_0$, AMP$_0$, ADP$_0$, ATP$_0$, dADP$_0$, dATP$_0$, dAMP$_0$, A$_0$, AdR$_0$, HxdR$_0$, Hx$_0$, HxR$_0$, XMP$_0$, GMP$_0$, GDP$_0$, GTP$_0$, dGDP$_0$, dGTP$_0$, dGMP$_0$, XR$_0$, GdR$_0$, X$_0$, G$_0$, GR$_0$, THF$_0$, f-THF$_0$, NADP$_0$, NADPH$_0$, NAD$_0$, NADH$_0$]$^T$

Before discussing the methods, features and details of rate equation derivation, let us consider what the static model will help us to learn about the system under study. It appears that the stoichiometric matrix allows one, firstly, to derive the relationships between steady-state fluxes and, secondly, to derive a number of conservation laws and to write them down in analytical form. Indeed, when solving the system of linear algebraic equations

$$\mathbf{N} \cdot \mathbf{v} = \mathbf{0} \qquad (4.2),$$

we find that any steady-state reaction rate (a steady-state flux), $v_i$, $i = 1, \ldots, n$, can be expressed as a linear combination of $s$ selected steady-state rates. The number $s$, which is equal to the dimension of the kernel of the matrix $\mathbf{N}$, and the coefficients of relationships, which express any steady-state rate in terms of $s$ selected rates, are fully determined by the stoichiometric matrix. As an example, we consider the relationships between steady-state rates of the purine biosynthesis pathway (see figure 4.1):

$$v_1 = v_2 + v_{23} - v_{24} + v_{29} - v_{50} - v_{51} - v_{52} - v_{53} + v_{55} + v_{57},$$

$$v_3 = v_6 = v_7 = v_8 = v_9 = v_{10} = v_{11} = v_{12} = v_{13} = v_2,$$

$$v_4 = v_{59} - v_2, \ v_5 = 2 \cdot v_{59} - v_2, \ v_4 = v_{59} - v_2,$$

$$v_{14} = v_{15} = v_2 - v_{24} + v_{29} - v_{52} - v_{53},$$

$$v_{16} = v_2 + v_{23} - v_{24} + v_{29} - v_{44} - v_{52} - v_{53} + v_{54} + v_{55} + v_{57},$$

$$v_{19} = v_{18}, \ v_{20} = -v_{18} + v_{51}, \ v_{21} = v_{41} = v_{46} = v_{47} = 0,$$

$$v_{22} = 2 \cdot v_2 + v_{23} - v_{24} + v_{29} - v_{50} - v_{51} - v_{52} - v_{53} + v_{57},$$

$$v_{25} = v_{26} = -v_{24}, \ v_{27} = -v_{29} + v_{50} + v_{51} - v_{55} - v_{56} - v_{57},$$

$$v_{28} = 2 \cdot v_2 + v_{29} - v_{50} - v_{51} - v_{52} - v_{53} + v_{57},$$

$$v_{30} = v_{42} + v_{50} + v_{51} + v_{52} + v_{53} + v_{56} - v_{57},$$

$$v_{32} = v_{24} - v_{29} - v_{31} + v_{42} + v_{50} + v_{51} + v_{52} + v_{53} - v_{55} + v_{56} - v_{57},$$

$$v_{33} = v_{57}, \ v_{35} = -v_{34} - v_{44} + v_{57}, \ v_{36} = v_{52} + v_{53},$$

$$v_{37} = v_2 - v_{24} + v_{29} - v_{39}, \ v_{38} = -v_{39} + v_{53}, \ v_{40} = v_{39}, \ v_{43} = v_{42}, \ v_{45} = v_{44},$$

$$v_{48} = v_{42} + v_{44}, \ v_{49} = v_{58} = -v_{52} - v_{53} + v_{57},$$

$$v_{32} = 10 \cdot v_2 + 2 \cdot v_{23} - 3 \cdot v_{24} + 3 \cdot v_{29} - 2 \cdot v_{44} + v_{50} + v_{51} + 2 \cdot v_{54} + v_{55}$$
$$+ v_{56} + v_{57} - v_{59} + v_{61}$$

It follows from these relationships that any steady-state rate can be expressed in terms of twenty-one rates $v_2$, $v_{17}$, $v_{18}$, $v_{23}$, $v_{24}$, $v_{29}$, $v_{31}$, $v_{34}$, $v_{39}$, $v_{42}$, $v_{44}$, $v_{50}$, $v_{51}$, $v_{52}$, $v_{53}$, $v_{54}$, $v_{55}$, $v_{56}$, $v_{57}$, $v_{59}$ and $v_{61}$; i.e., $s$ is equal to 21.

Conservation laws are the first linear integrals of the system of differential equations (4.1) describing the kinetics of the biochemical system under study. As the simplest example of conservation law valid for the purine biosynthesis pathway, we consider the following algebraic expression:

$$NAD + NADH = const_1 \qquad (4.3)$$

This relationship results from summation and integration of the differential equations that describe the variations of NAD and NADH concentrations with time. The meaning of equation (4.3) is that the sum of the concentrations of NAD and NADH does not change with time. It is easy to show that the number of conservation laws in the kinetic model, which describes the biochemical system consisting of $m$ intermediates connected with $n$ reactions, is given by the following formula:

$$(\text{The number of conservation laws}) = m - n + s \qquad (4.4)$$

In the case of the purine biosynthesis pathway, $m$, $n$ and $s$ are equal to 43, 61 and 21, respectively. In accordance with equation (4.4), this implies that the number of conservation laws in the kinetic model of purine biosynthesis pathway is equal to 3. The relationship (4.3) represents one of these three conservation laws. The other two are given by the following expressions:

$$NADP + NADPH = const_2$$

$$THF + f\text{-}THF = const_3 \qquad (4.5)$$

Since equations (4.3) and (4.5) are valid at any time point, including time zero, the values of parameters $const_i$, $i = 1$, 2 and 3, are completely determined by the initial conditions:

$$const_1 = NAD_0 + NADH_0,$$

$$const_2 = NADP_0 + NADPH_0,$$

$$const_3 = THF_0 + f\text{-}THF_0$$

## Derivation of Rate Law of Enzymatic Reactions

Once the appropriate static network has been chosen, the second stage of model construction is the generation of rate equations that describe the dependence of each reaction rate on the concentrations of intermediates involved in the studied pathway. In order to make the models scalable and comparable with a range of experimental data, we have developed detailed and reduced descriptions of every biochemical process involved in the model.

The detailed description of a biochemical reaction (i.e., enzyme catalytic cycle) implies the authentic view of the molecular mechanism of this reaction and takes into account all the possible states of the protein involved, including its possible inactivated states (because of phosphorylation) or dead-end inhibitor complexes. Usually, the detailed description comprises a combined set of ordinary differential equations and algebraic equations (if steady-state or conservation constraint assumptions are made).

A reduced description represents the reaction rate as an explicit analytic function of the concentrations of substrates, products, inhibitors, and activators, as well as the total protein concentration and all kinetic constants of the processes. Derivation of the rate equation is described in detail in chapter 5 and accomplished by the following processes:

Construction of the enzyme catalytic cycle using structural and kinetic data (in most cases, information on the mechanism of enzyme operation is available in the literature);

Derivation of the rate equation in terms of parameters of the catalytic cycle (rate constants and dissociation/equilibrium constants of elementary stages of the catalytic cycle) on the basis of quasi steady state or rapid equilibrium approaches (Cornish-Bowden, 2001);

Derivation of the equations that express parameters of the catalytic cycle in terms of kinetic parameters (Michaelis constants, inhibition constants, catalytic constants);

Derivation of the rate equation in terms of kinetic parameters of the enzymatic reaction.

In the framework of the kinetic modelling approach, the level of detailing the catalytic cycle and the complexity of the rate equation derived on the basis of this cycle are fully determined by availability of the experimental data on structural and functional organization of the enzyme. Indeed, once the catalytic cycle of the enzyme has been established and experimentally proved, one can use this information to derive the rate equation. For cases in which the mechanism of enzyme operation is unknown, we suggest to develop a minimal catalytic cycle, which: (a) satisfies all structural and stoichiometric data available in the literature, (b) allows derivation of the rate equation describing available kinetic experimental data, and (c) is the simplest of all possible catalytic cycles that satisfy the two preceding requirements.

## BASIC PRINCIPLES OF KINETIC MODEL VERIFICATION

The third stage of the model development is parameter evaluation. For evaluation of kinetic parameters, we use *in vitro* and *in vivo* experimental data from the following sources:

Literature data on the values of $K_m$, $K_i$, $K_d$, rate constants, etc.;

Electronic databases (only a few databases with specific kinetic content are available at present; in particular, the EMP database (Selkov et al., 1996) and BRENDA (Schomburg, Chang, and Schomburg 2002);

Experimentally measured dependencies of the initial rates of enzymatic reaction on the concentrations of substrates, products, inhibitors and activators;

Time series data obtained from kinetic experiments;

*In vitro* data obtained with cell-free extracts;

*In vivo* data that describe the intracellular kinetics of the cellular pathway under study.

### Verification of Kinetic Model Using *in Vitro* Experimental Data Measured for Purified Enzymes

The kinetic parameters of rate equations are usually estimated as follows. First, we search for all the *in vitro* experimental data on kinetics of the studied enzyme available in the literature. These data are usually presented as dependencies of either the initial reaction rate on the substrate/product concentration or as dependencies of the substrate/product concentration on time. A special technique has been elaborated for quantitative description of all available *in vitro* experiments. We use an explicit rate equation applicable to individual enzymes for fitting the initial reaction rate dependences on the substrate/product concentration. The reaction rate equation is also used for determination of inhibition parameters. In order to describe the time series experiments carried out with purified enzymes, we construct minimodels, i.e., systems of ordinary differential equations, which have the solutions corresponding to the measured time dependences. Rate laws and the concentrations of experimentally measured intermediates are used in the construction of these minimodels. The kinetic parameters of rate equations are evaluated by fitting the minimodel solutions against experimentally measured time series (see examples in chapter 5 of this approach).

## Verification of the Kinetic Model Using *in Vitro* and *in Vivo* Experimental Data Measured for a Biochemical System

The preceding sections show how one can reconstruct the catalytic cycle of an enzyme and derive the rate equation on the basis of this cycle using structural information and kinetic data obtained for purified enzyme preparations. Then, the parameters of the rate equation can be estimated using *in vitro* experimental dependencies characterizing the kinetic properties of this enzymatic reaction. However, it may also happen that the *in vitro* data characterizing the kinetics of the purified enzyme will be insufficient to estimate all parameters of the rate equation. For example, only the dependencies of initial rate on the concentration of substrate have been measured for many enzymes but the capacity of products to inhibit the enzymatic reaction has not yet been studied. There are two possible ways to estimate the values of these unknown parameters. This may be achieved by using: (a) *in vitro* kinetic data obtained for the same enzyme isolated from another (micro)organism; or (b) *in vitro* and *in vivo* data that characterise the kinetics of the entire metabolic pathway or the part containing the enzyme under study.

There are two types of data that characterise the kinetics of the entire metabolic pathway or the part containing the enzyme under study: (a) *in vitro* data obtained with cell-free extracts; and (b) *in vivo* data that describe the intracellular kinetics of the metabolic pathway under study.

Let us consider the features of experiments in which these data are generated and discuss various types of kinetic models that can use these data for verification.

The *in vitro* data obtained with cell-free extracts are usually either specific activities of individual enzymes or time dependencies of the concentrations of metabolites (Leung and Schramm 1980). The experiments are carried out in the following way: after disruption of cells and removal of membranes and DNA, the cell suspension yields a cell-free extract, which represents a soluble fraction of cellular proteins. Then, low-molecular-weight constituents representing intermediates of intracellular metabolism are removed from this protein fraction. In kinetic experiments, low-molecular-weight compounds serving as substrates of an individual enzyme or a fragment of a metabolic pathway are added into the purified protein fraction and the kinetics of consumption of these substrates by the cell-free extract are measured. In the simplest case, when the kinetics of a single enzyme are studied, only the substrates of this enzyme are added into the cell-free extract. If the concentrations of these added substrates are much higher than the value of corresponding Michaelis constants of the studied enzyme, the initial rate of their consumption represents a maximal rate of the enzymatic reaction and is defined as a specific activity of the enzyme. It follows from the description of experiments that the experimental data obtained with cell-free extracts describe functioning of only a metabolic part of the intracellular processes. In fact, in experiments with cell-free systems, the concentrations of enzymes of the metabolic pathway under study do not depend on the concentrations of intermediates that control the expression of corresponding genes.

Using the specific activities of individual enzymes measured in cell-free extracts, we can evaluate the maximal rates of enzymatic reactions or can calculate the intracellular concentrations of these enzymes provided that their turnover numbers are known. The maximal rate of enzymatic reaction ($V_{max}$) can be evaluated *in vivo* from the specific activity (SA) of the corresponding enzyme in a cell lysate using a relationship between the intracellular volume and dry cell weight (d.c.w) or protein weight. For example, in *E. coli* cells, proteins constitute 55 percent of the dry cell weight, 70 percent of the protoplasm is water, and the density of protoplasm is 1.1 (Neidhardt 1987). Taking all this into account, one can calculate that 1 mg d.c.w corresponds approximately to 3 μL of intracellular volume, 1 mg total cellular protein corresponds to 5.5 μL of intracellular volume, and $V_{max}$ can be approximated in terms of the specific activity as:

$$V_{max} = \text{SA μmol/min/(mg intracellular protein)}$$
$$= \text{SA μmol/min/(5.5 μL of intracellular volume)}$$
$$= \text{(SA/5.5) M/min} = 182 \cdot \text{SA mM/min}$$

The *in vivo* data characterizing intracellular kinetics of the metabolic pathway under study are usually measured as follows: a suspension of (bacterial) cells is cultivated under certain conditions. At some time point, these conditions are changed. For example, the concentration of substrate in the medium is changed, another substrate is added or the temperature is changed. Then, small amounts of cells are periodically taken from the suspension, all reactions in these samples are stopped, and the concentrations of intracellular metabolites of interest are measured. Finally, the time dependencies of concentrations of intracellular metabolites are obtained, which characterise the response of cells to the change of external conditions. It follows from this that the data obtained in these experiments describe the cofunctioning of both metabolic and gene segments of intracellular processes. Indeed, in these experiments, the concentrations of enzymes constituting the metabolic pathway under study depend on the concentration of intermediates of this pathway that control the expression of the corresponding genes.

It follows from the above that data obtained in experiments with cell-free extracts must be used for verification of models that describe only metabolism and do not take into account the dependency of expression of enzymes on the concentrations of intermediates formed in this pathway. On the contrary, metabolic models of this type cannot be used for interpretation of experimental data obtained in *in vivo* experiments. Data obtained in *in vivo* experiments must be used for verification of models that take into account the dependency of expression of enzymes on the concentration of intermediates formed in the pathway under study. These models describe both the metabolism and its interaction with transcription and translation as well as their mutual regulation.

## STUDY OF DYNAMIC AND REGULATORY PROPERTIES OF THE KINETIC MODEL

A system of ordinary differential equations describing any part of intracellular processes is numerically integrated using specialised software packages like DBSolve (Goryanin, Hodgman, Selkov 1999; Gizzatkulov et al. 2004). The dynamic and regulatory properties of the system are studied at this stage. Specifically, the number of stationary states in the system is determined. The steady-state concentrations of metabolites are obtained as a function of parameters of system inputs and outputs and the inhibition and activation parameters of various reactions. In addition, using the methods of metabolic control theory, the distribution of the control over steady-state fluxes between different enzymes is studied and the sensitivity of the system to variations of other kinetic parameters is analyzed.

## CHAPTER 5

# Introduction to DBSolve

There are many stand-alone software packages available for systems biology and kinetic modelling. Most are available as a result of the efforts of the SBML (Systems Biology Markup Language) community (Hucka et al. 2001). However, only a few contain a full range of tools to allow kinetic model creation, parameter fitting and analysis. DBSolve is one of these packages. DBSolve is a software environment for creation, analysis and visualisation of kinetic models of biological processes. A number of versions have been released during more than ten years of software development (Gizzatkulov et al. 2004; Goryanin 1996; Goryanin and Serdyuk 1994; Goryanin, Hodgman, and Selkov 1999). During this ten-year period, DBSolve has been extensively used in Moscow State University and by GlaxoSmithKline to create hundreds of kinetic models for both research and teaching. The package has built-in algorithms and tools for constructing models and fitting parameters to the experimental data. All the models are considered to be systems of nonlinear ordinary differential equations and/or nonlinear algebraic differential equations with arbitrary right-hand sides. These features allow modellers to expand the class of possible applications to include chemical, PK/PD, ecological or other biomedical systems.

DBSolve includes the following methods:

Generation of stoichiometric matrix based on the list of the reactions describing the system;

Automatic analysis of the stoichiometric matrix;

Automatic generation of the systems of ordinary differential equations and conservation laws based on the stoichiometric matrix;

Calculation of functional dependencies defined explicitly;

Numerical solution of nonlinear ODE systems and visualisation of the solution;

Calculation of functional dependencies defined implicitly as a system of algebraic equations (generally nonlinear);

Automatic search of optimal values of the parameters of a system based on the experimental data (fitting);

Analysis of stability of the dynamic system (bifurcation) and calculation of the control coefficients as defined in metabolic control analysis.

DBSolve is written in C++. The executable version (for Windows) is attached to this volume on CD together with example files that cover the most important DBSolve features. The software has an object-oriented structure containing a range of methods and associated data that can be viewed from the main application window. DBSolve has an internal language for building and storing models and an internal compiler for processing the models and running simulations. A derived model can be saved in the internal format (with the SLV extension) and/or exported to other systems biology software as an ASCII text file or as an SBML file. The SLV file (extended SBML) contains all the necessary information about the mathematical model: the stoichiometric matrix, system equations, initial values, experimental parameter values, information about the biological components together with the reactions/processes, and links to external databases, ontology and controlled vocabularies. The full description of the SLV files is available from CD or from the ISBSPb Web site (ISBSPb 2008). DBSolve supports both import and export to SBML 2.0 format.

## CREATION AND ANALYSIS OF THE MODELS USING DBSOLVE. FUNCTIONAL DESCRIPTION

In this chapter we present a DBSolve-based strategy for the development and analysis of kinetic models and describe the corresponding sections of the package. All steps of the strategy are illustrated with an example of a simple biochemical system.

This chapter contains only a minimal description of the menu, hot keys and visualisation tools available to create, run and analyse models.

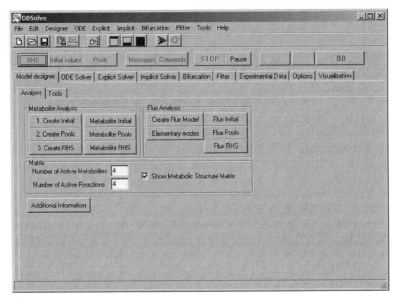

FIGURE 5.1   The Main Form 1.

For further details on the program interface, please refer to the DBSolve manual available on the book CD and on the Web site www.insysbio.ru.

## A General Look at the Interface

The main interface of DBSolve consists of two forms: the Main Form 1 and the Auxiliary Form 2, presented as independent pages. The main form is used to perform different operations with the model: entering parameters, editing the ODE system and so on. The form consists (figure 5.1) of the menu bar, two hot key bars and a number of tabs allowing switching to another interface within the workspace for managing various types of calculations.

Initially, most of the tabs and buttons of Form 1 are disabled. They become available only after the model has been created.

The Auxiliary Form 2 (figure 5.2) is used for visualisation purposes: drawing of charts, editing of the stoichiometric matrix, etc. It contains the same tabs as the Main Form 1, and they are automatically synchronised with the tabs on the main form.

## Description of the Example

Let's look at the scheme (figure 5.3) of the biochemical reactions for our example.

Letters A, B, C, D stand for substances (intermediates) participating in reactions. The arrows show the ongoing reactions. The rate of each

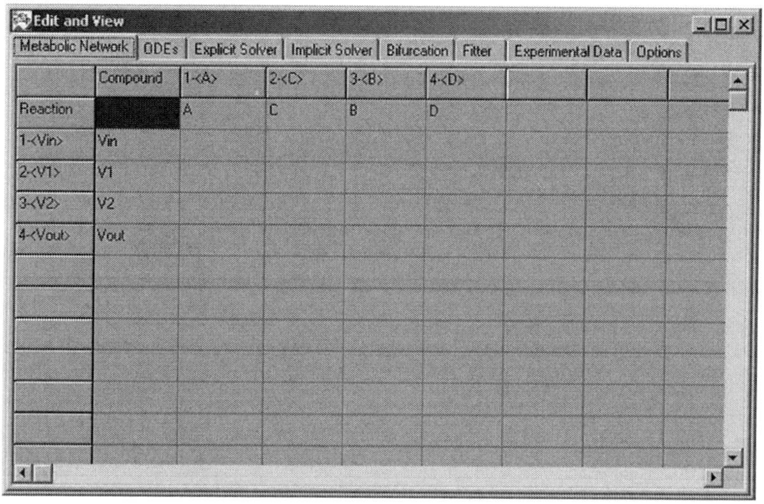

FIGURE 5.2    The Auxiliary Form 2.

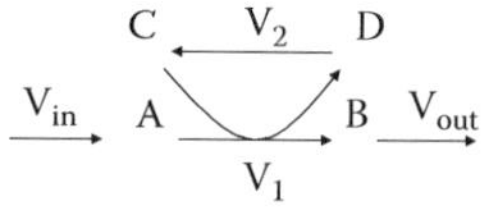

reaction is displayed above the corresponding arrow. The rate $V_{out}$ is the enzymatic reaction, i.e., it occurs by means of an enzyme, and the rest of the reactions are running spontaneously; i.e., in accordance with the law of mass action. $V_{in}$ and $V_{out}$ denote input of substance A into the system and output of the substance B from the system, respectively. For simplicity, we assume that all the reactions with the exception of $V_{in}$ are irreversible and their directions correspond to the arrows in the figure.

FIGURE 5.3    Schema of the biochemical reactions.

For the purposes of this introduction, a kinetic model is a description of a (biological) system which includes a set of components with attributes which include their functional dependencies, quantities, activity and interactions. A simulation is an electronic experiment with the model to discover consequences of interventions and changes in the initial conditions. The kinetic modelling process includes several steps: starting with the creation of the ODE system that corresponds to the stoichiometric matrix/reaction scheme, then introducing the functional dependencies of the reaction rates on concentrations of the participating substances, and finally the determination of the parameter values appearing in rate equations. The kinetic model allows us to study the system *in silico*; i.e., to build up concentrations and reaction rates, time and parameter dependencies, both transient and in steady-state. DBSolve can perform most of these

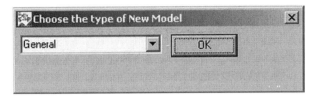

FIGURE 5.4   Opening dialog box.

and other tasks either automatically or with minimal intervention from the researcher.

## The 'Metabolic Network' Tab: Creation of ODE System (Simple Method)

In order to create the model, press the button 'New' located in the Main Form 1. Then, press the OK button in the opening dialog box (figure 5.4).

A standard 'Save As' dialog box appears on the screen. After entering the name and path of the file press the 'Save' button, following which the form in figure 5.5 is displayed, informing you that the new file has been created.

The next step is to complete the 'Metabolic Network' form (figure 5.6). Don't be confused by the term 'Metabolic Network'. Actually, one can model any biomedical processes in DBSolve and include a range of species such as RNA, RNAi, DNA, proteins, cells, hormones, cytokines, etc.

1. Enter the number of metabolites involved in the system (in our example it is four) in the 'Number of Active Metabolites field';

2. Enter the number of all the reactions, including influxes and out-fluxes of intermediaries (in our example it is four) in the 'Number of Active Reactions field';

3. Check the 'Show Metabolic Structure Matrix' check box. This allows you to edit the table 'Metabolic Network' in the Auxiliary Form 2 (figure 5.7).

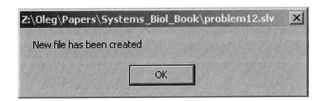

FIGURE 5.5   Message, New file has been created.

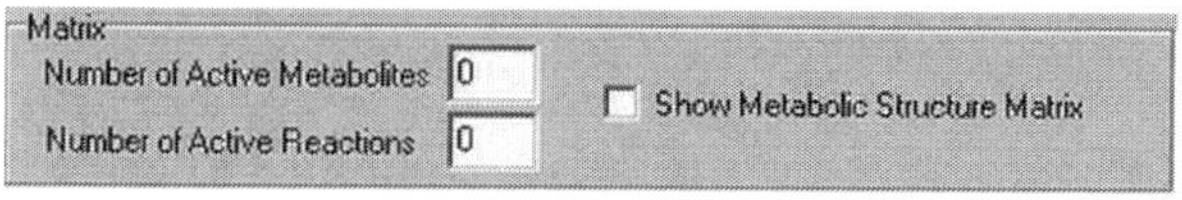


FIGURE 5.6 Metabolic Network' form.

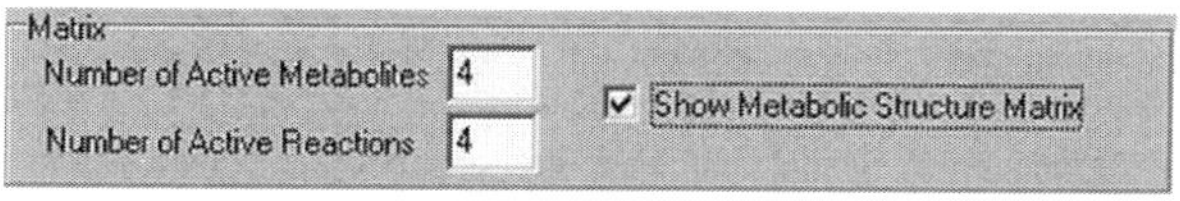


FIGURE 5.7 'Show Metabolic Structure Matrix' check box.

The next step is to complete the table on Auxiliary Form 2 in accordance with the stoichiometric matrix of the system under investigation. In the first row you should enter the names of the metabolites, and in the first column the conventional names of the reactions in the system (figure 5.8). At the next stage, it is necessary to generate the model (system of ODE and corresponding algebraic equations) in accordance with the stoichiometric matrix. To proceed one should use Main Form 1 and press buttons in the following sequence (figure 5.9):

Create Initial: This corresponds to the creation of initial values of system variables and parameters by a random-number generator.

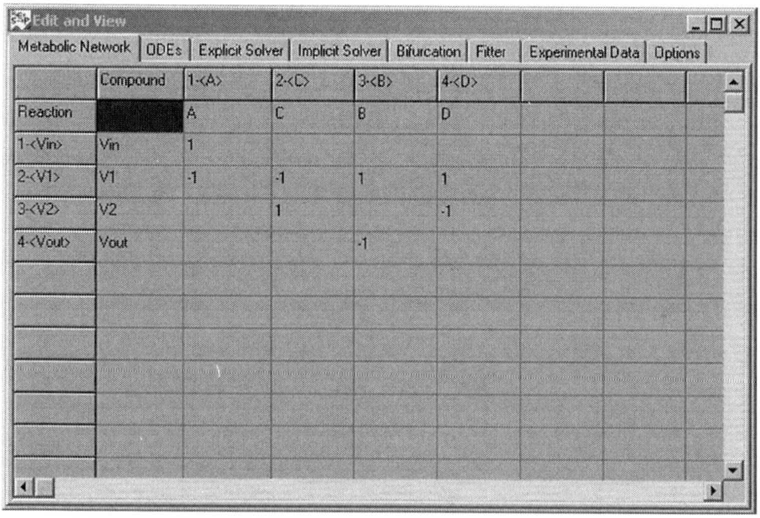


| | Compound | 1-<A> | 2-<C> | 3-<B> | 4-<D> |
|---|---|---|---|---|---|
| Reaction | | A | C | B | D |
| 1-<Vin> | Vin | 1 | | | |
| 2-<V1> | V1 | -1 | -1 | 1 | 1 |
| 3-<V2> | V2 | | 1 | | -1 |
| 4-<Vout> | Vout | | | -1 | |

FIGURE 5.8 Table on Auxiliary Form 2.

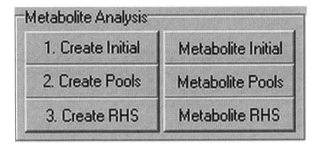

FIGURE 5.9    Main Form 1 Create buttons.

Create Pools: This corresponds to the recalculation of conservation laws (pools) in accordance with the initial values.

Create RHS: This corresponds to creation of the right-hand sides of the differential equations.

This completes the model creation stage.

## Creation of the ODE System Using RCT Format (The Alternative Method)

There is a more convenient input procedure for the large-scale 'Metabolic Network' matrix. The special RCT (ReaCTions) format has been developed. This format allows the automatic generation of the stoichiometric matrix using the reactions of the system. The RCT format is an ASCII text file, where each reaction from the 'Metabolic Network' can be represented with the following syntax:

```
(reaction name): (substance)+...+(substance)=(substance)+...+(substance);
// single-line comment
/*
Multiple line comment
A semicolon is mandatory at the end of the equations.
*/
(reaction name): (substance)+...+(substance)=(substance)+...+(substance);
```

For example, the RCT ASCII text file corresponding to the example considered above would be written as follows:

```
V_in: =A;// in the case of an influx, the left part of the equation is empty
V1: A+C=B+D;
V2: D=C;
V_out: B=;// in the case of an outflux, the right part of the equation is empty
```

An RCT file may be edited in any text editor and must be saved with the RCT extension. The next step is to run the DBSolve tool, File → Import

RCT (ASCII) from the menu and upload the RCT file. DBSolve will then automatically create a new model including the stoichiometric matrix. The next stage of the modelling process is the creation of the ODE system as described in the previous section.

## DBSolve Editors: RHS, Initial Values, Pools

There are three buttons to edit the model automatically generated from the stoichiometric matrix, located on Main Form 1: RHS (right-hand sides), Initial values and Pools.

### RHS Editor

The window that appears after pressing the RHS button allows us to edit the ODE system, rate laws and conservation laws (Pools) as well as any arbitrary algebraic function. The RHS editor is shown in figure 5.10.

There are four sections in the RHS window: Pools, Rate Laws, Differential Equations and Explicit Functions.

In the section 'Pools', the conservation laws can be written down. DBSolve derives the functions for rate laws from the analysis of the stoichiometric matrix. In our example, DBSolve has found only one conservation law $D + C = Pool[1]$. The value of Pool[1] will remain constant during

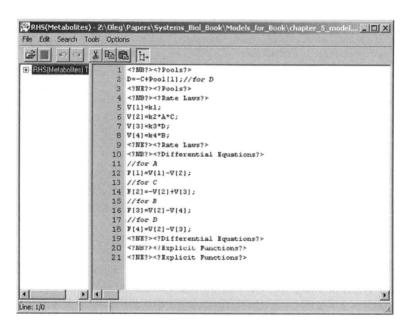

FIGURE 5.10   The RHS editor.

integration and is considered as a parameter for the ODE solver. Concentration D has been chosen as a dependent variable.

In the section 'Rate Laws', the reaction rates as functions of the concentrations of metabolites and kinetic parameters are entered. By default, DBSolve provides the simplest mechanism for all the rates; i.e., irreversible mass action law. If necessary, the function for any reaction rate can be changed. In our example, the reaction, corresponding to the rate $V_{out}$ (V[4] in DBSolve RHS notation) is enzyme catalyzed and reaction corresponding to $V_{in}$ (V[1] in DBSolve RHS notation) is reversible. To take this into account one should manually replace the mass action law expression (V[4]) with the Michaelis-Menten equation and add a term reflecting reversibility in expression for V[1]:

```
V[1]=k1-k_1*A;
V[2]=k2*A*C;
V[3]=k3*D;
V[4]=Vmax*B/(Km+B);
```

In the section 'Differential Equations', the equations are written down. F[i] stands for the corresponding time derivatives for the ith metabolite according to the order in the stoichiometric matrix. The section 'Explicit Function' is designed to manually input various complementary functions and remains empty when a model is generated automatically.

### Initial Values

The initial values window appears by pressing the button of the same name and allows values to be assigned to the parameters of the rate equations and the values of the pools and initial values for concentrations to be set. These initial values will be used to start the process of integration. The parameters are initialised automatically by assigning some random values. These values may be changed to known kinetic constants obtained from the literature or from databases, or one can use the procedure for automatic fitting of kinetic constants (see figure 5.11).

### Pools

This window is designed for viewing and editing conservation laws.

### ODE Tab: Solving the ODE System. Model Integration or *in Silico* Experiments

This window (figure 5.12) allows us to solve the ODE system, to specify parameters of the calculation (figure 5.13), to choose the output variables (figure 5.14) and to visualise time dependencies (figure 5.15). DBSolve uses numerical

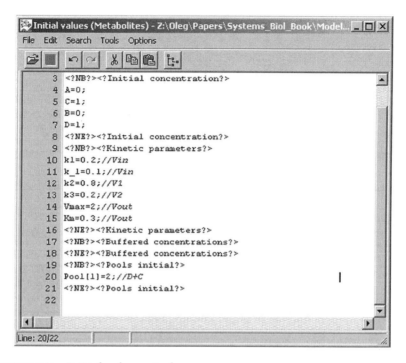

FIGURE 5.11 Initial values window.

FIGURE 5.12 ODE window.

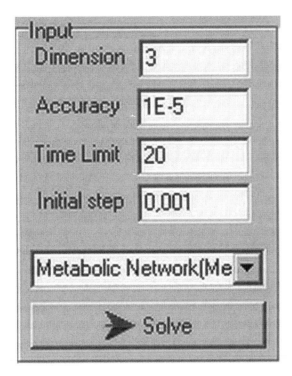

FIGURE 5.13   Description of the parameters for solving ODE system.

procedures for integration of ordinary differential equations (ODEs), or mixed ODE/nonlinear algebraic equations (NAEs), to describe the dynamics of these models. In particular, it has an original implicit integration algorithm, with the Newton prediction-correction procedure at every step of integration with a step size control similar to that published in Gear and Petzold 1984. A popular LSODE algorithm (Hindmarsh 1983) has also been implemented. Both methods have special subroutines for getting output for user-defined time points, which is essential for fitting algorithms.

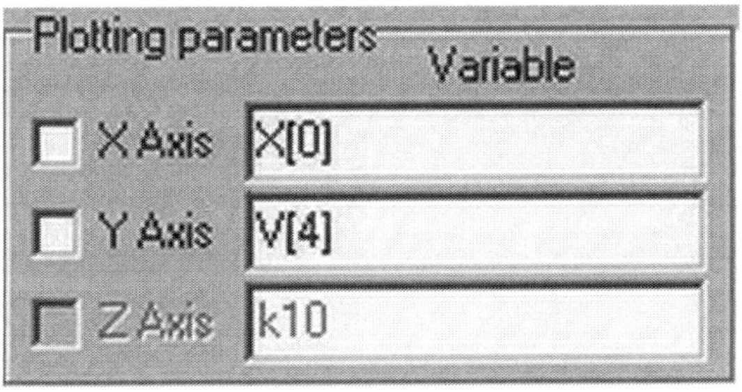

FIGURE 5.14   Parameters for plotting calculated data.

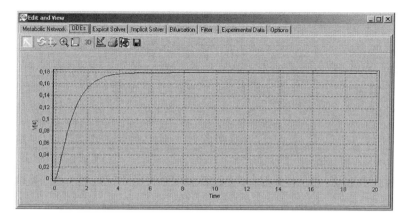

FIGURE 5.15   The dependency of V[4] on X[0].

Using the ODE solver we have calculated the time-dependency of the V[4] rate. The result of the calculation can be seen on the following graph on Auxiliary Form 2. Similarly, one can calculate the dependencies of any concentrations, rates, or complementary functions on time.

By pressing the 'Messages' button the following information appears in the 'Messages' window.

1. Entered initial values for variables;

2. Initial values of the right hand sides;

3. Right-hand sides and variables values on the last time step.

For our example the 'Messages' window looks as follows:

```
Time Limit=20
EPS Relative=1e-05
EPS Absolute=1e-05
Initial Step=0.001
Method=Stiff BDF, chord iterations, full jac
Initial Point
<?NB?><?initials:date:24.04.2008 time:13:18:41?>
A=0.000000e+00;
C=1.000000e+00;
B=0.000000e+00;
D=1.000000e+00;
<?NE?><?initials:date:24.04.2008 time:13:18:41?>
Initial RHS
<?NB?><?initials RHS:date:24.04.2008 time:13:18:41?>
F[1]=2.000000e-01;//A
F[2]=2.000000e-01;//C
F[3]=0.000000e+00;//B
F4=0.000000e+00;//D
```

```
<?NE?><?initials RHS:date:24.04.2008 time:13:18:41?>
V[1]=0.200000
V[2]=0.000000
V[3]=0.200000
V[4]=0.000000
fcn=97 jac=10 step=55
Last Point
A=2.032424e-01;
C=1.104457e+00;
B=2.959365e-02;
D=8.955428e-01;
Last RHS
F[1]=9.775019e-05;//A
F[2]=-4.694430e-04;//C
F[3]=1.451424e-06;//B
F4=0.000000e+00;//D
V[1]=0.179676
V[2]=0.179578
V[3]=0.179109
V[4]=0.179577
```

## Explicit Tabbed Page. Calculating Dependencies Determined Explicitly

Users may have their own particular equations which require solving and wish to be applied to a set of experimental data. DBSolve offers the facility to encode and solve such 'explicitly stated' dependencies. Using this tabbed page (figure 5.16), one can calculate any algebraic function

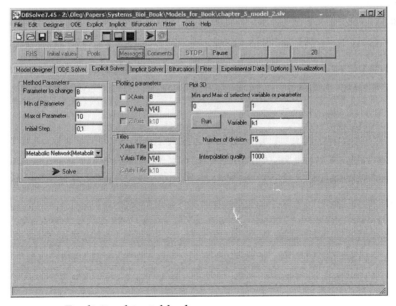

FIGURE 5.16   Explicit solver tabbed page.

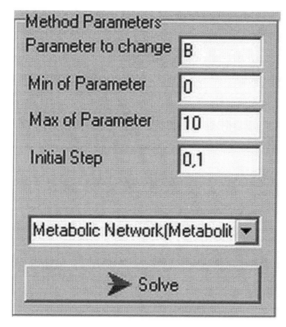

FIGURE 5.17   Description of parameters for calculation.

defined explicitly. For example, let us calculate the dependency of the rate of enzymatic reaction ($V[4]$) on the variable concentration of metabolite B. To do this we should open the Explicit solver and enter parameters for calculation.(figure 5.17).

Please, note that the value B specified in initial values should be assigned to 0, since the initial value of the variable should not exceed the lower limit of the range of B variation (figure 5.18).

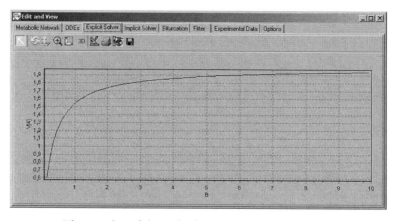

FIGURE 5.18   The results of the calculation.

## The Implicit Solver Tabbed Page. The Study of the System in a Steady State

The method allows the user to trace the changes in the system as a result of changing one or more of its parameters. The Implicit Solver window (figure 5.19) allows us to study the system under steady-state conditions. The tool available on this page enables us to calculate dependencies of steady-state concentrations and rates on parameter values. The tool uses the original parameter continuation procedure. This procedure is very useful for determining any functional dependencies (such as overall steady-state flux, control coefficients, $IC_{50}$) against any external (substrate concentrations) or internal (enzyme concentrations) model parameters. It is especially useful in the case of nonlinear algebraic systems which have no explicit solution or have multiple and unstable solutions.

Let us consider our example and calculate the dependency of the steady-state concentration of A on the value of the parameter $k_1$. To do this one can open the Implicit Solver page and set the following parameters for calculation (figure 5.20).

This method of analysis needs some preliminary remarks. Prior to proceeding with the calculation of the steady-state dependencies, sometimes it is necessary to find out the steady-state concentrations by ODE solver.

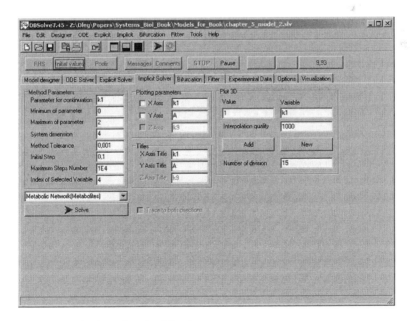

FIGURE 5.19   The 'Implicit' tabbed page.

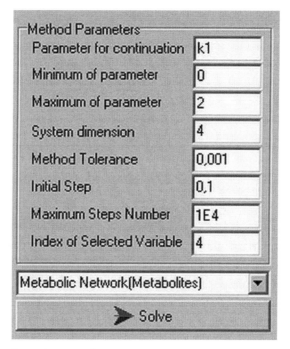

FIGURE 5.20   Description of parameters.

In the initial values the editor enters the value of the parameter in question equal to the value of the minimum of parameter. Then he chooses the ODE tab and, in the 'Time Limit' field, enters a time value sufficient for the system to reach the steady state. On the graph it looks like a curve in saturation. After having built the time dependency in the 'Messages' window, you can see the last calculated point, which is close to the steady state (if the time period was sufficient). These steady-state values of the variables must be copied from the 'Messages' window to the bottom of the 'initial values editor' window. Now the initial values for the concentrations of metabolites correspond to the stationary state for the minimum value of the continuation parameter.

In the 'System Dimension' field of the 'Implicit' page it is necessary to specify the algebraic system dimension. It is usually one equation greater than the ODE system that corresponds to the model (ODE window). Sometimes it is useful to perform this analysis to identify the sensitivity of the steady state to the dimension reduction.

When first filling in the 'Index of Selected Variable' field you should enter the value corresponding to the value from the 'System Dimension' field, because the varied parameter is assigned the greatest number (index).

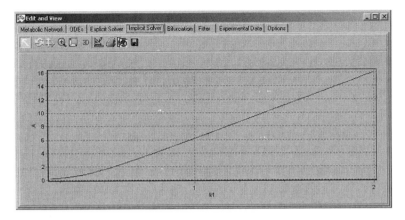

FIGURE 5.21 Implicit solution. Steady-state concentration of A is a non-linear function of the parameter $k_1$.

After completing the settings, press the 'Run' button to start the calculation. The results will be shown as in figure 5.21.

Experimental Data Tab: Creation of the Table with Experimental Data

The tab 'Experimental Data', together with the 'Fitter' tab, is designed to implement the procedure of parameter identification from experiments. The data are entered in the form of a table. For the purpose of parameter identification, any of the above-mentioned methods and corresponding experimental data dependencies (ODE Solver, Explicit Solver and Implicit Solver) may be used.

Let us assume that we have experimental data for the reaction V[4]; for example, the dependency of the rate of the enzyme-catalyzed reaction on time. This type of experimental information can be represented by a system of differential equation corresponding to our model. It means that we should use the ODE Solver to identify the parameters of the model on the basis of available experimental dependencies.

The experimentally measured dependency of V[4] on time (X[0]) is given by 6 six columns of the table (figure 5.22):

X#1: value and name of the argument;

Y#1: value and name of the function;

Weight#1: weight of the experimentally measured points; the parameter appears in the formula of discrepancy (see the example of identification);

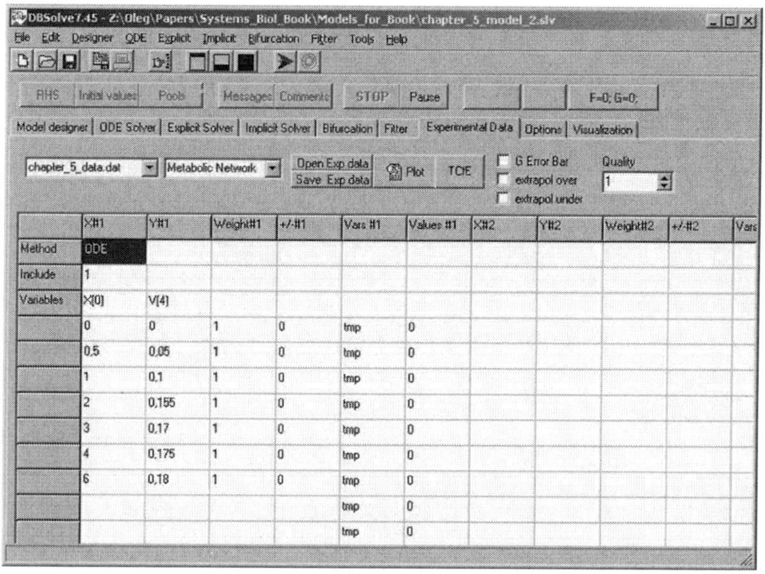

FIGURE 5.22 The experimental data is entered into the table in the 'Experimental Data' tab.

Numbers in the fourth column represent technical information which does not affect the process of identification.

The fifth column (the parameter) and the sixth column (the value) are designed for specifying parameters for different experimental conditions.

In the 'Method' row of the table you should also specify the method to be used for simulation of this set of experimental data (Explicit, Implicit, ODE). The 'Include' row may contain logical values. Specifically, either 0 or 1 depending on whether or not this set is used for fitting parameters. In the 'Variables' row you should specify the name of the function and the name of the argument that describes the given experimental dependency.

To plot one set of the experimental data (figure 5.23) and the corresponding model generated curve you can use the 'Single Plot' tab (figure 5.24).

FIGURE 5.23 Experimental data file.

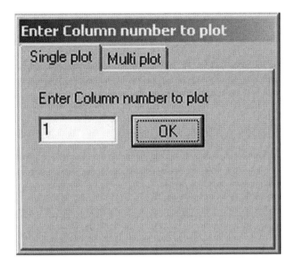

FIGURE 5.24 By pressing 'Plot' button the following two-tabs dialog appears.

To plot these curves you should enter the sequential number of the set in the table and press the OK button.

From figure 5.25 one can observe a substantial deviation of the theoretical calculations from the experimental data. This indicates that the current values of the kinetic parameters for the rate laws (we have changed values of $k_1$ and $k_3$ from the values indicated in previous sections) are not appropriate for this experimental data. To adjust the parameter values

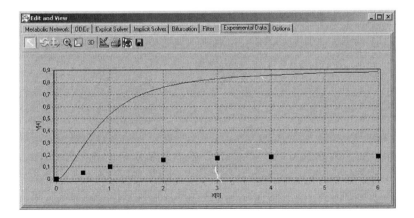

FIGURE 5.25 Experimental data tab. The 'Multi plot' tab is designed to display several sets of experimental data and DBSolve will generate curves on the same graph in accordance with the user's choice.

in order to match the theoretical curve with the experimental data one should carry out the procedure of parameter fitting, described in the next section.

The table with the experimental data should be saved in a separate file with the '.dat' extension. The experimental data file is not included in the 'slv' file of the model, so, when exporting the model to another PC, one should also copy all the experimental data files.

## The Fitter Tabbed Page: Automatic Parameter Fitting

This tabbed page (figure 5.26) is designed for setting up and executing the algorithms that automatically fit the parameters to the experimental data specified on the 'Experimental Data' page. The mechanism of selection of optimal parameters is based on the following. In the first stage, DBSolve calculates the objective function based on the deviation between theoretical and experimental data and a penalty function (to take into account parameter constraints). The parameters specified by the user are changed and another calculation of objective function takes place until the objective function reaches a specified minimum value. Thus, one can find the values of the parameters which allow the model to describe the experimental data in the best possible way. DBSolve allows the user to choose how

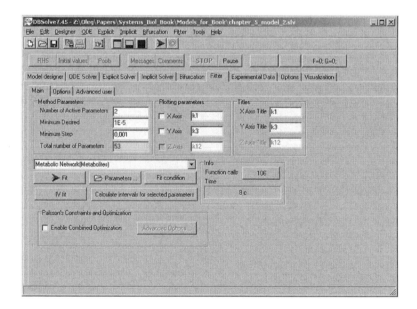

FIGURE 5.26   Fitter main window consists of three tabs.

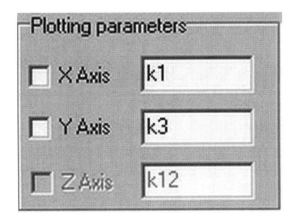

FIGURE 5.27    Parameters for plotting.

the objective function should be calculated and the algorithm to be used. Sometimes the 'best' fit is not easily found. However, to check the quality of the procedure, the standard deviation and confidence intervals for every active parameter as well as ANOVA table are shown in the 'Message' window to help users to make their assessment.

In the next section, 'Plotting Parameters', one can specify parameters which are used for monitoring the fitting process as a graph in Auxiliary Form 2 (figure 5.27).

The parameters (figure 5.28) that are being fitted and the limits they vary between are specified in the table invoked by pressing the 'Parameters...' button. In the first column there are parameter names. In the second, third and fourth columns there are values, defining the current value of

| Parameters for fitter |  |  |  |  |  |  |  |  |  | _|□|x| |
|---|---|---|---|---|---|---|---|---|---|---|
| OK | Steps | Permutation | Number of enabled Parameters | 2 |  |  |  |  |  |  |
| Parameters |  |  |  |  |  |  |  |  |  |  |

| OK | Parameter | Value | Minimum | Maximum | Initial Step | Network | Process ID | Rate Law # | Rate Law # | ON | ▲ |
|---|---|---|---|---|---|---|---|---|---|---|---|
| 1 | k1 | 1 | 0 | 3 | 0,05 | 0 | 0 | 0 | 0 | 1 |  |
| 2 | k3 | 1 | 0 | 5 | 0,05 | 0 | 0 | 0 | 0 | 1 |  |
| 3 | k2 | 0,8 | 0 | 100 | 0,04 | 0 | 0 | 0 | 0 | 0 |  |
| 4 | k4 | 0 | 0 | 100 | 0 | 0 | 0 | 0 | 0 | 0 |  |
| 5 | k5 | 0 | 0 | 100 | 0 | 0 | 0 | 0 | 0 | 0 |  |
| 6 | k6 | 0 | 0 | 100 | 0 | 0 | 0 | 0 | 0 | 0 |  |
| 7 | P[7] | 0 | 0 | 0 | 0 | 0 | 0 | 0 | 0 | 0 |  |
| 8 | P[8] | 0 | 0 | 0 | 0 | 0 | 0 | 0 | 0 | 0 |  |
| 9 | P[9] | 0 | 0 | 0 | 0 | 0 | 0 | 0 | 0 | 0 |  |
| 10 | P[10] | 0 | 0 | 0 | 0 | 0 | 0 | 0 | 0 | 0 |  |
| 11 | P[11] | 0 | 0 | 0 | 0 | 0 | 0 | 0 | 0 | 0 |  |
| 12 | P[12] | 0 | 0 | 0 | 0 | 0 | 0 | 0 | 0 | 0 |  |
| 13 | P[13] | 0 | 0 | 0 | 0 | 0 | 0 | 0 | 0 | 0 | ▼ |

FIGURE 5.28    Parameters for 'Fitter' window.

the parameter and the lower and upper limits. In the 'Initial Step' column the initial step for changing the parameter is defined. In order to define the initial step automatically, press the 'Steps' button located in the upper part of the panel. The rightmost field (ON) is 1 if the parameter is used for the process of fitting and 0 if it is not used. In order to exclude some parameters from the fitting procedure, you need to specify 0 in the 'ON' field and press the 'Permutation' button. At the same time, in the 'Number of Enabled Parameters' field, a new number of used parameters will be displayed, and the table rows will be rearranged accordingly. The rest of the columns do not allow adjustments by the user.

The 'Fit Conditions' button of the 'Fitter' tabbed page invokes the page which allows the user to define special conditions on parameters chosen for fitting, such as inequalities.

### Options Tab

The parameter Lambda is used only for the Marquardt-Levenberg algorithm (figure 5.29). Depending on the items checked in the 'Method' section (figure 5.30), different mathematical functions are used to calculate deviations. Deviation is defined by the following relation:

$$\text{GlobalFmin} = \sum_k \text{Convergency}_k$$

The next two values are not used for fitting and are calculated only for output to 'Messages.'

$$\text{ResidualSS} = \sum_k (v_k - u_k)^2$$

$$\text{TotalSS} = \sum_k (u_k)^2$$

FIGURE 5.29   Selection of algorithm for parameters fitting.

The value $cValue_k$ is an intermediary auxiliary one, necessary for determining the step used in calculation. 'Convergency' and 'cValue' are calculated on the basis of the following cases:

1. If AbsoluteFitting = 0 and LeastFitting = 1,

$$\text{Convergency}_k = weight_k \frac{(v_k - u_k)^2}{v_k^2}$$

2. If AbsoluteFitting = 1 and LeastFitting = 1,

$$\text{Convergency}_k = weight_k * (v_k - u_k)^2 \rightarrow$$

3. If AbsoluteFitting = 0 and LeastFitting = 0,

$$\text{Convergency}_k = weight_k * \frac{|(v_k - u_k)|}{v_k}$$

4. If AbsoluteFitting = 1 and LeastFitting = 0,

$$\text{Convergency}_k = weight_k |v_k - u_k|$$

*Advanced User Tab*

The 'Advanced User' tab is designed to allow advanced users to adjust fitting methods and parameters.

Steady state via ODE: steady state will be determined by integration until Time to steady state (SS);

Use Experimental data: usage of experimental data; otherwise the objective function F[0] will be minimised.

Use ODE's Solver: the option includes use of the ODE system;

Use Implicit Solver: the option includes use of the Implicit method;

See the user manual on the CD for more details.

*Example of Fitting*

As a simple example, let us consider the sequence of operations needed to determine the optimal parameters for our model. As shown above, the model with changed values of $k_1$ and $k_3$, $k_1 = k_3 = 1$ cannot describe the experimental data well (figure 5.25).

To prepare the fitting process one needs to set values in 'Experimental Data' and 'Fitting' pages as described above.

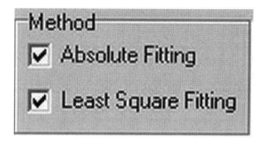

FIGURE 5.30   The 'Method' tab defines the procedure used for discrepancy calculation.

Step 1.   Choose the method for calculation of the objective function as shown on figure 5.30.

Step 2.   Check that the 'Parameters …' window contains the $k_1$ and $k_3$ coefficients of interest and are assigned the correct intervals and step (see figure 5.28).

Step 3.   Press the 'Fit' button to start (figure 5.31)

Step 4.   Observe the process of parameters fitting on the screen. The process of calculation stops either when the 'Stop' button is pressed or when the stop criteria defined in the 'Method' parameters are reached (figure 5.32).

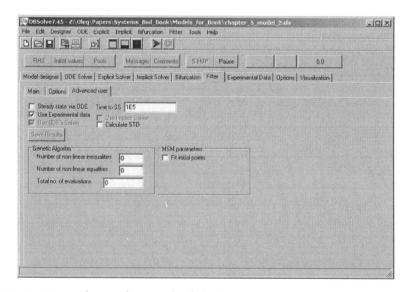

FIGURE 5.31   Advanced user tab of the fitter window.

FIGURE 5.32   Message.

By pressing the 'Messages' button the following information is displayed:

```
Fitting Started: File:chapter_5_data.dat
Number of Active Parameters=2
k1=1.000000e+00;
k3=1.000000e+00;
Step too small #Calls 220
Least square Minimum 8.188729e-06
Fitting Finished: Last call exit code=0
The last observed at call 220 objective function value: 0
Used experimental points #7
Degrees of Freedom (Residual df) 5
Residual SS 1.744230e+01
Total SS 1.836835e+01
Model SS 9.260462e-01
Model mean square 4.630231e-01
Residual mean square 3.488461e+00
F Statistics 1.327299e-01
k1=1.999987e-01; +/-0.000000e+00 CI 1.999987e-01 1.999987e-01
k3=1.994909e-01; +/-0.000000e+00 CI 1.994909e-01 1.994909e-01
P#1-1 V[4]=0.0000 SS=0.000000e+00 0.000000e+00 0.000000e+00 0.000000e+00
0.000000e+00
P#1-2 V[4]=0.0500 SS=1.604970e-08 0.000000e+00 0.000000e+00 4.987331e-02
0.000000e+00
P#1-3 V[4]=0.1000 SS=5.708518e-07 0.000000e+00 0.000000e+00 1.007555e-01
0.000000e+00
P#1-4 V[4]=0.1550 SS=3.197579e-06 0.000000e+00 0.000000e+00 1.532118e-01
0.000000e+00
P#1-5 V[4]=0.1700 SS=8.397432e-07 0.000000e+00 0.000000e+00 1.709164e-01
0.000000e+00
P#1-6 V[4]=0.1750 SS=2.183339e-06 0.000000e+00 0.000000e+00 1.764776e-01
0.000000e+00
P#1-7 V[4]=0.1800 SS=1.381168e-06 0.000000e+00 0.000000e+00 1.788248e-01
0.000000e+00
k1=1.999987e-01;
k3=1.994909e-01;
```

The last two lines contain the new optimised values for the parameters. The new values should be copied to the clipboard and then pasted to the

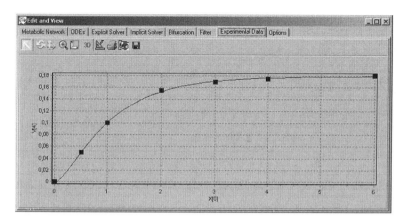

FIGURE 5.33    Fitted curve.

bottom of the initial values editor, in order to assign them as new optimised values for further simulations. One can plot the graph figure 5.33 on the Experimental data page for the new values.

The 'Options' Tabbed Page

The 'Options' tabbed page (figure 5.34, figure 5.35 and figure 5.36) allows additional parameters to be set for visualisation, reporting and debugging. There are three tabs on this page.

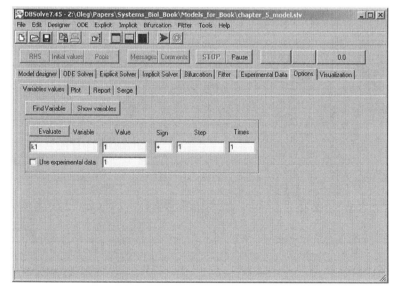

FIGURE 5.34    'Variables' page.

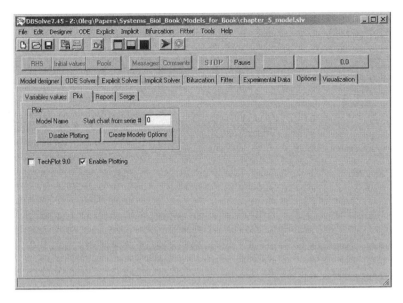

FIGURE 5.35 'Plot' page.

The window in figure 5.34 allows the user to retrieve the value of all model variables and parameters. If the 'Use Experimental Data' field is checked, then the values will be corrected based on the experimental data. This window allows the user to change parameters or variables specified in the previous form and performs a series of calculations. Several curves with different values of the parameter will be plotted. 'Sign' defines the operation on the parameter; 'Step' defines the numerical step for the operation, and 'Times' defines how many curves are to be drawn with the changed parameter. Check the 'Family of curves' check box in the 'Report' page to display them.

The Report page (figure 5.36) allow a user to produce various output files and plots. The option "Start Chart from Series #" (figure 5.35) allows us to draw curves on the graph without erasing the previous ones. By default, the value of this option is 0. It means that plotting of a new graph erases previous ones. By using numbers 1, 2, etc., for each plot all lines can be preserved.

Disable Plotting: this button disables the display of graphs;

Zeros Report: warns of 'division by zero' errors.

Message Window: switches on reports in 'Messages' window.

Change Initial Point: the last values for variables are automatically copied from 'Messages' to 'Initial Values' by pressing the 'Solve' button.

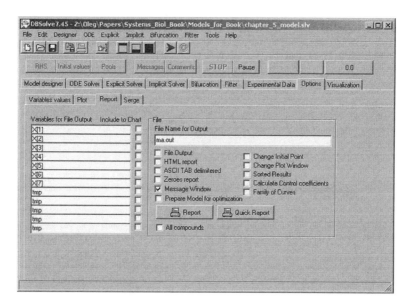

FIGURE 5.36 The 'Report' page is designed for the composition of reports and for storing results of simulations into files.

Change Plot Window: refreshes the page of graph displayed by pressing the 'Solve' button.

Calculate Control Coefficient: automatic calculation of control coefficients.

Family of Curves: switches to plot the family of curves, described in section 'Variables Values Page'. Note: Make sure this option is unchecked when drawing a single curve and when fitting the parameters; otherwise, the data of the calculation may be wrong.

## Some Examples from the CD

In the CD attached the reader can find a great number of examples of DBSolve applications. Below we would like to illustrate some of them, including calculation of trajectories of the ODE system with chaotic attractors and oscillatory behaviour (figure 5.37).

- Modelling of Chaos: file ODE_peroxidase-oxidase_reaction.slv

    Models of peroxidase-oxidase reaction taken from (Fed'kina and Bronnikova 1995) are considered. The model represents chaotic behaviour near the chaotic attractor. The black area with

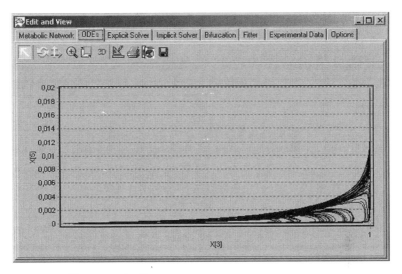

FIGURE 5.37    Phase portrait.

trajectories is expected to be more filled with longer integration times. To run the example, go to the ODE solver page.

- Modelling of damped oscillations: file Damped_oscillations.slv

Modelling of oscillations (Goldstein and Goryanin 1996) for two activities of 6-phosphofructo-2-kinase/fructose-2, 6-bis-phosphatase in rat liver has been performed. Figure 5.38 shows damped oscillation calculated on the basis of the model.

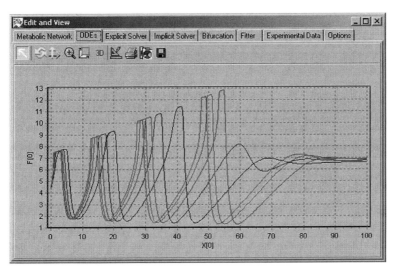

FIGURE 5.38    Damped oscillations.

## CHAPTER 6

# Enzyme Kinetics Modelling

## INTRODUCTION

Cellular metabolism, the integrated interconversion of thousands of metabolic substrates through enzyme-catalyzed biochemical reactions, is the most investigated system of intracellular molecular interactions. Activities of most if not all of the enzymes involved in cellular metabolism are regulated by end products and intermediates of corresponding pathways. This complex network of positive and negative feedbacks as well as genetic regulation of expression level provides flexible adaptation of the metabolic network to fast and slow changes in the external environment. It is the overall dynamic nature of the cell that determines not only its present properties but its future ones as well. The cellular regulatory system is responsible for maintenance of homeostasis and for transitions between different physiological states, so it is extremely important to include regulatory properties (effects) of metabolic pathways into cell models. This can be achieved by using the kinetic modelling approach.

In order for these metabolic regulations to be described properly, models should be constructed for each individual enzyme entered into a system. In this chapter, we describe a strategy to develop these models of individual enzymes. Our approach is based upon utilization of all available experimental data characterizing structural peculiarities and kinetics of the enzyme considered. To demonstrate how different types of experimental data can be incorporated into kinetic models, we present kinetic models of the adenine nucleotide translocator from mitochondria (Metelkin, Goryanin, and Demin 2006) and the following *Escherichia coli* enzymes: histidinol dehydrogenase (Demin et al. 2005), imidazologlycerol-phosphate synthetase (Demin et al. 2004), isocitrate dehydrogenase (Mogilevskaya et al. 2007), isocitrate dehydrogenase kinase/phosphatase (Mogilevskaya et al. 2007), phosphofructokinase-1 (*Peskov, Goryanin, and Demin* 2008) and galactosidase (Metelkin et al. 2008). Development of kinetic models of these enzymes illustrates the basic principles of kinetic description of enzymatic reactions cited above, namely, how kinetic data measured under different conditions (pH, temperature and others) can be amalgamated in order to construct a quantitative description predicting the kinetic behaviour of the enzyme under a range of conditions. The method presented in this chapter allows us to predict the kinetic behaviour of the enzymes under any set of *in vivo* or *in vitro* conditions.

## BASIC PRINCIPLES OF MODELLING OF INDIVIDUAL ENZYMES AND TRANSPORTERS

In the framework of our strategy to make the models scalable and comparable with different kinds of experimental data we have developed both detailed and reduced descriptions for every enzyme. The detailed reaction description includes the exact molecular mechanism of the reaction (i.e., the enzyme catalytic cycle). Usually, the detailed description comprises a set of differential algebraic equations.

The reduced description represents the reaction rates as an explicit analytic function of the substrates and products. To derive the corresponding rate equations from the catalytic cycle, we have used quasi-steady-state and rapid equilibrium approaches. The catalytic cycle of each enzyme is described by linear differential equations. Initially, concentrations of substrates, products and effectors (inhibitors and activators) are assumed to be buffered; i.e., they do not change with time.

In the framework of the approach suggested in this chapter, the level of detailed elaboration of the catalytic cycles of selected enzymes and subsequent derivation of rate equations are fully determined by available experimental data on the structural and functional organization of the enzyme. Indeed, if the catalytic cycle of the enzyme is established and proved experimentally, then we use it to derive the rate equation. If the mechanism underlying enzyme operation is unknown we suggest a 'minimal' catalytic cycle which:

(1) Satisfies all structural and stoichiometric data available from the literature.

(2) Allows us to derive a rate equation describing the available kinetic experimental data.

(3) Is the simplest catalytic cycle of all possible ones satisfying clauses 1 and 2.

Another important feature of the development of the enzyme kinetic description based on *in vitro* data is that kinetic experimental data published in different literature sources are measured under different conditions (pH, temperature). This means that we should construct a catalytic cycle and derive a rate equation which:

(4) Satisfies available experimental data describing dependence of enzyme operation on pH, temperature and other experimental conditions.

(5) Allows the mechanism describing dependence of reaction rate on pH and temperature in the catalytic cycle of the enzyme to be accounted for in the simplest of all possible ways.

To estimate the kinetic parameter values we use the following sources:

1) Literature data on the values of $K_m$, $K_i$, $K_d$, rate constants, pH optimum, etc.;

2) Electronic databases; only a few databases with specific kinetic content are available at the moment, in particular EMP (Selkov et al. 1996) and BRENDA (Shomburg, Chang, and Shomburg 2002);

3) Experimentally measured dependencies of the initial reaction rates on concentrations of substrates, products, inhibitors and activators;

4) Time series data from enzyme kinetics.

## Methods to Derive Rate Equation on the Basis of Enzyme Catalytic Cycle

In this section we present various techniques of derivation of rate equations. These techniques are based on application of different assumptions to describe the function of individual processes of the catalytic cycle and enable us to take into account the diversity of dynamic and regulatory properties of different enzymes. We consider three approaches to derive the rate equation: quasi-equilibrium, quasi-steady-state and combined quasi-equilibrium, quasi-steady-state approaches. The main features of these approaches are exemplified by consideration of various catalytic cycles and derivation of corresponding rate equations.

### Quasi-Equilibrium Approach

This approach can be applied to derive rate equations for the catalytic cycles with the following characteristic features:

(i)  All stages of the catalytic cycle can be subdivided into a group of fast reactions and a group of slow reactions.

(ii)  Fast reactions can be considered at quasi-equilibrium in comparison with slow reactions.

(iii)  All concentrations of enzyme states can be expressed in terms of parameters of the catalytic cycle and substrate/product/effector concentrations on the basis of equilibrium relationships valid for fast reactions.

We exemplify application of the quasi-equilibrium approach by derivation of the reaction rate of the enzyme functioning in accordance with the random bi bi mechanism (Cleland 1963). The catalytic cycle of the enzyme is depicted in figure 6.1.

To derive rate equations which describe the dependence of the reaction rate on parameters of the catalytic cycle and concentrations of the substrates, products and effectors, we assume that the rates of all reactions of the substrate binding and dissociation of the products are significantly higher than the rates of the catalytic reaction designated as 1 in figure 6.1. This means that each of these 'fast' reactions can be considered as a quasi-equilibrium one (the dissociation constants are given near the corresponding reactions in figure 6.1); thus, we can write the following relationships:

$$K_A = \frac{E \cdot A}{E \bullet A}, \quad K_P = \frac{E \cdot P}{E \bullet P}$$

$$K_B = \frac{E \cdot B}{B \bullet E}, \quad K_Q = \frac{E \cdot Q}{Q \bullet E} \tag{6.1}$$

$$K_{AB} = \frac{B \cdot E \bullet A}{B \bullet E \bullet A}, \quad K_{PQ} = \frac{Q \cdot E \bullet P}{Q \bullet E \bullet P}$$

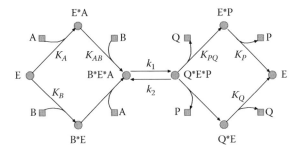

FIGURE 6.1 Catalytic cycle of an enzyme functioning in accordance to random bi bi mechanism.

For concentrations of the enzyme states, the following conservation law is also satisfied:

$$E + E{\bullet}A + B{\bullet}E + B{\bullet}E{\bullet}A + E{\bullet}P + Q{\bullet}E + Q{\bullet}E{\bullet}P = e_0 \qquad (6.2)$$

where $e_0$ is the total concentration of the enzyme. Solving the system of linear (relative to concentrations of the enzyme states) algebraic equations (6.1) and (6.2), we obtain the following expressions for the concentrations of the enzyme states:

$$B{\bullet}E{\bullet}A = \frac{A}{K_A} \frac{B}{K_{AB}} \frac{e_0}{\Delta}, \quad Q{\bullet}E{\bullet}P = \frac{P}{K_P} \frac{Q}{K_{PQ}} \frac{e_0}{\Delta}$$

$$\Delta = 1 + \frac{A}{K_A} + \frac{B}{K_B} + \frac{A}{K_A} \frac{B}{K_{AB}} + \frac{P}{K_P} + \frac{Q}{K_Q} + \frac{P}{K_P} \frac{Q}{K_{PQ}} \qquad (6.3)$$

According to the scheme of the catalytic cycle presented in figure 6.1, the rate equation for the reaction can be written as follows:

$$v = k_1{\bullet}B{\bullet}E{\bullet}A - k_{-1}{\cdot}Q{\bullet}E{\bullet}P \qquad (6.4)$$

Substitution of (6.3) into equation (6.4) gives the following equation for the reaction rate:

$$v = \frac{e_0}{\Delta} \cdot \frac{k_1}{K_A} \frac{1}{K_{AB}} \left( A \cdot B - P \cdot Q \cdot \frac{k_{-1}}{k_1} \frac{K_A}{K_P} \frac{K_{AB}}{K_{PQ}} \right) \qquad (6.5)$$

where $K_A$, $K_B$, $K_P$ and $K_Q$ are the dissociation constants of the substrates A, B and products P, Q from free enzyme; $K_{AB}$ and $K_{PQ}$ are the dissociation constants of the substrates B and Q from the enzyme complex with the substrates A and P, respectively; $k_1$, $k_{-1}$ are the rate constants of the catalytic stage of the enzyme cycle. The dissociation and rate constants are presented in figure 6.1 near the corresponding reactions.

### Quasi-Steady-State Approach

This approach can be applied to derive rate equations for catalytic cycles with the following characteristic features:

(i) No stages of the catalytic cycle can be subdivided into a group of fast reactions and a group of slow reactions.

(ii) All concentrations of enzyme states can be expressed in terms of parameters of the catalytic cycle and substrate/product/effector concentrations on the basis of a steady-state solution of a system of ordinary differential equations describing the catalytic cycle.

We exemplify the application of the quasi-steady-state approach by derivation of the reaction rate of an enzyme functioning in accordance with the ordered uni bi mechanism (Cleland 1963). The catalytic cycle of the enzyme is depicted in figure 6.2 and the dissociation and rate constants are given near the corresponding stages.

To derive equations which describe the dependence of the reaction rate on parameters of the catalytic cycle and concentrations of the substrates, products and effectors, we assume that there are no fast and slow stages in the catalytic cycle but all reaction rates are of the same order of magnitude. This means that the dynamics of the catalytic cycle are described by

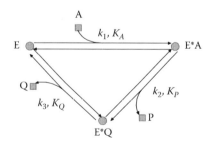

FIGURE 6.2 Catalytic cycle of an enzyme functioning in accordance to ordered uni bi mechanism.

the following system of differential equations:

$$\frac{dE \bullet A}{dt} = v_1 - v_2$$

$$\frac{dE \bullet Q}{dt} = v_2 - v_3 \tag{6.6}$$

$$E + E \bullet A + E \bullet Q = e_0$$

where $e_0$ is the total concentration of the enzyme. Rate equations of individual stages of the catalytic cycle are expressed in the following manner:

$$
\begin{aligned}
v_1 &= k_1 \cdot (E \cdot A / K_A - E \bullet A), \\
v_2 &= k_2 \cdot (E \bullet A - P \cdot E \bullet Q / K_P), \\
v_3 &= k_3 \cdot (E \bullet Q - Q \cdot E / K_Q)
\end{aligned}
\tag{6.7}
$$

Solving the system of (6.6) and (6.7) at steady state, we obtain the following expressions for the concentrations of the enzyme states:

$$E = \frac{e_0}{\Delta} \cdot \left( k_3 \cdot (k_1 + k_2) + k_1 \cdot k_2 \frac{P}{K_P} \right),$$

$$E \bullet A = \frac{e_0}{\Delta} \cdot \left( k_1 \cdot k_3 \frac{A}{K_A} + k_1 \cdot k_2 \frac{A}{K_A} \frac{P}{K_P} + k_2 \cdot k_3 \frac{Q}{K_Q} \frac{P}{K_P} \right) \tag{6.8}$$

$$\Delta = k_3 \cdot (k_1 + k_2) + k_1 \cdot (k_2 + k_3) \frac{A}{K_A} + k_1 \cdot k_2 \frac{P}{K_P} + k_3 \cdot (k_1 + k_2) \frac{Q}{K_Q}$$

$$+ k_1 \cdot k_2 \frac{A}{K_A} \frac{P}{K_P} + k_2 \cdot k_3 \frac{Q}{K_Q} \frac{P}{K_P}$$

Substitution of (6.8) into the first equation of (6.7) gives the following expression for the reaction rate:

$$v = \frac{e_0}{\Delta} \cdot \frac{k_1 \cdot k_2 \cdot k_3}{K_A} \left( A \cdot B - P \cdot Q \cdot \frac{K_A}{K_P} \frac{1}{K_Q} \right) \tag{6.9}$$

where $K_A$, $K_P$ and $K_Q$ are the dissociation constants of the substrate A and products P, Q; $k_i$, $i = 1, 2, 3$, are the rate constants of the corresponding stages of the catalytic cycle.

*Combined Quasi-Equilibrium, Quasi-Steady-State Approach*

This approach can be applied to derive rate equations for catalytic cycles with the following characteristic features:

(i) All stages of the catalytic cycle can be subdivided into a group of fast reactions and a group of slow reactions.

(ii) Fast reactions can be considered at quasi-equilibrium in comparison with slow reactions.

(iii) The initial catalytic cycle of the enzyme can be reduced to a catalytic cycle including slow processes only.

(iv) All concentrations of enzyme states can be expressed in terms of parameters of the catalytic cycle and substrate/product/effector concentrations on the basis of both equilibrium relationships valid for fast reactions and a steady-state solution of a system of ordinary differential equations describing the reduced catalytic cycle.

We exemplify an application of the combined quasi-equilibrium, quasi-steady-state approach by derivation of the reaction rate of an enzyme functioning in accordance with the ping pong bi bi mechanism (Cleland 1963). The catalytic cycle of the enzyme is depicted in figure 6.3 and the dissociation and rate constants are given near the corresponding stages.

The dynamics of the catalytic cycle are described by the following system of differential equations:

$$\frac{dE}{dt} = v_6 - v_3$$

$$\frac{dE \bullet A}{dt} = v_3 - v_1$$

$$\frac{dQ \bullet E}{dt} = v_2 - v_6 \tag{6.10}$$

$$\frac{dE^*}{dt} = v_4 - v_5$$

$$\frac{dE^* \bullet P}{dt} = v_1 - v_4$$

$$\frac{dB \bullet E^*}{dt} = v_5 - v_2$$

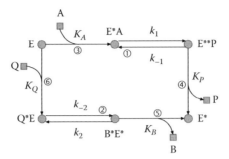

FIGURE 6.3 Catalytic cycle functioning in accordance to ping pong bi bi mechanism.

where rate equations of individual stages of the catalytic cycle are given by the following expressions:

$$v_1 = k_1 \cdot E \bullet A - k_{-1} \cdot E^* \bullet P$$
$$v_2 = k_2 \cdot B \bullet E^* - k_{-2} \cdot Q \bullet E$$
$$v_3 = k_3 \cdot w_3 = k_3 \cdot (E \cdot A / K_A - E \bullet A) \qquad (6.11)$$
$$v_4 = k_4 \cdot w_4 = k_4 \cdot (E^* \bullet P - E^* \cdot P / K_P)$$
$$v_5 = k_5 \cdot w_5 = k_5 \cdot (E^* \cdot B / K_B - B \bullet E^*)$$
$$v_6 = k_6 \cdot w_6 = k_6 \cdot (Q \bullet E - E \cdot Q / K_Q)$$

To derive equations which describe the dependence of the reaction rate on parameters of the catalytic cycle and concentrations of the substrates, products and effectors, we assume that the rates of all reactions of the substrate binding and dissociation of the products are significantly higher than the rates of catalytic reactions designated as 1 and 2 in figure 6.3; i.e., $k_i \gg k_{\pm j}$, $i = 3, 4, 5, 6$ and $j = 1, 2$. This allows us to subdivide all stages of the catalytic cycle into fast (reactions 3, 4, 5, 6) and slow (reactions 1, 2) processes.

Let us transform the system of differential equations (6.10) in such a way that, in the resulting system, we have two differential equations which the right-hand sides consist of linear combinations of the rates of slow reactions only. In order to do this we proceed with the following linear transformations of the ODE system:

(a) Add up the first three differential equations of the system (6.10) and substitute the resulting differential equation for the first equation of the system (6.10).

(b) Add up the last three differential equations of the system (6.10) and substitute the resulting differential equation for the fourth equation of the system (6.10).

The transformed system of differential equations can be presented in the following way:

$$\frac{d(E+Q\bullet E+E\bullet A)}{dt}=v_2-v_1$$

$$\frac{dE\bullet A}{dt}=v_3-v_1$$

$$\frac{dQ\bullet E}{dt}=v_2-v_6$$

$$\frac{d(E^*+E^*\bullet P+B\bullet E^*)}{dt}=v_1-v_2 \qquad (6.12)$$

$$\frac{dE^*\bullet P}{dt}=v_1-v_4$$

$$\frac{dB\bullet E^*}{dt}=v_5-v_2$$

Let X and Y stand for the following sums of concentrations of the enzyme states:

$$X=E+Q\bullet E+E\bullet A$$
$$Y=E^*+E^*\bullet P+B\bullet E^* \qquad (6.13)$$

Substituting relationship (6.11) into system (6.12) and tending $k_i$, $i=$ 3, 4, 5, 6 to infinity, one obtains:

$$\frac{dX}{dt}=v_2-v_1$$

$$\qquad (6.14)$$

$$\frac{dY}{dt}=v_1-v_2$$

$$w_3=0,\ w_4=0,\ w_5=0,\ w_6=0 \qquad (6.15)$$

The system of differential equation (6.14) corresponds to the reduced catalytic cycle depicted in figure 6.4a. To derive the rate equation we should solve the system at steady state. To proceed with this, we should express variables of the catalytic cycle entering rate equations $v_1$ and $v_2$

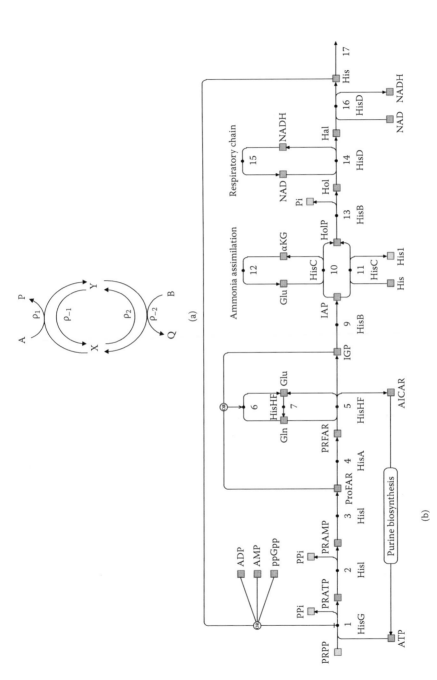

FIGURE 6.4 (a) Reduced catalytic cycle of the enzyme operation in accordance with bi bi ping pong mechanism. (b) Scheme of histidine biosynthesis pathway in *Escherichia coli.*

(E•A, E*•P, B•E*, Q•E) in terms of new variables $X$ and $Y$. Solving the system of the algebraic equations (6.13) and (6.15), one obtains:

$$E \bullet A = \frac{X}{\Delta_1} \cdot \frac{A}{K_A}, \quad Q \bullet E = \frac{X}{\Delta_1} \cdot \frac{Q}{K_Q}$$

$$E^* \bullet P = \frac{Y}{\Delta_2} \cdot \frac{P}{K_P}, \quad B \bullet E^* = \frac{Y}{\Delta_2} \cdot \frac{B}{K_B} \tag{6.16}$$

$$\Delta_1 = 1 + \frac{A}{K_A} + \frac{Q}{K_Q}, \quad \Delta_2 = 1 + \frac{B}{K_B} + \frac{P}{K_P}$$

Substituting expression (6.16) into the rate equations for $v_1$ and $v_2$ (6.11), one arrives at the following rate equations in terms of variables of the reduced catalytic cycle $X$ and $Y$:

$$v_1 = \rho_1 \cdot A \cdot X - \rho_{-1} \cdot P \cdot Y$$
$$v_2 = \rho_2 \cdot B \cdot Y - \rho_{-2} \cdot Q \cdot X \tag{6.17}$$

where the apparent rate constant can be expressed in terms of parameters of the catalytic cycle and substrate/product concentrations in the following manner:

$$\rho_1 = \frac{1}{\Delta_1} \cdot \frac{k_1}{K_A}, \quad \rho_{-1} = \frac{1}{\Delta_2} \cdot \frac{k_{-1}}{K_P}$$

$$\rho_2 = \frac{1}{\Delta_2} \cdot \frac{k_2}{K_B}, \quad \rho_{-2} = \frac{1}{\Delta_1} \cdot \frac{k_{-2}}{K_Q} \tag{6.18}$$

Solving the system of (6.14), (6.17) and (6.18) at steady state, we obtain the following expressions for steady-state concentrations of $X$ and $Y$:

$$X = \frac{e_0 \cdot (\rho_{-1} \cdot P + \rho_2 \cdot B)}{\rho_1 \cdot A + \rho_{-1} \cdot P + \rho_2 \cdot B + \rho_{-2} \cdot Q}, Y = \frac{e_0 \cdot (\rho_1 \cdot A + \rho_{-2} \cdot Q)}{\rho_1 \cdot A + \rho_{-1} \cdot P + \rho_2 \cdot B + \rho_{-2} \cdot Q} \tag{6.19}$$

Substitution of (6.19) into the first equation (6.17) results in the following expression for the reaction rate:

$$v = \frac{e_0 \cdot \dfrac{k_1 \cdot k_2}{K_A \cdot K_B} \left( A \cdot B - P \cdot Q \cdot \dfrac{k_{-1} \cdot k_{-2}}{k_1 \cdot k_2} \dfrac{K_A}{K_P} \dfrac{K_B}{K_Q} \right)}{\left( k_1 \cdot \dfrac{A}{K_A} + k_{-2} \cdot \dfrac{Q}{K_Q} \right) \cdot \left( 1 + \dfrac{B}{K_B} + \dfrac{P}{K_P} \right) + \left( k_2 \cdot \dfrac{B}{K_B} + k_{-1} \cdot \dfrac{P}{K_P} \right) \cdot \left( 1 + \dfrac{A}{K_A} + \dfrac{Q}{K_Q} \right)}$$

$$\tag{6.20}$$

where $K_A$, $K_B$, $K_P$ and $K_Q$ are the dissociation constants of the substrates A, B and products P, Q from enzyme states E and E*; $k_i$, $k_{-i}$, $i = 1, 2$ are the rate constants of catalytic stages of the enzyme cycle.

## How to Express Parameters of the Catalytic Cycle in Terms of Kinetic Parameters

In accordance with the approach described in this chapter, we derive a rate equation for the enzyme on the basis of its catalytic cycle by applying one of the techniques described in the previous sections. The resulting rate equation represents fractionally rational functions of the concentrations of substrates, products, effectors and parameters of the catalytic cycle such as rate and dissociation constants of its individual reactions. However, in enzyme kinetics, rate equations are usually written using parameters which characterize kinetic properties of the enzyme as a whole. The Michaelis constants, the turnover number of the enzyme, and the equilibrium constant are used as such kinetic parameters (Cornish-Bowden 2001). In this section we present a method allowing us to express parameters of the catalytic cycle in terms of kinetic parameters.

Let (6.21) be the rate equation of the enzymatic process, in which parameters of the catalytic cycle (the rate and dissociation constants of certain stages) are the terms:

$$v = [E]_{tot} \cdot f(S_1, \ldots, S_i, \ldots, S_n, P_1, \ldots, P_j, \ldots, P_m, M_1, \ldots, M_h, \ldots, M_q) \quad (6.21)$$

where $[E]_{tot}$ is the total enzyme concentration and $S_i$ $(i = 1, \ldots, n)$, $P_j$ $(j = 1, \ldots, m)$ and $M_h$ $(h = 1, \ldots, q)$ are concentrations of the substrates, products and modifiers (inhibitors and activators), respectively. Using biochemical definitions of conventional parameters of enzymatic kinetics (the Michaelis constants of the substrates and products, the equilibrium constants, the turnover number of enzyme in the forward reaction in the presence and in the absence of activators and inhibitors) lets us find how to express the kinetic parameters via parameters of the catalytic cycle. By definition, the turnover number of the enzyme in the forward reaction is the ratio of the maximal rate of enzyme functioning to the total enzyme concentration at saturating concentrations of all substrates and zero concentrations of all products and modifiers. This means that for calculation of the turnover number of enzyme in the forward reaction, the following expression should be used:

$$k_{cat}^f = \lim_{\substack{S_i \to \infty, i = 1, \ldots, n \\ P_j = 0, j = 1, \ldots, m \\ M_h = 0, h = 1, \ldots, q}} f(S_1, \ldots, S_n, P_1, \ldots, P_m, M_1, \ldots, M_q) \quad (6.22)$$

Analogously, the turnover number of enzyme at the saturating concentration of modifier (inhibitor or activator) $M_r$ can be calculated:

$$k_{cat}^{f,M_r} = \lim_{\substack{S_i \to \infty\, i=1,\ldots,n \\ M_r \to \infty \\ P_j=0,\, j=1,\ldots,m \\ M_h=0,\, h=1,\ldots r-1,\, r+1,\ldots,q}} f(S_1,\ldots,S_n,P_1,\ldots,P_m,M_1,\ldots,M_q) \qquad (6.23)$$

The equilibrium constant can be found from the following expression:

$$K_{eq} = \prod_{j=1}^{m} P_j^{eq} \Big/ \prod_{i=1}^{n} S_i^{eq} \qquad (6.24)$$

where the equilibrium concentrations of the substrates $S_i^{eq}$ ($i = 1,\ldots,n$) and products $P_j^{eq}$ ($j = 1,\ldots,m$) are solutions of the following equation:

$$f(S_1^{eq},\ldots,S_i^{eq},\ldots,S_n^{eq},\, P_1^{eq},\ldots,P_j^{eq},\ldots,P_m^{eq},\, M_1,\ldots,M_h,\ldots,M_q) = 0 \quad (6.25)$$

By definition, the Michaelis constant of the enzyme with respect to a some substrate is the concentration of that substrate at which the rate of the enzymatic process is a half of its maximal rate under the conditions when the products and modifiers (inhibitors and activators) are absent and concentrations of all other substrates are saturating. In accord with this definition, $K_{m,S_t}$, the Michaelis constant of substrate, $S_t$, is a solution of the following equation:

$$\frac{k_{cat}^f}{2} = F_{S_t}(K_{m,S_t}) \qquad (6.26)$$

where

$$F_{S_t}(S_t) = \lim_{\substack{S_i \to \infty\, i=1,\ldots,t-1,\,t+1,\ldots,n \\ P_j=0,\, j=1,\ldots,m \\ M_h=0,\, h=1,\ldots,q}} f(S_1,\ldots,S_n,\, P_1,\ldots,P_m,\, M_1,\ldots,M_q)$$

Analogously, the Michaelis constant of substrate $S_t$ at the saturating concentration of modifier (inhibitor or activator) $M_r$ is a solution of the following equation:

$$\frac{k_{cat}^{f,M_r}}{2} = F_{S_t}^{M_r}\!\left(K_{m,S_t}^{M_r}\right) \qquad (6.27)$$

where

$$F_{S_t}^{M_r}(S_t) = \lim_{\substack{S_i \to \infty\, i=1,\ldots,t-1,t+1,\ldots,n \\ M_r \to \infty \\ P_j=0,\, j=1,\ldots,m \\ M_h=0,\, h=1,\ldots,r-1,r+1,\ldots,q}} f(S_1,\ldots,S_n,P_1,\ldots,P_m,M_1,\ldots,M_q)$$

and the Michaelis constant of product $P_t$ are calculated:

$$\frac{k_{cat}^b}{2} = F_{P_t}(K_{m,P_t}) \tag{6.28}$$

where

$$F_{P_t}(P_t) = \lim_{\substack{P_j \to \infty\, j=1,\ldots,t-1,t+1,\ldots,m \\ S_i=0,\, i=1,\ldots,n \\ M_h=0,\, h=1,\ldots,q}} (-f(S_1,\ldots,S_n,P_1,\ldots,P_m,M_1,\ldots,M_q))$$

In this expression, the turnover number of enzyme in the reverse reaction, $k_{cat}^b$, is calculated as follows:

$$k_{cat}^b = \lim_{\substack{P_j \to \infty\, j=1,\ldots,m \\ S_i=0,\, i=1,\ldots,n \\ M_h=0,\, h=1,\ldots,q}} (-f(S_1,\ldots,S_n,P_1,\ldots,P_m,M_1,\ldots,M_q)) \tag{6.29}$$

## Examples of Rate Equations Expressed in Terms of Kinetic Parameters

In this section we apply methods described in the previous section to express parameters of the catalytic cycles presented in the previous sections in terms of kinetic parameters. Corresponding rate equations in terms of the kinetic parameters are then constructed.

### Random Bi Bi Mechanism

Applying formulae (6.21)–(6.29) to the rate equation (6.5), we derive the following relationships between the parameters of the catalytic cycle and the conventional parameters of enzyme kinetics:

$$k_{cat}^f = k_1$$

$$K_{m,P} = K_P \cdot \frac{K_{PQ}}{K_Q}, \quad K_{eq} = \frac{k_1}{k_{-1}} \cdot \frac{K_P \cdot K_{PQ}}{K_A \cdot K_{AB}}, \quad K_{m,A} = K_A \cdot \frac{K_{AB}}{K_B}, \quad K_{m,Q} = K_{PQ} \tag{6.30}$$

$$K_{m,B} = K_{AB}$$

These relationships allow expression of parameters of the catalytic cycle in terms of kinetic parameters:

$$k_1 = k_{cat}^f$$

$$K_{AB} = K_{m,B}$$

$$K_{PQ} = K_{m,Q} \tag{6.31}$$

$$K_B = K_A \cdot \frac{K_{m,B}}{K_{m,A}}, \quad k_{-1} = \frac{k_{cat}^f \cdot K_P \cdot K_{m,Q}}{K_{eq} \cdot K_A \cdot K_{m,B}}, \quad K_Q = K_P \frac{K_{m,Q}}{K_{m,P}}$$

Substitution of these relationships into equation (6.5) gives the following rate equation in terms of kinetic parameters:

$$v = \frac{e_0 \cdot k_{cat}^f}{\Delta} \cdot (A \cdot B - P \cdot Q / K_{eq}) \tag{6.32}$$

where

$$\Delta = K_A \cdot K_{m,B} + K_{m,B} \cdot A + K_{m,A} \cdot A + A \cdot B + \frac{K_A \cdot K_{m,B}}{K_P} \cdot P$$

$$+ \frac{K_A \cdot K_{m,B}}{K_P} \cdot \frac{K_{m,P}}{K_{m,Q}} \cdot Q + \frac{K_A \cdot K_{m,B}}{K_P \cdot K_{m,Q}} \cdot P \cdot Q$$

*Ordered Uni Bi Mechanism*

Applying formulae (6.21)–(6.29) to the rate equation (6.9), we derive the following relationships between the parameters of the catalytic cycle and the conventional parameters of enzyme kinetics:

$$k_{cat}^f = \frac{k_2 \cdot k_3}{k_2 + k_3}$$

$$K_{eq} = \frac{K_P \cdot K_Q}{K_A}$$

$$k_{cat}^b = k_1$$

$$K_{m,A} = K_A \cdot \frac{k_3 \cdot (k_1 + k_2)}{k_1 \cdot (k_2 + k_3)}$$

$$K_{m,P} = K_P \cdot \frac{k_1 + k_2}{k_2}$$

$$K_{m,Q} = K_Q \cdot \frac{k_1}{k_3} \qquad (6.33)$$

These relationships allow expression of parameters of the catalytic cycle in terms of kinetic parameters:

$$k_1 = k_{cat}^b$$

$$k_2 = \frac{k_{cat}^b \cdot \alpha}{\alpha^2 \cdot \delta - 1}, \quad \alpha = \frac{k_{cat}^b}{k_{cat}^f}, \quad \delta = \frac{K_{m,A} \cdot K_{eq}}{K_{m,P} \cdot K_{m,Q}}, \quad \alpha^2 \cdot \delta > 1$$

$$K_3 = k_{cat}^b \cdot \alpha \cdot \delta$$

$$K_A = \frac{\alpha^2 \cdot K_{m,A} \cdot \delta}{\alpha \cdot (1+\alpha) \cdot \delta - 1} \quad K_P = \frac{\alpha \cdot K_{m,P} \cdot \delta}{\alpha \cdot (1+\alpha) \cdot \delta - 1} \quad K_Q = \alpha \cdot K_{m,Q} \cdot \delta \quad (6.34)$$

Substitution of these relationships into equations (6.9) gives following rate equation in terms of kinetic parameters:

$$v = \frac{e_0 \cdot k_{cat}^f}{\Delta} \cdot (A - P \cdot Q / K_{eq}) \qquad (6.35)$$

where

$$\Delta = K_{m,A} + A + \frac{K_{m,A}}{\alpha \cdot \delta \cdot K_{m,P}} \cdot P + \frac{K_{m,A}}{\alpha \cdot \delta \cdot K_{m,Q}} \cdot Q + \frac{1 + \frac{1}{\alpha} - \frac{1}{\alpha^2 \delta}}{\alpha \cdot K_{m,P}} \cdot A \cdot P$$

$$+ \frac{1}{\alpha \cdot K_{eq}} \cdot P \cdot Q$$

## Ping Pong Bi Bi Mechanism

Applying formulae (6.21)–(6.29) to the rate equation (6.20), we derive the following relationships between the parameters of the catalytic cycle and the conventional parameters of enzyme kinetics:

$$k_{cat}^f = \frac{k_1 \cdot k_2}{k_1 + k_2}$$

$$K_{m,B} = K_B \cdot \frac{k_1}{k_1 + k_2}, \quad K_{eq} = \frac{k_1 \cdot k_2}{k_{-1} \cdot k_{-2}} \frac{K_P \cdot K_Q}{K_A \cdot K_B}, \quad K_{m,A} = K_A \cdot \frac{k_2}{k_1 + k_2}$$

$$K_{m,Q} = K_Q \cdot \frac{k_{-1}}{k_{-1} + k_{-2}} \tag{6.36}$$

These relationships allow expression of parameters of the catalytic cycle in terms of kinetic parameters:

$$k_1 = k_{cat}^f \cdot \frac{K_A}{K_{m,A}}$$

$$k_2 = k_{cat}^f \cdot \frac{K_A}{K_A - K_{m,A}}, \quad K_A > K_{m,A}$$

$$k_{-1} = k_{cat}^f \cdot \frac{K_P}{K_{m,P}} \cdot \phi, \quad \phi = \sqrt{\frac{K_{m,P} \cdot K_{m,Q}}{K_{m,A} \cdot K_{m,B} \cdot K_{eq}}}$$

$$k_{-2} = k_{cat}^f \cdot \frac{K_P}{K_P - K_{m,P}} \cdot \phi, \quad K_P > K_{m,P}$$

$$K_Q = K_{m,Q} \cdot \frac{K_P}{K_P - K_{m,P}} \quad K_B = K_{m,B} \cdot \frac{K_A}{K_A - K_{m,A}} \tag{6.37}$$

Substitution of these relationships into equation (6.20) gives following rate equations in terms of kinetic parameters:

$$v = \frac{e_0 \cdot k_{cat}^f}{\Delta} \cdot (A \cdot B - P \cdot Q / K_{eq}) \tag{6.38}$$

where

$$\Delta = \left( A + \frac{K_{m,A}}{K_{m,Q}} \cdot \phi \cdot Q \right) \cdot \left( K_{m,B} + \frac{K_A - K_{m,A}}{K_A} \cdot B + \frac{K_{m,B}}{K_P} \cdot P \right)$$

$$+ \left( B + \frac{K_{m,B}}{K_{m,P}} \cdot \phi \cdot P \right) \cdot \left( K_{m,A} + \frac{K_{m,A}}{K_A} \cdot A + \frac{K_{m,B}}{K_{m,Q}} \cdot \frac{K_P - K_{m,P}}{K_P} \cdot Q \right)$$

## 'HYPERBOLIC' ENZYMES

'Hyperbolic' enzymes are enzymes that display hyperbolic dependencies of reaction rate on concentration of substrates/products. In this section we describe several examples of hyperbolic enzymes operating in *E. coli* in accordance to different catalytic mechanisms.

### Kinetic Model of Histidinol Dehydrogenase from *Escherichia coli*

To illustrate basic principles of catalytic cycle construction and derivation of the rate equation we consider an enzyme catalyzing two successive reactions of the histidine biosynthesis pathway of *E. coli* (see figure 6.4B). This enzyme (EC 1.1.1.23), encoded by gene *hisD*, is able to catalyze two separate reactions (Umbarger 1996): oxidation of histidinol to histidinal (histidinoldehydrogenase activity):

$$Hol + NAD = Hal + NADH \qquad (6.39)$$

and oxidation of histidinal to histidine (histidinaldehydrogenase activity):

$$Hal + NAD = His + NADH \qquad (6.40)$$

*Available Experimental Data*

To construct the catalytic cycle of histidinol dehydrogenase the following data on structural and functional features of the enzyme were used:

i. When histidinol is used as a substrate of the enzyme, one histidine molecule is formed and two NAD molecules are reduced per one histidinol molecule consumed, but histidinal accumulation is not experimentally detected (Adams 1955ab; Loper and Adams 1965).

ii. When histidinal is used as a substrate of the enzyme, one histidine molecule is formed and one NAD molecule is reduced per one histidinal molecule consumed (Loper and Adams 1965).

iii. The enzyme has one catalytic site (Adams 1955ab; Loper and Adams 1965), in other words, different substrates of the histidinol dehydrogenase and histidinal dehydrogenase reactions compete with each other.

iv. The binding of substrates as well as dissociation of products proceeds in random order (Adams 1955ab; Loper and Adams 1965; Umbarger 1996).

v. The pH optimum of the histidinoldehydrogenase reaction differs from that of the histidinaldehydrogenase reaction by two units of pH (Loper and Adams 1965).

Kinetic properties of the enzyme operating as histidinoldehydrogenase, i.e., catalyzing reaction (6.39) only, have been partly studied in a paper (Loper and Adams 1965) where turnover number and Michaelis constants for Hol and NAD have been estimated at different pH values:

$$K_{m,Hol}^{pH=7.7} = 0.029\,mM,\ pH = 7.7; \quad K_{m,Hol}^{pH=9.3} = 0.012\,mM,\ pH = 9.3$$

$$K_{m,NAD}^{Hol} = 1.53\,mM,\ pH = 7.5; \quad K_{m,NAD}^{Hol} = 1.26\,mM,\ pH = 9.3$$

$$k_{cat,Hol} = 1073\,min^{-1}; \quad pH = 9.4 \tag{6.41}$$

From equation (6.41) it follows that an increase in pH from 7.7 to 9.3 decreases the Michaelis constant for histidinol by almost three times but changes the Michaelis constant for NAD ($K_{m,NAD}^{Hol}$) by less than twenty percentage points; i.e., in the range of experimental error. From this one can conclude that the Michaelis constant for NAD does not depend on pH.

Kinetic properties of the enzyme operating as histidinaldehydrogenase, i.e., catalyzing reaction (6.40) only, have been partly studied in a paper (Loper and Adams 1965) where turnover number and Michaelis constants for Hal and NAD were estimated at different pH values:

$$K_{m,Hal} = 0.0078\,mM,\ pH = 7.7;$$

$$K_{m,NAD}^{Hal} = 0.21\,mM,\ pH = 7.5;$$

$$k_{cat,Hal} = 1073\,min^{-1}; \quad pH = 7.6 \tag{6.42}$$

In addition, dependencies of the maximal activity of histidinoldehydrogenase and histidinaldehydrogenase on pH have been measured in (Loper and Adams 1965) and the time dependence of NADH accumulation in the presence of 2.7 μM of histidinol dehydrogenase at pH 8.9 has been described in (Adams 1955).

## Construction of the Catalytic Cycle

To construct the catalytic cycle of histidinol dehydrogenase (see figure 6.5a) we have used experimental data described in clauses (i)-(iv) of the previous section. We have assumed that the enzyme operation follows the random bi bi mechanism of Cleland's classification (Cleland 1963). This has been

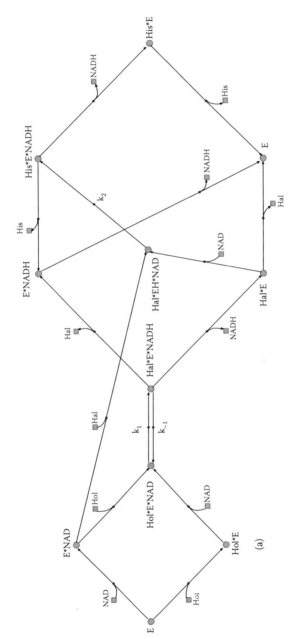

FIGURE 6.5 Catalytic cycle of histidinol dehydrogenase without (a) and with (b) mechanism assigning pH dependence.

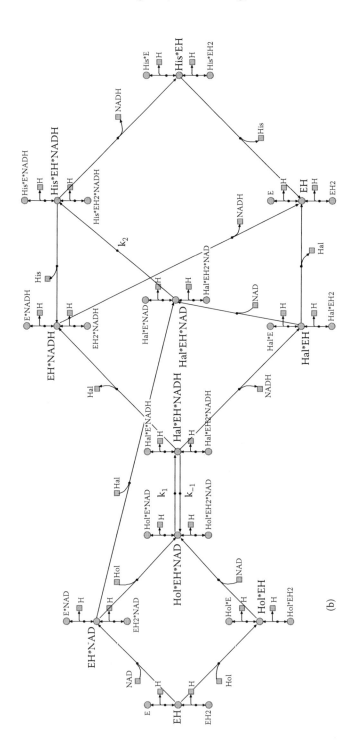

FIGURE 6.5  (Continued).

confirmed by statement (iv) of the previous section. The suggested mechanism has four ternary complexes Hol°E°NAD, Hal°E°NADH, Hal°E°NAD, His°E°NADH. The transition of ternary enzyme-substrate complex Hol°E°NAD to the complex of enzyme with products Hal°E° NADH corresponds to the histidinoldehydrogenase activity of the enzyme but the irreversible (i.e., far-from-equilibrium) transition (Loper and Adams 1965) of ternary complex Hal°E°NAD to complex His°E°NADH corresponds to the histidinaldehydrogenase activity of the enzyme.

In accordance with statement (iii), histidinol should compete with histidinal for the catalytic site of the enzyme. To take this into account, we have introduced an additional reaction of binding of histidinal to the enzyme with NAD bound in the catalytic site (E°NAD → Hal°E°NAD).

If histidinol and NAD are used as initial substrates then in accordance with the suggested catalytic cycle one molecule of histidine is formed and two molecules of NAD are reduced per one histidinol molecule consumed. This stoichiometry corresponds to successive transitions via the following stages of the catalytic cycle: E → E°NAD → Hol°E°NAD → Hal°E°NADH → Hal°E → Hal°E°NAD → His°E°NADH → His°E → E. Note that the histidinal molecule, being an intermediate in the pathway of histidine formation from histidinol, is not released during cycling through these stages of the catalytic cycle but remains in the catalytic site of the enzyme and is oxidized to histidine. This means that it is not possible to detect free histidinal experimentally if the enzyme mainly operates via the selected cycle of stages of catalytic cycle. This corresponds to statement (i) of the previous section. If histidinal and NAD are used as initial substrates, then one molecule of histidine is formed and one molecule of NAD is reduced per one histidinal molecule consumed. This stoichiometry corresponds to successive transitions via the following stages of catalytic cycle: E → E°NAD → Hal°E°NAD → His E°NADH → His°E → E and is in agreement with statement (ii) of the previous section.

The next stage of catalytic cycle construction consists of taking into account the possibility of proton binding to different states of the enzyme. The catalytic cycle (figure 6.5a) should be modified and a rate equation should be derived according to the following experimental observations/constraints on pH:

A. pH optima of histidinoldehydrogenase activity differs from that of histidinaldehydrogenase activity (this is in agreement with clause v of the previous section).

B. The Michaelis constant with respect to histidinol depends on pH.

C. The Michaelis constant with respect to NAD of the histidinoldehydrogenase reaction does not depend on pH.

In order to introduce the pH dependence of enzyme operation and to fit the requirements formulated above, the approach described in Cornish-Bowden (2001) was used.

The enzyme (or, in other words, one or several amino acid residues of the catalytic site directly participating in catalysis) can be deprotonated, singly protonated or doubly protonated. The catalytic cycle has been modified under the assumption that singly protonated enzyme is catalytically active (figure 6.5b.)

*Derivation of Rate Equations*

Using the scheme of the catalytic cycle (figure 6.5b), rates of the histidinoldehydrogenase and histidinaldehydrogenase reactions can be obtained:

$$v_{Hol} = k_1 \cdot Hol°EH°NAD - k_{-1} \cdot Hal°EH°NADH \qquad (6.43)$$

$$v_{Hal} = k_2 \cdot Hal°EH°NAD \qquad (6.44)$$

In order to derive rate equations describing the dependence of rates of the histidinoldehydrogenase and histidinaldehydrogenase reactions on concentrations of substrates, products and effectors we assume that the reactions of substrate binding, product dissociation and enzyme protonation are much faster than the catalytic reactions designated by numbers 1 and 2 in figure 6.5b. Consequently, we consider all fast reactions as quasi-equilibrium and obtain analytical expressions of concentrations of enzyme states (Hol°EH°NAD, Hal°EH°NADH, Hal°EH°NAD) included in the right-hand sides of equations (6.43) and (6.44). This enables us to derive dependencies of the rates of the histidinoldehydrogenase and histidinaldehydrogenase reactions on substrates, products and proton concentrations:

$$v_{Hol} = \frac{HisD}{\Delta} \cdot \left( k_1 \cdot \frac{Hol}{K_{NAD,Hol}} \cdot \frac{NAD}{K_{NAD}} - k_{-1} \cdot \frac{NADH}{K_{Hal,NADH}} \cdot \frac{Hal}{K_{Hal}} \right)$$

$$v_{Hal} = \frac{HisD \cdot k_2 \cdot NAD \cdot Hal/K_{Hal,NAD}/K_{Hal}}{\Delta}$$

$$\Delta = h_E + h_{Hol\circ E} \cdot \frac{Hol}{K_{Hol}} + h_{E\circ NAD} \cdot \frac{NAD}{K_{NAD}} + h_{Hol\circ E\circ NAD} \cdot \frac{Hol}{K_{NAD,Hol}} \cdot \frac{NAD}{K_{NAD}}$$

$$+ h_{Hal\circ E} \cdot \frac{Hal}{K_{Hal}} + h_{Hal\circ E\circ NAD} \cdot \frac{NAD}{K_{Hal,NAD}} \cdot \frac{Hal}{K_{Hal}} + h_{E\circ NADH} \cdot \frac{NADH}{K_{NADH}}$$

$$+ h_{Hal\circ E\circ NADH} \cdot \frac{NADH}{K_{Hak,NADH}} \cdot \frac{Hal}{K_{Hal}} + h_{His\circ E} \cdot \frac{His}{K_{His}}$$

$$+ h_{His\circ E\circ NADH} \cdot \frac{His}{K_{NADH,His}} \cdot \frac{NADH}{K_{NADH}} \tag{6.45}$$

Here the functions $h_X$ define the level of singly protonated enzyme state X, $X \in \{E, E°NAD, Hol°E°NAD \text{ etc.}\}$:

$$h_X = h_X(pH) = 1 + \frac{K_X^1}{H} + \frac{H}{K_X^2} \tag{6.46}$$

Here, H is proton concentration; $K_X^1$ and $K_X^2$ are dissociation constants describing proton dissociation from the singly and doubly protonated enzyme state X, respectively. HisD is the total enzyme concentration; $K_A$ is the dissociation constant of substrate (or product) A from free enzyme; $K_{BA}$ is the dissociation constant of substrate (or product) A from the complex of the enzyme with substrate (or product) B; and $k_1$, $k_{-1}$, $k_2$ are rate constants of catalytic steps of the catalytic cycle. Dissociation constants of substrates/products and rate constants are shown in figure 6.5b near corresponding reactions.

Equation (6.45) involves thirty-two parameters: twelve parameters are rate and dissociation constants describing kinetic properties of individual steps of the catalytic cycle (figure 6.5b), and twenty parameters characterize proton binding to different enzyme states. In accordance with our approach we introduce several assumptions, which do not conflict with experimental data cited above on the one hand and simplify mechanism of pH dependence of enzyme operation on the other hand. This enables us to reduce the number of unknown parameters. Let us assume:

1. Ten enzyme states of catalytic cycle, depicted in figure 6.5a, can be subdivided into three groups:

   Group 1: E, E°NAD, E°NADH

   Group 2: Hol°E, Hol°E°NAD

   Group 3: Hal°E°NADH, Hal°E°NAD, His°E°NADH, Hal°E, His°E

2. Protonation of amino acid residues of catalytic site of enzyme states included in one group are described by identical dissociation constants:

$$h_E = h_{E \circ NAD} = h_{E \circ NADH} = h_0 = 1 + \frac{K_0^1}{H} + \frac{H}{K_0^2} \tag{6.47}$$

$$h_{Hol \circ E} = h_{Hol \circ E \circ NAD} = h_{Hol} = 1 + \frac{K_{Hol}^1}{H} + \frac{H}{K_{Hol}^2} \tag{6.48}$$

$$h_{Hal \circ E} = h_{Hal \circ E \circ NAD} = h_{Hal \circ E \circ NADH} = h_{His \circ E} = h_{His \circ E \circ NADH}$$

$$= h_{Hal,His} = 1 + \frac{K_{Hal,His}^1}{H} + \frac{H}{K_{Hal,His}^2} \tag{6.49}$$

These assumptions allow us to reduce the number of unknown parameters for pH dependence of enzyme operation from twenty to six and to rewrite equation (6.45) as:

$$v_{Hol} = \frac{HisD}{\Delta} \left( k_1 \cdot \frac{Hol}{K_{NAD,Hol}} \cdot \frac{NAD}{K_{NAD}} - k_{-1} \cdot \frac{NADH}{K_{Hal,NADH}} \cdot \frac{Hal}{K_{Hal}} \right)$$

$$v_{Hal} = \frac{HisD \cdot k_2 \cdot NAD \cdot Hal / K_{Hal,NAD} / K_{Hal}}{\Delta}$$

$$\Delta = h_0 \cdot \left( 1 + \frac{NAD}{K_{NAD}} + \frac{NADH}{K_{NADH}} \right) + h_{Hol} \left( \frac{Hol}{K_{Hol}} + \frac{Hol}{K_{NAD,Hol}} \cdot \frac{NAD}{K_{NAD}} \right) + h_{Hal,His}$$

$$\left( \frac{Hal}{K_{Hal}} + \frac{NAD}{K_{Hal,NAD}} \cdot \frac{Hal}{K_{Hal}} + \frac{NADH}{K_{Hal,NADH}} \cdot \frac{Hal}{K_{Hal}} + \frac{His}{K_{His}} + \frac{His}{K_{NADH,His}} \cdot \frac{NADH}{K_{NADH}} \right)$$

$$\tag{6.50}$$

It is easy to see that this simplified mechanism of pH dependence completely satisfies experimentally established facts (A), (B) and (C) cited in the previous section. Moreover, it is not difficult to prove that further simplification of the mechanism of pH dependence (such as subdivision of all enzyme states to two groups instead of three) does not allow us to derive rate equations describing these experimental facts correctly.

So, the suggested mechanism of pH dependence is the minimal one of all possible mechanisms able to describe experimental facts (A), (B) and (C).

## Estimation of Kinetic Parameters of the Rate Equations Using in Vitro Experimental Data

To reduce the number of unknown parameters of equation (6.50) we have used values of kinetic parameters (6.41) and (6.42) measured experimentally. Loper and Adams (1965) have estimated turnover numbers and Michaelis constants with respect to substrates of the histidinoldehydrogenase and histidinaldehydrogenase reactions at different pH values. We have obtained analytical expressions for these kinetic parameters via parameters of the catalytic cycle; i.e., dissociation and rate constants. Then, using these relationships between kinetic parameters and parameters of the catalytic cycle we have obtained an expression for seven parameters in equation (6.50) via experimentally measured kinetic parameters (6.41) and (6.42) and eleven remaining parameters of the catalytic cycle:

$$k_1 = k_{cat,Hol} \cdot h_{Hol}(pH=9.4), \ k_2 = k_{cat,Hal} \cdot h_{Hal,His}(pH=7.6), \ K_{Hal,NAD} = K_{m,NAD}^{Hal}$$

$$K_{NAD} = \frac{K_{Hol} \cdot K_{m,NAD}^{Hol}}{K_{m,Hol}^{pH=9.3}} \cdot \frac{h_0(pH=9.3)}{h_{Hol}(pH=9.3)}, \ K_{NAD,Hol} = K_{m,Hol}^{pH=9.3} \cdot \frac{h_{Hol}(pH=9.3)}{h_0(pH=9.3)}$$

$$K_0^1 = \frac{\dfrac{K_{m,Hol}^{pH=9.3}}{K_{m,Hol}^{pH=7.7}} \cdot \dfrac{h_{Hol}(pH=9.3)}{h_{Hol}(pH=7.7)} \cdot \left(1 + \dfrac{10^{-4.7}}{K_0^2}\right) - \left(1 + \dfrac{10^{-6.3}}{K_0^2}\right)}{\dfrac{1}{10^{-6.3}} - \dfrac{1}{10^{-4.7}} \cdot \dfrac{K_{m,Hol}^{pH=9.3}}{K_{m,Hol}^{pH=7.7}} \cdot \dfrac{h_{Hol}(pH=9.3)}{h_{Hol}(pH=7.7)}},$$

$$K_{Hal} = \frac{K_{Hol} \cdot K_{m,NAD}^{Hol} \cdot K_{m,Hal}}{K_{m,NAD}^{Hal} \cdot K_{m,Hol}^{pH=9.3}} \cdot \frac{h_0(pH=9.3)}{h_{Hol}(pH=9.3)} \cdot \frac{h_{Hal,His}(pH=7.7)}{h_0(pH=7.7)} \quad (6.51)$$

where,

$$h_0(pH) = 1 + \frac{K_0^1}{10^{-3+pH}} + \frac{10^{-3+pH}}{K_0^2}, \ h_{Hol}(pH) = 1 + \frac{K_{Hol}^1}{10^{-3+pH}} + \frac{10^{-3+pH}}{K_{Hol}^2}$$

$$h_{Hal,His}(pH) = 1 + \frac{K_{Hal,His}^1}{10^{-3+pH}} + \frac{10^{-3+pH}}{K_{Hal,His}^2} \quad (6.52)$$

To estimate values of the remaining eleven parameters, experimental data (Loper and Adams 1965) on the pH dependence of the maximal velocity of histidinoldehydrogenase and histidinaldehydrogenase have been used. Data (Adams 1955) on time dependence of NADH accumulation in the presence of 2.7 μM of histidinol dehydrogenase at pH 8.9 has also been used.

pH dependence of the maximal velocity of histidinoldehydrogenase activity of the enzyme normalized to its value at optimal pH is determined:

$$V_{max,norm}^{Hol}(pH) = \frac{1 + 2 \cdot \sqrt{\frac{K_{Hol}^1}{K_{Hol}^2}}}{h_{Hol}(pH)} \tag{6.53}$$

In a similar way, the pH dependence of the maximal velocity of histidinaldehydrogenase activity normalized to its value at optimal pH is determined:

$$V_{max,norm}^{Hal}(pH) = \frac{1 + 2 \cdot \sqrt{\frac{K_{Hal,His}^1}{K_{Hal,His}^2}}}{h_{Hal,His}(pH)} \tag{6.54}$$

Values of the four parameters from equations (6.53) and (6.54) have been chosen so that pH dependencies of the maximal velocities of histidinol dehydrogenase have coincided with experiments measured in Loper and Adams (1965). Figure 6.6 shows pH dependencies of maximal velocities calculated from equations (6.53) and (6.54) (solid lines) and experimentally measured ones (symbols). The values of estimated parameters are listed in table 6.1.

The remaining seven parameters of equation (6.50) have been estimated from experimental data (Adams 1955). The kinetic experiment has been conducted as follows: the reaction was started by adding 20 μM of histidinol to a cuvette containing 500 μM of NAD and 2.7 μM of histidinol dehydrogenase at pH 8.9. Accumulation of NADH was followed with time. The kinetic model was developed to incorporate this experiment (figure 6.7):

$$dNADH/dt = v_{Hol} + v_{Hal}$$
$$NAD + NADH = 500 \text{ μM}$$
$$Hol + Hal + His = 20 \text{ μM}$$
$$NADH + 2 \cdot Hol + Hal = 40 \text{ μM} \tag{6.55}$$

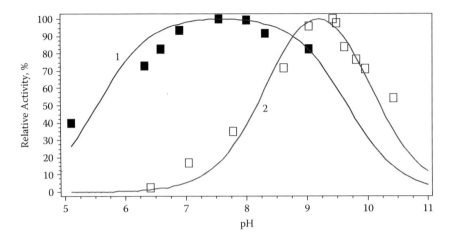

FIGURE 6.6 Dependencies of experimentally measured (symbols) and calculated from model (solid lines) histidinoldehydrogenase (empty squares and line 2) and histidinaldehydrogenase (filled squares and line 1) activities of histidinol dehydrogenase on pH.

TABLE 6.1 Values of Kinetic Parameters of Histidinol Dehydrogenase

| Parameter | Values Resulted from Model Identification Against Experimental Data Published in Adams (1955ab), Loper and Adams (1965) | Parameter | Values Resulted from Model Identification Against Experimental Data Published in Adams (1955ab), Loper and Adams (1965) |
|---|---|---|---|
| $k_1$ | 1073 min$^{-1}$ | $K_{NAD}$ | 47.2 mM |
| $k_{-1}$ | 645 min$^{-1}$ | $K_{Hol}$ | 0.208 mM |
| $k_2$ | 1073 min$^{-1}$ | $K_{NAD,Hol}$ | $6.7 \cdot 10^{-3}$ mM |
| $K_0^1$ | $3.8 \cdot 10^{-7}$ mM | $K_{Hol}$ | 1.8 mM |
| $K_0^2$ | $8.5 \cdot 10^{-7}$ mM | $K_{Hol, NAD}$ | 0.21 mM |
| $K_{Hol}^1$ | $10^{-7}$ mM | $K_{NADH}$ | 0.63 mM |
| $K_{Hol}^2$ | $4.2 \cdot 10^{-6}$ mM | $K_{His}$ | 0.2 mM |
| $K_{Hol, His}^2$ | $2.2 \cdot 10^{-7}$ mM | $K_{Hol,NADH}$ | $2.8 \cdot 10^{-4}$ mM |
| $K_{Hol, His}^2$ | $2.8 \cdot 10^{-3}$ mM | $K_{NADH, His}$ | 2.1 mM |

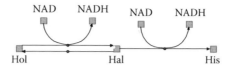

FIGURE 6.7  Kinetic scheme of experiment.

The initial values of model variables corresponded to initial concentrations of substrates in experiment:

$$Hol = 20 \ \mu M,$$
$$NAD = 500 \ \mu M$$
$$NADH = Hal = His = 0 \tag{6.56}$$

Seven unknown parameters for rate equations of the histidinoldehydrogenase and histidinadehydrogenase reactions were chosen so that the time dependence of NADH accumulation resulting from the numerical solution of system (6.55), (6.56) (solid line in figure 6.8) coincided with the corresponding time dependence measured experimentally (symbols in figure 6.8) in Adams (1955ab). Values of estimated parameters are listed in table 6.1.

## Kinetic Model of *Escherichia coli* Isocitrate Dehydrogenase and Its Regulation by Isocitrate Dehydrogenase Kinase/Phosphatase

*E. coli* isocitrate dehydrogenase (IDH; EC 1.1.1.42) encoded by the *icd* gene is a Krebs cycle enzyme that catalyzes isocitrate (iCit) decarboxylation

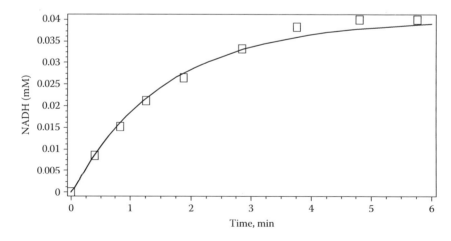

FIGURE 6.8  Experimentally measured (empty squares) and calculated from model (solid line) time dependence of NADH accumulation.

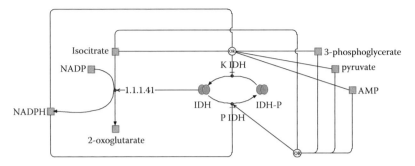

FIGURE 6.9  Regulation of isocitrate dehydrogenase by kinase/phosphatase. IDH and IDHP are the free and the phosphorylated forms of isocitrate dehydrogenase, respectively; K_IDH is isocitrate dehydrogenase kinase; and P_IDH is isocitrate dehydrogenase phosphatase. The dashed arrows with the plus and minus signs indicate the activating and inhibiting influences of metabolites, respectively.

to form 2-ketoglutarate (KG), with concomitant NADP reduction (Neidhardt 1987): iCit + NADP = KG + NADPH + $CO_2$. The activity of the enzyme is regulated by its phosphorylation/dephosphorylation catalyzed by IDH kinase/phosphatase (*aceK* gene; figure 6.9), which is expressed together with glyoxylate shunt enzymes in aerobic *E. coli* growth on acetate. Rerouting of iCit to the glyoxylate shunt prevents complete oxidation of acetate in decarboxylating reactions of the Krebs cycle (Kornberg and Krebs 1957). When IDH is in the inactive phosphorylated state, iCit is consumed by isocitrate lyase of the glyoxylate shunt to supplement the stores of succinate and malate, which are spent in biosynthetic processes. If energy consumption in the cell rises, IDH is dephosphorylated and works actively to consume iCit, thus increasing the energy function of the Krebs cycle. The indices of the energy state of the cell are the metabolites pyruvate, 3-phosphoglycerate and AMP: if their concentrations change, the ratio between the kinase and phosphatase activities of IDH kinase/phosphatase also changes.

In this section we propose catalytic mechanisms for IDH and IDH kinase/phosphatase that are consistent with the data on the structural organization of the active sites of these enzymes (Miller et al. 2000) and agree with the measured kinetic dependences (Dean and Koshland 1993; Miller et al. 2000; Nimmo 1986). Moreover, the mechanism for IDH kinase/phosphatase takes into account the regulation by two groups of effectors: IDH substrates and catabolic intermediates (Miller et al. 2000) pyruvate, 3-phosphoglycerate and AMP. Based on the mechanisms proposed, we derive equations for the rates of reaction of these enzymes and, using published experimental data, determine the kinetic parameters in these equations. The equations are used to predict the dependence of the IDH activity on the energy load on the *E. coli* cell.

### Available Experimental Data

To describe the mechanism of enzyme functioning, we used the published experimental data obtained by an *in vitro* study of isolated *E. coli* isocitrate dehydrogenase (Dean and Koshland 1993; Nimmo 1986) and also the results on the operation of IDH kinase/phosphatase in the presence of its protein substrates (phosphorylated and free IDH) and various effectors, including IDH substrates (Miller et al. 2000). An equation for the enzyme reaction rate was derived according to principles described previously. The equation parameters were chosen to ensure the best agreement between the experimental data and the results of numerically solving the rate equations using the program package DBSolve (see chapter 5 for description).

### Kinetic Model of Isocitrate Dehydrogenase

We propose a rapid equilibrium random bi ter mechanism involving the formation of two dead-end enzyme complexes as the simplest mechanism that is consistent with the data on the enzyme structure and can describe the experimental data published in Dean et al. (1993), Nimmo (1986) and Miller et al. (2000). A scheme of the catalytic cycle is presented in figure 6.10. We assume that the catalytic step (figure 6.10, step 1) is the limiting step, whereas all the steps of binding of substrates and products to the enzyme are in quasi-equilibrium and are characterized by corresponding dissociation constants.

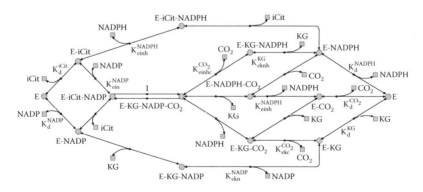

FIGURE 6.10 Scheme of the catalytic cycle of isocitrate dehydrogenase.

Not all dissociation constants are independent; some of them can be expressed through the other dissociation constants. In the scheme (figure 6.10), we indicate only the selected independent dissociation constants that were used in deriving the rate equation. It is seen that nine steps are characterized by dependent dissociation constants. For example, the constant of iCit dissociation from the E–iCit–NADP complex is expressed through independent dissociation constants as:

$$K(E - iCit - NADP \rightarrow E - NADP + iCit) = \frac{K_d^{iCit} * K_{ein}^{NADP}}{K_d^{NADP}}$$

Using the above assumptions, we derive the reaction equation and expressed parameters of the catalytic cycle in terms of kinetic parameters (see section 'Methods to Derive Rate Equation on the Basis of Enzyme Catalytic Cycle'):

$$V_{IDH} = IDH \frac{\frac{k_f}{K_m^{iCit} K_d^{NADP}} (iCit * NADP - NADPH * KG * CO_2 / K_{eq}^{IDH})}{1 + \frac{iCit * K_m^{NADP}}{K_m^{iCit} K_d^{NADP}} + \frac{NADP}{K_d^{NADP}} + \frac{iCit * NADP}{K_m^{iCit} K_d^{NADP}} + \frac{iCit}{K_d^{iCit}} + \frac{NADPH * K_m^{NADP}}{K_m^{iCit} K_d^{NADP} K_{einh}^{NADPH}} + \frac{NADPH * K_{eknh}^{KG} K_m^{CO_2}}{K_m^{KG} K_d^{NADPH} K_d^{CO_2}}}$$

$$+ \frac{CO_2}{K_d^{CO_2}} + \frac{KG * K_m^{NADPH} K_{ekc}^{CO_2}}{K_m^{KG} K_{enhc}^{NADPH} K_d^{CO_2}} + \frac{KG}{K_m^{KG}} \frac{NADPH * K_m^{CO_2}}{K_{enhc}^{NADPH} K_d^{CO_2}} + \frac{NADPH}{K_{enhc}^{NADPH}} \frac{CO_2}{K_d^{CO_2}} + \frac{KG * CO_2 * K_m^{NADPH}}{K_m^{KG} K_{enhc}^{NADPH} K_d^{CO_2}}$$

$$+ \frac{KG * K_m^{NADPH} K_{ekc}^{CO_2}}{K_m^{KG} K_{enhc}^{NADPH} K_d^{CO_2}} \frac{NADP}{K_{ekn}^{NADP}} + \frac{NADPH}{K_{enhc}^{NADPH}} \frac{KG}{K_m^{KG}} \frac{CO_2}{K_d^{CO_2}} \quad (6.57)$$

Here, $IDH$ is the isocitrate dehydrogenase concentration, $K_{eq}$ is the equilibrium constant, and $k_f$ is the enzyme turnover number. $K_m^{iCit}, K_m^{NADP}, K_m^{CO_2}$, $K_m^{KG}$, $K_m^{NADPH}$ are the Michaelis constants for the respective substrates and products, which are combinations of dissociation constants and were obtained by an algorithm described in the section 'Methods to Derive Rate Equation on the Basis of Enzyme Catalytic Cycle' and Demin et al. (2004a). The equilibrium constant and the enzyme turnover number were estimated by Dean and Koshland (1993; table 6.2). The Michaelis constants for iCit and NADP and the dissociation constants for NADPH were found so that rate equation (6.57) best fits the experimental data (Dean and Koshland 1993; figure 6.11a, table 6.2). The parameters $K_{ekn}^{NADP}$, $K_{einh}^{NADPH}$, $K_m^{NADPH}$, $K_{eknh}^{KG}$, $K_d^{CO2}$, $K_{ekc}^{CO2}$, $K_{enhc}^{NADPH}$ were estimated from the experimental data on product inhibition (Nimmo 1986; figure 6.11b, figure 6.11c, table 6.2).

TABLE 6.2 Kinetic Parameters of *E. coli* Isocitrate Dehydrogenase and IDH Kinase/Phosphatase That Are Known from the Literature or Obtained by Identification of the Model Using Experimental Data (Michaelis Constants and Dissociation Constants Are Expressed in mM and Rate Constants Are Expressed in 1/min)

| | Kinetic Parameter Values | |
|---|---|---|
| Enzyme | Experimentally Measured (ref) | Estimated in Model Identification from Experimental Data (ref) |
| Isocitrate dehydrogenase | $K_{eq} = 1000\ mM$ (Dean, Koshland 1993) | $K_m^{iCit} = 0.0059$; $K_d^{NADP} = 0.0013$; $K_m^{NADP} = 0.0227$ (Dean, Koshland 1993) |
| | $k_f = 4830$ (Dean, Koshland 1993) | $K_d^{NADPH} = 1.4e\text{-}4$; $K_d^{iCit} = 3e\text{-}4$ (Dean, Koshland 1993, t = 21°C) |
| | $K_m^{iCit} = 0.011$ (Dean, Koshland 1993) | $K_d^{NADPH} = 0.12$; $K_d^{iCit} = 0.03$ (Miller et al, 2000, t = 37°C) |
| | $K_d^{NADP} = 0.006$ (Dean, Koshland 1993) | $K_{einh}^{NADPH} = 7e\text{-}3$; $K_{eknh}^{KG} = 5.5$; $K_d^{CO2} = 1.6$; $K_{ekc}^{CO2} = 1.6$; |
| | $K_m^{NADP} = 0.017$ (Dean, Koshland 1993) | $K_{ekn}^{NADP} = 1.6e\text{-}4$; $K_m^{NADPH} = 3.6e\text{-}3$; $K_{enhc}^{NADPH} = 0.028$ (Dean, Koshland 1993) |
| | $K_m^{KG} = 0.038$ (Uhr, Thompson, Cleland,1974) | |
| | $K_m^{CO2} = 2.2$ (Uhr, Thompson, Cleland,1974) | |
| IDH kinase | $k_1 = 3640$ (Miller et al, 2000) | $K_{eq} = 0.98$ (7); $K_m^{ATP} = 0.6$; $K_m^{IDH} = 1.2e\text{-}3$ (Miller et al, 2000) |
| | $k_2 = 27{,}000$ (Miller et al, 2000) | $K_m^{ADP} = 0.6$; $K_m^{P} = 0.6$; $K_m^{IDHP} = 0.001$ (Stueland, Gorden., LaPorte, 1988) |
| | | $K_{eq}^1 = 100$; $K_{eq}^2 = 100$ (Stueland, Gorden., LaPorte, 1988) |
| IDH phosphatase | $K_m^{IDHP} = 0.001$ (Stueland et al, 1987) | $K_d^{AMP} = 0.01$; $K_d^{3PG} = 0.41$; $K_d^{Pyr} = 0.16$; $K_{d,IDHP}^{NADPH} = 0.22$ (Miller et al, 2000) |
| | $k_p = 918$ (Miller et al, 2000) | $K_m^{P} = 1.5e\text{-}2$; $K_{eq}^{p} = 10$; $K_m^{IDH} = 0.1$ (Stueland, Gorden., LaPorte, 1988) |

The Michaelis constants for 2-ketoglutarate and $CO_2$ were taken for the enzyme isolated from pig heart mitochondria (Uhr, Thompson, and Cleland 1974) because of the unavailability of the values of these parameters for *E. coli* enzyme (table 6.2).

## Kinetic Model of IDH Kinase/Phosphatase

The functioning of IDH is regulated by its reversible phosphorylation by IDH kinase/phosphatase. The phosphorylation of Ser113 at the active site of the enzyme blocks the binding of iCit to it; i.e., IDH becomes inactive

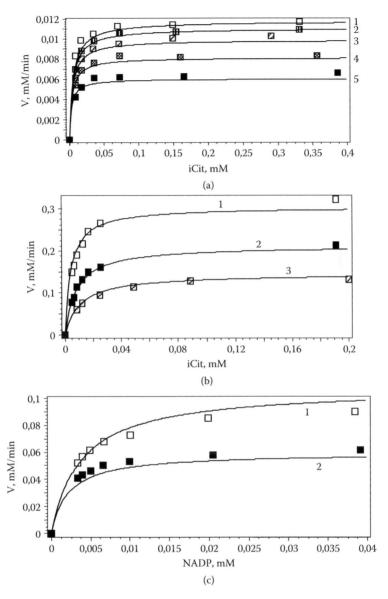

FIGURE 6.11 Dependence of the initial rate of the isocitrate dehydrogenase reaction on the substrate concentration. The points represent the experimental data (a) and (b, c); the lines were calculated by equation (6.1). (a) The NADP concentration is (1) 0.336, (2) 0.168, (3) 0.084, (4) 0.042, and (5) 0.021 mM; T = 21°C; and pH 7.3. (b) The NADP concentration is 0.05 mM; the NADPH concentration is (1) 0, (2) 0.01, and (3) 0.025 mM; T = 25°C; and pH 6.2. (c) The iCit concentration is 0.03 mM; the KG concentration is (1) 1 and (2) 2 mM; T = 25°C; and pH 6.2.

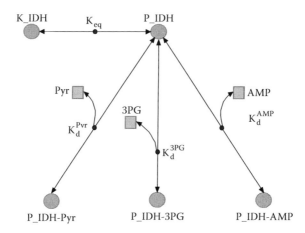

FIGURE 6.12  Regulation of IDH kinase/phosphatase by pyruvate, 3-phosphoglycerate and AMP. K_IDH is IDH kinase; P_IDH is IDH phosphatase; Pyr is pyruvate; 3PG is 3-phosphoglycerate; P_IDH–Pyr, P_IDH–3PG, and P_IDH–AMP are complexes of IDH phosphatase with effectors.

(Miller et al. 2000). The two forms of the regulatory enzyme—the kinase and the phosphatase—are in equilibrium. Pyruvate, 3-phosphoglycerate and AMP shift the equilibrium toward phosphatase (Miller et al. 2000) by binding to the free phosphatase to form complexes that also exhibit phosphatase activity (figure 6.12). The concentrations of these effectors increase with an increase in the energy demand of the cell, which leads to an increase in the fraction of active IDH and, hence, in the Krebs cycle flux. We assume that the steps of effector binding (figure 6.12) are in equilibrium. Then, the concentrations of IDH kinase (K_IDH) and all the enzyme forms that exhibit phosphatase activity (P_IDHtot) are given by the expressions

$$K\_IDH = \frac{KP\_IDH}{K_{eq}\left(1 + \frac{1}{K_{eq}} + \frac{Pyr}{K_d^{Pyr}} + \frac{3PG}{K_d^{3PG}} + \frac{AMP}{K_d^{AMP}}\right)};$$

$$P\_IDH_{tot} = KP\_IDH \frac{1 + \frac{Pyr}{K_d^{Pyr}} + \frac{3PG}{K_d^{3PG}} + \frac{AMP}{K_d^{AMP}}}{\left(1 + \frac{1}{K_{eq}} + \frac{Pyr}{K_d^{Pyr}} + \frac{3PG}{K_d^{3PG}} + \frac{AMP}{K_d^{AMP}}\right)} \qquad (6.58)$$

Here, $KP\_IDH$ is the total concentration of IDH kinase/phosphatase: $KP\_IDH = K\_IDH + P\_IDH + P\_IDH\_Pyr + P\_IDH\_3PG + P\_IDH\_AMP;$

$K_{eq}$ is the equilibrium constant between IDH kinase and IDH phosphatase; and $K_d^{Pyr}$, $K_d^{3PG}$, $K_d^{AMP}$ are the constants of dissociation of effectors from complexes with IDH phosphatase. The values of these parameters were found so that equations (6.58) best fit the experimental data obtained by Miller et al. (2000), who measured the maximal activities of the enzymes in the presence and absence of effectors (table 6.2).

IDH kinase catalyzes the reaction IDH + ATP = ADP + IDHP. The rate equation was derived under the assumption that here the enzyme operates by the random bi bi mechanism (figure 6.13a):

$$V_{K\_IDH} = \frac{K\_IDH * k_1(ATP * IDH - ADP * IDHP / K_{eq}^1)}{K_m^{ATP} K_m^{IDH} + ATP * K_m^{IDH} + IDH * K_m^{ATP} + ATP * IDH + K_m^{ATP} K_m^{IDH}}$$

$$\left( \frac{ADP}{K_m^{ADP}} + \frac{IDHP}{K_m^{IDHP}} + \frac{ADP}{K_m^{ADP}} \frac{IDHP}{K_m^{IDHP}} + \frac{P}{K_m^P} + \frac{ADP}{K_m^{ADP}} \frac{P}{K_m^P} \right)$$

(6.59)

Here, $k_1$ is the enzyme turnover number; $K_{eq}^1$ is the kinase reaction equilibrium constant; and $K_m^{ATP}, K_m^{IDH}, K_m^{ADP}, K_m^{IDHP}, K_m^P$ are the Michaelis constants for the respective substrates. The Michaelis constants for IDH and ATP were found by fitting equation (6.59) to the experimental data (Miller et al. 2000; figure 6.15b; table 6.2). The equilibrium constant and the Michaelis constants for ADP, IDHP and P could not be estimated from these experimental data.

IDH kinase also exhibits ATPase activity (figure 6.13a), described under the assumption that the enzyme functions by the random uni bi mechanism:

$$V_{ATPase} = \frac{K\_IDH * k_2(ATP - ADP * P / K_{eq}^2)}{K_m^{ATP} + ATP + \frac{IDH * K_m^{ATP}}{K_m^{IDH}} + \frac{ATP * IDH}{K_m^{IDH}} + K_m^{ATP}}$$

$$\left( \frac{ADP}{K_m^{ADP}} + \frac{IDHP}{K_m^{IDHP}} + \frac{ADP}{K_m^{ADP}} \frac{IDHP}{K_m^{IDHP}} + \frac{P}{K_m^P} + \frac{ADP}{K_m^{ADP}} \frac{P}{K_m^P} \right)$$

Here, $k_2$ is the turnover number and $K_{eq}^2$ is the equilibrium constant in the ATPase reaction (table 6.2).

IDH phosphatase (P_IDH) dephosphorylates IDHP and converts it into the active state: IDHP = IDH + P. The reaction rate equation was derived

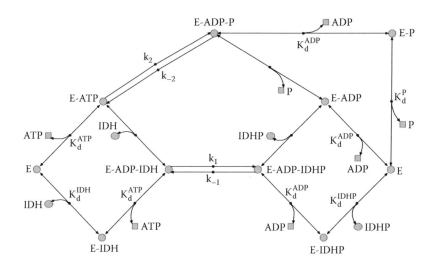

(a)

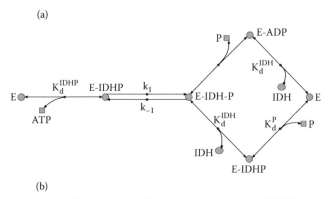

(b)

FIGURE 6.13  Schemes of the catalytic cycles of IDH kinase/phosphatase: (a) kinase and ATPase activities and (b) phosphatase activity. IDH is the dephosphorylated (active) form, IDHP is the phosphorylated (inactive) form and E is IDH kinase/phosphatase.

under the assumption that the enzyme functions in accordance with the random uni bi mechanism (figure 6.13b):

$$V_{P\_IDH} = \frac{P\_IDH * k_p (IDHP - P * IDH / K_{eq}^p)}{K_m^{IDHP} + IDHP + K_m^{IDHP} \frac{IDH}{K_m^{IDH}} + K_m^{IDHP} \frac{P}{K_m^P} + K_m^{IDHP} \frac{IDH}{K_m^{IDH}} \frac{P}{K_m^P}} \quad (6.60)$$

Here, $k_p$ is the phosphatase turnover number; $K_{eq}^p$ is the phosphatase reaction equilibrium constant; and $K_m^{IDH}, K_m^{IDHP}, K_m^P$ are the Michaelis constants for the respective substrates. The Michaelis constant for IDHP is available in the literature (Stueland et al. 1987; table 6.2). The Michaelis constants for the reaction products were not known.

The IDH kinase turnover numbers $k_1$ and $k_2$ and the IDH phosphatase turnover number $k_p$ were found from the experimental data obtained by Miller et al. (2000), who measured all the three maximal activities of IDH kinase/phosphatase in the absence of effectors. Knowing the enzyme concentration and the previously estimated K_IDH/P_IDH equilibrium constant, we calculated the rate constants (table 6.2).

IDH kinase/phosphatase is also subject to regulation by the IDH substrates (Miller et al. 2000) iCit and NADPH, which bind to IDH and IDHP and, thus, decrease the concentrations of substrates accessible for kinase/phosphatase (figure 6.14). In this case, we assume that IDH is protected from phosphorylation only in its complexes with iCit and NADPH, whereas complexes of the enzyme with its other substrates and products (figure 6.10) can be phosphorylated. Assuming all the steps of substrate and product binding to IDH to be quasi-equilibrium (see 'Kinetic Model of Isocitrate Dehydrogenase'), we expressed the concentrations of the enzyme forms that are subject to phosphorylation. Then, the fraction of the enzyme that is subject to phosphorylation is defined as the ratio of the total concentration of these forms of the enzyme to the total concentration of all forms of the enzyme (figure 6.10):

$$
IDH = \frac{IDH_{tot}\left(1 + \frac{NADP}{K_d^{NADP}} + \frac{KG^* K_m^{NADPH} K_{ekc}^{CO_2}}{K_m^{KG} K_{enhc}^{NADPH} K_d^{CO_2}} + \frac{NADPH}{K_{enhc}^{NADPH}} \frac{CO_2}{K_d^{CO_2}} + \frac{KG^* K_m^{NADPH} K_{ekc}^{CO_2}}{K_m^{KG} K_{enhc}^{NADPH} K_d^{CO_2}} \frac{NADP}{K_{ekn}^{NADP}} + \frac{KG^* CO_2^* K_m^{NADPH}}{K_m^{KG} K_{enhc}^{NADPH} K_d^{CO_2}}\right)}{1 + \frac{iCit^* K_m^{NADP}}{K_m^{iCit} K_d^{NADP}} + \frac{NADP}{K_d^{NADP}} + \frac{iCit^* NADP}{K_m^{iCit} K_d^{NADP}} + \frac{iCit}{K_d^{iCit}} \frac{NADPH^* K_m^{NADP}}{K_m^{iCit} K_d^{NADP} K_{einh}^{NADPH}} + \frac{NADPH^* K_{eknh}^{KG} K_m^{CO_2}}{K_m^{KG} K_{enhc}^{NADPH} K_d^{CO_2}}}
$$

$$
+ \frac{CO_2}{K_d^{CO_2}} + \frac{KG^* K_m^{NADPH} K_{ekc}^{CO_2}}{K_m^{KG} K_{enhc}^{NADPH} K_d^{CO_2}} + \frac{KG}{K_m^{KG}} \frac{NADPH^* K_m^{CO_2}}{K_{enhc}^{NADPH} K_d^{CO_2}} + \frac{NADPH}{K_{enhc}^{NADPH}} \frac{CO_2}{K_d^{CO_2}} + \frac{KG^* CO_2^* K_m^{NADPH}}{K_m^{KG} K_{enhc}^{NADPH} K_d^{CO_2}}
$$

$$
+ \frac{KG^* K_m^{NADPH} K_{ekc}^{CO_2}}{K_m^{KG} K_{enhc}^{NADPH} K_d^{CO_2}} \frac{NADP}{K_{ekn}^{NADP}} + \frac{NADPH}{K_{enhc}^{NADPH}} \frac{KG}{K_m^{KG}} \frac{CO_2}{K_d^{CO_2}}
$$

$$
\tag{6.61}
$$

IDH phosphatase affects only the fraction of IDHP that is not bound to NADPH (figure 6.14):

$$
IDHP = \frac{IDHP_{tot}}{1 + \frac{NADPH}{K_{IDHP}^{NADPH}}}
\tag{6.62}
$$

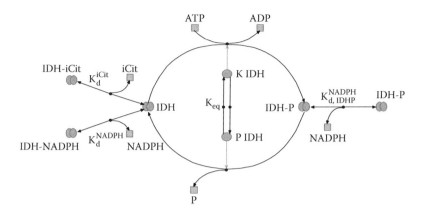

FIGURE 6.14 Scheme of regulation of IDH kinase/phosphatase by isocitrate and NADPH: K_IDH is IDH kinase and P_IDH is IDH phosphatase.

Here, $K_{IDHP}^{NADPH}$ is the NADPH–IDHP dissociation constant. The constants of iCit and NADPH dissociation from IDH were found from the experimental data (Miller et al. 2000; figure 6.15a and figure 6.15b). The constant of dissociation of NADPH from IDHP was also determined from the experimental data (Miller et al. 2000; figure 6.15c). One can see that the values of the constants of dissociation of iCit and NADPH from the complex with IDH found in different experiments (Dean and Koshland 1993; Miller et al. 2000; Nimmo 1986) differ significantly (table 6.2). These differences could be due to differences in experimental conditions, namely, temperature: 21°C versus 37°C (Miller et al. 2000). Hence, we can find the enthalpy of binding of iCit and NADPH to IDH from the van't Hoff equation:

$$\Delta H = -\frac{RT_1T_2}{T_1 - T_2} \ln \frac{K_d(T_1)}{K_d(T_2)}$$

From the two values of the dissociation constants for iCit, we have

$$\Delta H = -51.5 \, \text{kcal/mol}$$

From the two values of the dissociation constants for NADPH, we have

$$\Delta H = -76.2 \, \text{kcal/mol}$$

Further, we studied the dependence of the degree of IDH phosphorylation on the concentrations of various effectors. For this purpose, we

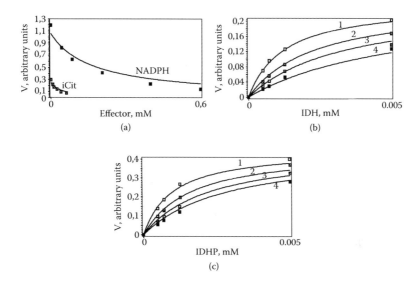

FIGURE 6.15 Dependence of the initial reaction rate for (a, b) IDH kinase and (c) IDH phosphatase on the substrate concentration. The points represent the experimental data; the lines were calculated by equations (6.3) and (6.4). (a) The IDH concentration is 0.0005 mM and the ATP concentration is 0.05 mM. (b) The ATP concentration is 0.05 mM and the iCit concentration is (1) 0, (2) 0.025, (3) 0.05, and (4) 0.1 mM. (c) The NADPH concentration is (1) 0, (2) 0.14, (3) 0.28, and (4) 0.42 mM.

constructed a general model of IDH regulation (figure 6.9) by including the rate equations (6.59) and (6.60) and the parameter values found. Thus, our model is described by a system of ordinary differential equations:

$$\frac{dIDH_{tot}}{dt} = V_{P\_IDH} - V_{K\_IDH}$$

$$\frac{dIDHP_{tot}}{dt} = V_{K\_IDH} - V_{P\_IDH} \tag{6.63}$$

Here, $IDH_{tot}$ and $IDPH_{tot}$ are model variables, $V_{P\_IDH}$ is the IDH phosphatase reaction rate described by equation (6.60), and $V_{K\_IDH}$ is the IDH kinase reaction rate described by equation (6.59), where the IDH kinase and IDH phosphatase reaction rate equations involve the fractions of $IDH_{tot}$ and $IDPH_{tot}$ that are determined by equations (6.61) and (6.62), respectively. The concentrations of the effectors—iCit, NADPH, pyruvate, 3-phosphoglycerate and AMP—are model parameters. It is known from the literature that, in *E. coli* cell growth on acetate, the ratio between the phosphorylated and free forms of IDH is 0.75 (Stueland, Gorden, and LaPorte 1988). The concentrations of the metabolites that are parameters of our model in the *E. coli* cell growing on acetate are also available in the literature (table 6.3). Obviously, these data are insufficient to uniquely identify the eight other parameters that remain undetermined. In view of the lack of other experimental data, we searched for a possible set of values of the undetermined parameters of IDH kinase/phosphatase within physiological ranges. Three parameters were chosen as follows.

The Michaelis constants of IDH kinase for ADP and P were chosen to be equal to the Michaelis constant of this enzyme for ATP, and the equilibrium constant $K_{eq}^2$ of the ATPase reaction of IDH kinase was taken to be equal to the equilibrium constant $K_{eq}^1$ of the kinase reaction (table 6.2). The values of the other five parameters were chosen so that the steady-state ratio IDHP/IDH = 0.75 (table 6.2). These parameters were checked for sensitivity and proved to be sensitive; i.e., even an insignificant change in each parameter led to a shift of the steady-state ratio between the phosphorylated and free IDH forms.

## Model Predictions

Using the model, we studied the effect of the described effectors on IDH phosphorylation. We investigated the dependence of the fraction of phosphorylated IDH (i.e., the IDHP/(IDH + IDHP) ratio) on the concentration

TABLE 6.3 Published Values of the Metabolite Concentrations in the *E. coli* Cell Growing on Acetate under Aerobic Conditions

| Metabolite | Concentration (mM) |
| --- | --- |
| iCit | 0.13 (Lowry et al, 1971) |
| NADP$_{tot}$ | 0.05 (Walsh , Koshland, 1984) |
| NADPH/NADP | 1 (Andersen, von Meyenburg, 1977) |
| KG | 0.07 (Lowry et al, 1971) |
| 3PG | 2.5 (LaPorte, Koshland, 1983) |
| Pyr | 0.01 (Lowry et al, 1971) |
| AMP | 0.29 (Lowry et al, 1971) |
| P | 5(HYPERLINK "http://redpoll.pharmacy.ualberta.ca/CCDB/cgi-bin/STAT_NEW.cgi" ) |
| ATP | 1.45 (Lowry et al, 1971) |
| ADP | 0.54 (Lowry et al, 1971) |

of each effector, with the concentrations of the other effectors fixed (figure 6.16). The most noticeable effect is produced by AMP: an increase in the AMP concentration causes a decrease in the IDHP fraction. With an increase in the NADPH concentration, the fraction of IDHP also decreases: NADPH can bind to both forms of the IDH kinase/phosphatase protein substrate, and the inhibition of IDH kinase is stronger. It will be seen that we tracked the increase in the NADPH concentration only to 0.05 mM. This is because the NADP and NADPH concentrations in our model constitute a common pool and are related by the condition NADP + NADPH = 0.05 (the NADP pool in *E. coli*; Walsh and Koshland 1984; figure 6.16). The second strongest effect on the ratio between the two IDH forms is exerted by iCit, which binds to IDH (an IDH kinase substrate and an IDH phosphatase product), inhibits IDH kinase and activates IDH phosphatase. Thus, with an increase in the iCit concentration, the concentration of active IDH increases (figure 6.16). Pyruvate and 3-phosphoglycerate increase the phosphatase activity of IDH kinase/phosphatase, which also leads to a decrease in the fraction of phosphorylated IDH.

Then, we showed how, with a shift in the energy and biosynthetic demands of the *E. coli* cell, the distribution of fluxes between the glyoxylate shunt and IDH changes because of a shift in the degree of IDH phosphorylation. The energy demand of the *E. coli* cell is characterized by the concentrations of ATP, ADP and AMP: with an increase in the energy consumption, the ATP concentration decreases and the ADP and AMP concentrations increase, and vice versa. The biosynthetic load on the *E. coli* cell may vary with the growth conditions and the composition

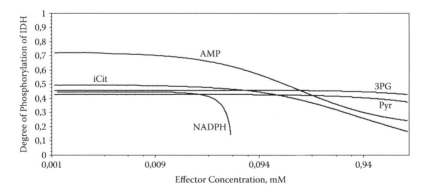

FIGURE 6.16 Dependence of the degree of isocitrate dehydrogenase phosphorylation on the effector concentrations.

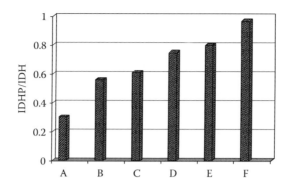

FIGURE 6.17 Degree of isocitrate dehydrogenase phosphorylation under various physiological conditions in the *E. coli* cell: (A) Increase in energy consumption ([AMP] = 1 mM, [ADP] = 1 mM, [ATP] = 0.3 mM), (B) Decrease in biosynthetic load ([NADPH] = 0.04 mM), (C) Addition of glucose ([Pyr] = 3 mM, [3PG] = 3 mM, [ATP] = 2 mM, [ADP] = 0.86 mM, [AMP] = 0.4 mM, [iCit] = 0.09 mM, and [KG] = 0.1 mM [15]), (D) Metabolites in *E. coli* growth on acetate (see metabolite concentrations in table 6.2), (E) Increase in biosynthetic load ([NADPH] = 0.01 mM), and (F) Decrease in energy consumption ([AMP] = 0.24 mM, [ADP] = 0.3 mM, [ATP] = 1.75 mM).

of the medium. Reactions of amino acid biosynthesis consume NADPH; therefore, to describe changes in the biosynthetic demand of the cell, we used this index. Figure 6.17 presents the histogram that characterizes the change in the degree of IDH phosphorylation with a change in the concentrations of a number of metabolites (model parameters). Initially, on acetate at published parameters values (table 6.3), the ratio between the phosphorylated and free IDH forms is 0.75 (figure 6.17D). With an increase in the energy consumption by the cell (when the ADP and AMP concentrations increase and the ATP concentration decreases), the degree of IDH phosphorylation decreases (figure 6.17A), which leads to an increase in the flux through IDH; i.e., in the energy function of the Krebs cycle. If the biosynthetic load on the *E. coli* cell (characterized by the NADPH concentration) decreases, the degree of IDH phosphorylation decreases (figure 6.17B), which decreases the flux through the glyoxylate shunt. After addition of glucose to the medium, in the *E. coli* cell, the concentrations of pyruvate, 3-phosphoglycerate, ATP, ADP and AMP increase and the iCit concentration decreases (Lowry et al. 1971); in this case, the degree of IDH phosphorylation decreases (figure 6.17C). That is, the energy function of the Krebs cycle increases since, under these conditions, it is unnecessary to synthesize glucose from Krebs cycle metabolites. With an increase in the biosynthetic demand of the *E. coli* cell, the NADPH concentration decreases, which causes an increase in the degree of IDH phosphorylation above the initial value (figure 6.17E); i.e., the glyoxylate shunt flux rises and the biosynthetic function of the Krebs cycle increases. With a decrease in the energy consumption by the *E. coli* cell (when the ATP concentration increases and the ADP and AMP concentrations decrease), the degree of IDH phosphorylation increases (figure 6.17F), which characterizes a decrease in the energy function of the Krebs cycle.

## Kinetic Model of β-Galactosidase from *Escherichia coli* Cells

β-galactosidase (EC 3.2.1.23) is an important enzyme of *E. coli* involved in sugar utilization. Together with lac-permease, this protein is encoded in the region of the lac-operon, which is the most popular model system for the study of transcription regulation in prokaryotes.

The enzyme has a complex catalytic cycle and catalyzes several reactions in a single catalytic site. As has been shown previously, the main catalytic activity of β-galactosidase following addition of lactose to the medium is hydrolysis of this substrate with the formation of glucose and galactose monosaccharides (Huber et al. 1976), as well as isomerization with the formation of allolactose (Burstein et al. 1965, Jobe and Bourgeois 1972). This work (Huber, Kurz, and Wallenfels 1976) has also shown that a number of tri- and tetrasaccharides appear in the medium.

In this section we have constructed a kinetic model of β-galactosidase, which describes both hydrolysis and transgalactosidase activity of the enzyme. On the basis of experimental data available, kinetic parameters of the model have been found. Using the constructed model, we have attempted to find a correlation between different enzyme activities under different conditions.

### Catalytic Cycle of β-Galactosidase Construction

The catalytic cycle of *E. coli* β-galactosidase (figure 6.18) was constructed based on the scheme proposed in Huber, Kurz, and Wallenfels (1976), extended by the addition of trisaccharide formation stages. The scheme

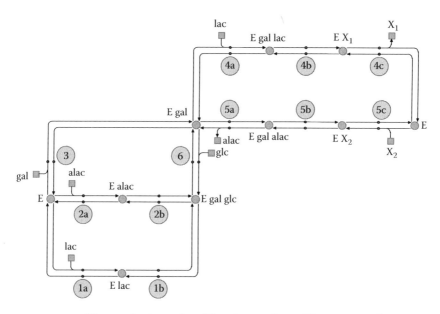

FIGURE 6.18 The catalytic cycle of β-galactosidase. The enzyme forms are at the nodes of the scheme. Every reaction is accompanied with the number of elementary stage. (Detailed description is in the text.)

describes both hydrolase and transgalactosidase activities of β-galactosidase. Stages 1a and 2a show the binding of lactose (*lac*) and allolactose (*alac*), respectively, with the catalytic center of free enzyme (*E*) with the formation of enzyme-substrate complexes (*E_lac* and *E_alac*, respectively). Stages 1b and 2b describe decomposition of disaccharides to glucose and galactose in the catalytic center of the enzyme, which results in the formation of a ternary complex (*E_gal_glc*). According to the literature data (Huber Gaunt, and Hurlburt 1984; Huber, Kurz, and Wallenfels 1976), the process of glucose (*glc*) and galactose (*gal*) dissociation occurs in succession (Stages 6, 3), via an intermediate enzyme form bound with galactose (*E_gal*). It follows from the scheme that the processes of lactose and allolactose hydrolysis are described by stages 1a-1b-6-3 and 2a-2b-6-3, and the transgalactosidase reaction passes through stages 1a-1b-2b-2a.

The most active of the reactions of oligosaccharide formation is galactosylation of lactose (4a-4b-4c) and allolactose (5a-5b-5c) as the most available substrates in a cell (Huber, Kurz, and Wallenfels 1976). It is supposed that, by analogy with glucose galactosylation (stages 6-1b-1a), these processes are realized through binding of lactose (stage 4a) or allolactose (stage 5a) to the enzyme form $E\_gal$ with the formation of trisaccharides $X_1$ and $X_2$ (specifying two types of trisaccharides: $gal + lac = X_1$, $gal + alac = X_2$).

*Derivation of the Rate Equation of β-Galactosidase*

To simplify derivation and analysis of the equation for the rate of beta-galactosidase functioning, each process of trisaccharide formation (4a-4b-4c and 5a-5b-5c) was described by a single integral stage, neglecting intermediate enzyme forms (figure 6.19). It was an approximation, because at present rather little is known about rate constants of these elementary stages, and the lack of kinetic data prevented us from assessing the contribution of each of them.

The simplified scheme of the catalytic cycle for the enzyme is given in figure 6.19. The corresponding rate as well as the rate or dissociation constant is indicated near each arrow.

With approximation of quasi stationary concentrations (Cornish-Bowden 2001), and entering a condition of the stationary for all the forms

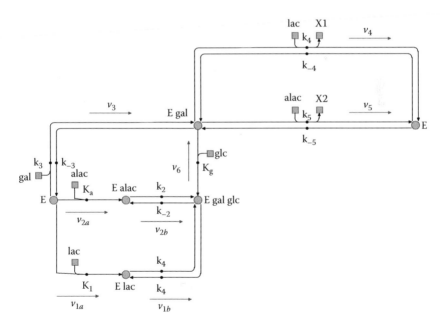

FIGURE 6.19 The reduced catalytic cycle for derivation of the rate equation of β-galactosidase. Every stage is accompanied with the rate constant or with the equilibrium constant. The additional arrows show the rate directions and its value. (Detailed description is in the text.)

of the enzyme, the following system of equations is obtained:

$$\begin{cases} (E\_lac)' = v_{1a} - v_{1b} = 0, \\ (E\_alac)' = v_{2a} - v_{2b} = 0, \\ (E\_gal\_glc)' = v_{1b} + v_{2b} - v_6 = 0, \\ (E)' = v_4 + v_5 - v_{1a} - v_{2a} - v_3 = 0, \\ (E\_gal)' = v_3 + v_6 - v_4 - v_5 = 0, \end{cases} \qquad (6.64)$$

Solving this system with respect to the rate of reactions, we find that in a stationary (steady) state the rates of the separate stages of the catalytic cycle correlate with each other as follows:

$$\begin{cases} v_{1a} = v_{1b}, \\ v_{2a} = v_{2b}, \\ v_6 = v_{1b} + v_{2b}, \\ v_3 = -v_{1b} - v_{2b} + v_4 + v_5. \end{cases} \qquad (6.65)$$

Thus, it turns out that all rates of elementary stages can be expressed as four velocities $v_{1b}$, $v_{2b}$, $v_4$ and $v_5$, hereinafter called 'independent'.

To derive the equation of the velocity of β-galactosidase, we write down the velocities of the elementary stages of the catalytic cycle using the mass action law (Cornish-Bowden 2001).

$$\begin{cases} v_{1b} = k_1 \cdot E\_lac - k_{-1} \cdot E\_gal\_glc, \\ v_{2b} = k_2 \cdot E\_alac - k_{-2} \cdot E\_gal\_glc, \\ v_3 = k_3 \cdot E \cdot gal - k_{-3} \cdot E\_gal, \\ v_4 = k_4 \cdot E\_gal \cdot lac - k_{-4} \cdot E \cdot X_1, \\ v_5 = k_5 \cdot E\_gal \cdot alac - k_{-5} \cdot E \cdot X_2. \end{cases} \qquad (6.66)$$

Substituting the expressions of the velocities (6.66) in the fourth equation of the system (6.65), one obtains the following equation:

$$k_3 \cdot gal \cdot E + k_1 \cdot E\_lac + k_2 \cdot E\_alac + k_{-4} \cdot X_1 \cdot E + k_{-5} \cdot X_2 \cdot E -$$

$$- k_{-3} \cdot E\_gal - k_{-1} \cdot E\_gal\_glc - k_{-2} \cdot E\_gal\_glc - \qquad (6.67)$$

$$- k_4 \cdot lac \cdot E\_gal - k_5 \cdot alac \cdot E\_gal = 0.$$

Concentrations of the enzyme states obey the following conservation law:

$$E + E\_lac + E\_alac + E\_gal\_glc + E\_gal = E_0. \qquad (6.68)$$

According to Huber, Gaunt, and Hurlburt (1984), the velocities of combination of lactose, allolactose and glucose are well above the rates of catalysis. So, we use an approximation of fast equilibrium for stages 1a, 2a and 6. This allows us, using the ratio of the constants of equilibrium for all quasi-equilibrium stages, to express concentrations of different forms of the enzyme as concentrations of the free form of the enzyme E and the complex of the enzyme and galactose E_gal:

$$E\_lac = \frac{E \cdot lac}{K_l},$$

$$E\_alac = \frac{E \cdot alac}{K_a}, \qquad (6.69)$$

$$E\_gal\_glc = \frac{E\_gal \cdot glc}{K_g}.$$

Substitution of the expression (6.69) in the formulae (6.67, 6.68) gives a system of two linear equations with respect to $E$ and $E\_gal$

$$\begin{cases} E\left(1 + \dfrac{lac}{K_l} + \dfrac{alac}{K_a}\right) + E\_gal\left(1 + \dfrac{glc}{K_g}\right) = E_0, \\[2mm] E\left(k_3 gal + k_1 \dfrac{lac}{K_l} + k_2 \dfrac{alac}{K_a} + k_{-4}X_1 + k_{-5}X_2\right) \\[2mm] \quad - E\_gal\left(k_{-3} + k_{-1}\dfrac{glc}{K_g} + k_{-2}\dfrac{glc}{K_g} + k_4 lac + k_5 alac\right) = 0. \end{cases}$$

Solving the system, the expressions for the stationary concentrations of the enzyme states $E$ and $E\_gal$ are obtained

$$E = \frac{E_0}{\Delta}\left\{k_{-3} + \frac{glc}{K_g}\left(k_{-1} + k_{-2}\right) + k_4 lac + k_5 alac\right\},$$

$$\qquad (6.70)$$

$$E\_gal = \frac{E_0}{\Delta}\left\{k_3 gal + k_1 \frac{lac}{K_l} + k_2 \frac{alac}{K_a} + k_{-4}X_1 + k_{-5}X_2\right\},$$

where

$$
\Delta = \left\{ k_3 + \frac{glc}{K_g}(k_{-1} + k_{-2}) + k_4 lac + k_5 alac \right\} \left( 1 + \frac{lac}{K_l} + \frac{alac}{K_a} \right)
$$
$$
+ \left\{ k_3 gal + k_1 \frac{lac}{K_l} + k_2 \frac{alac}{K_a} + k_{-4}X_1 + k_{-5}X_2 \right\} \left( 1 + \frac{glc}{K_g} \right).
$$

The concentrations of other stationary forms of the enzyme can be expressed using the formula (6.69). With knowledge of the stationary concentrations of all the forms of the enzyme, it is possible to calculate the velocity of any elementary stage. Using (6.66), (6.69) and (6.70) we obtain the following:

$$
v_{1b} = \frac{E_0}{\Delta} \left\{ k_1 \frac{lac}{K_l} \left( k_{-3} + (k_{-1} + k_{-2})\frac{glc}{K_g} + k_4 lac + k_5 alac \right) \right.
$$
$$
\left. - k_{-1} \frac{glc}{K_g} \left( k_3 gal + k_1 \frac{lac}{K_l} + k_2 \frac{alac}{K_a} + k_{-4}X_1 + k_{-5}X_2 \right) \right\} \quad (6.71)
$$

in much the same manner for other key velocities

$$
v_{2b} = \frac{E_0}{\Delta} \left\{ k_2 \frac{alac}{K_a} \left( k_{-3} + \left(k_{-1} + k_{-2}\right)\frac{glc}{K_g} + k_4 lac + k_5 alac \right) \right.
$$
$$
\left. - k_{-2} \frac{glc}{K_g} \left( k_3 gal + k_1 \frac{lac}{K_l} + k_2 \frac{alac}{K_a} + k_{-4}X_1 + k_{-5}X_2 \right) \right\} \quad (6.72)
$$

$$
v_4 = \frac{E_0}{\Delta} \left\{ k_4 lac \left( k_3 gal + k_1 \frac{lac}{K_l} + k_2 \frac{alac}{K_a} + k_{-4}X_1 + k_{-5}X_2 \right) \right.
$$
$$
\left. - k_{-4}X_1 \left( k_{-3} + (k_{-1} + k_{-2})\frac{glc}{K_g} + k_4 lac + k_5 alac \right) \right\} \quad (6.73)
$$

$$v_5 = \frac{E_0}{\Delta} \left\{ k_5 alac \left( k_3 gal + k_1 \frac{lac}{K_l} + k_2 \frac{alac}{K_a} + k_{-4} X_1 + k_{-5} X_2 \right) \right.$$

$$\left. - k_{-5} X_2 \left( k_{-3} + (k_{-1} + k_{-2}) \frac{glc}{K_g} + k_4 lac + k_5 alac \right) \right\} \qquad (6.74)$$

In the foregoing we have written equations of the velocities for all the elementary stages. These expressions include the constants of velocity and dissociation. In order to describe the behaviour of the real enzyme, the parameters of equations should be determined based on experimental data. The results of the study by Burstein et al. (1965) have been used as experimental data. The data were produced in the following experiment. To a solution containing 0.5 M of lactose, at zero point of time, β-galactosidase was added in such a way that its final concentration was equal to 130 μg/mL. Time series of the concentrations of lactose, allolactose, galactose, glucose and total concentration of oligosaccharides were measured.

Since the system described contains no influxes and effluxes of material, then, in accordance with the kinetic scheme depicted in figure 6.19, the change of concentrations of the metabolites with time is determined only by the activity of the enzyme and the following can be written:

$$\begin{cases} (lac)' = -v_{1a} - v_4, \\ (alac)' = -v_{2a} - v_5, \\ (gal)' = -v_3, \\ (glc)' = -v_6, \\ (X_1)' = v_4, \\ (X_2)' = v_5. \end{cases} \qquad (6.75)$$

In terms of the expression (6.65), which is the result of the use of an approximation of quasi-stationary concentrations for all the forms of the enzyme, the system transforms into the following one:

$$\begin{cases} (lac)' = -v_{1b} - v_4, \\ (alac)' = -v_{2b} - v_5, \\ (gal)' = -v_{1b} - v_{2b} - v_4 - v_5, \\ (glc)' = v_{1b} + v_{2b}, \\ (X_1)' = v_4, \\ (X_2)' = v_5. \end{cases} \tag{6.76}$$

The given system is a system of differential equations with concentrations of metabolites as variables. The system of equations (6.76) has two first linear integrals which correspond to the laws of conservation of monosaccharides glucose and galactose:

$$\begin{aligned} glc + lac + alac + (X_1 + X_2) = const_1, \\ gal + lac + alac + 2(X_1 + X_2) = const_2. \end{aligned} \tag{6.77}$$

The correctness of the first expression is supported if, based on (6.76), one observes that

$$(glc)' + (lac)' + (alac)' + (X_1)' + (X_2)' = 0,$$

and then integrates this expression. In a similar manner it is possible to demonstrate the correctness of the second expression of the system (6.77).

With the values of concentrations of metabolites in the system under study at zero point of time, it is possible to determine the values of the parameters $const_1$ and $const_2$. So, quantitative description of the above experiment reduces to a solution of the Cauchy problem:

$$\begin{cases} (lac)' = -v_1 - v_4, \\ (alac)' = -v_2 - v_5, \\ (X_1)' = v_4, \\ (X_2)' = v_5, \\ glc = 0,5M - lac - alac - (X_1 + X_2), \\ gal = 0,5M - lac - alac - 2(X_1 + X_2), \end{cases} \tag{6.78}$$

with initial terms

$$lac_0 = 0,5M,$$

$$alac_0 = glc_0 = gal_0 = X_{10} = X_{20} = 0M.$$

*Identification of the Parameters of the β-Galactosidase Rate Equation*

In order to describe the behaviour of the real system, it is necessary to identify the parameters included in the equation of the rate of an enzyme based on experimental data. In our work we used as a criterion of adequacy of the model constructed to the action of the real enzyme a sum of quadratic deviations of theoretical values, the results of modelling from the experimental points from the work of Huber, Kurz, and Wallenfels (1976). In this connection a search for optimal values of the model parameters determined by the system of equations (6.78) was used in designating a set of parameter values such that the criterion reaches a minimum. Minimization of deviation was performed according to the Hooke-Jeeves technique within a wide range of possible values of the parameters of the rate and dissociation. Using these procedures we estimated the parameters given in table 6.4. Experimental data of some constants (which can be found in the literature) are given in brackets. Figure 6.20 gives us the possibility to compare how well the experimental data are described.

*Model Predictions*

We used the model constructed by us for studying the ratio between different activities of an enzyme. β-galactosidase has, indeed, several activities: formation of glucose and galactose, transformation of lactose into

TABLE 6.4   Values of the Kinetic Parameters of β-galactosidase Estimated from Experimental Data

| | |
|---|---|
| $k_1 = 1,0 \cdot 10^4 \text{ min}^{-1}$ | $k_{-1} = 0,8 \cdot 10^3 \text{ min}^{-1}$ |
| $k_2 = 4 \cdot 10^4 \text{ min}^{-1}$ | $k_{-2} = 1,0 \cdot 10^4 \text{ min}^{-1}$ ($2,3\ 10^4 \text{ min}^{-1}$ measured experimentally) |
| $k_3 = 3 \cdot 10^1 \text{ min}^{-1}\text{mM}^{-1}$ | $k_{-3} = 1,6 \cdot 10^4 \text{ min}^{-1}$ |
| $k_4 = 2 \cdot 10^1 \text{ min}^{-1}\text{mM}^{-1}$ | $k_{-4} = 0,8 \cdot 10^3 \text{ min}^{-1}\text{mM}^{-1}$ |
| $k_5 = 2 \cdot 10^1 \text{ min}^{-1}\text{mM}^{-1}$ | $k_{-5} = 0,8 \cdot 10^3 \text{ min}^{-1}\text{mM}^{-1}$ |
| $K_l = 0,7 \text{ мM}$ (1,3 mM measured experimentally) | |
| $K_a = 0,8 \text{ mM}$ | |
| $K_g = 14 \text{ mM}$ (17 mM measured experimentally) | |

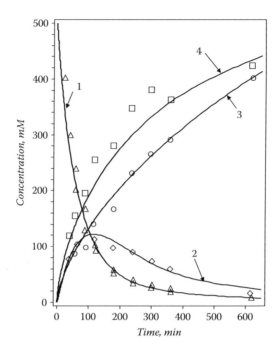

FIGURE 6.20 The comparison of experimental data and model results. Circles, squares and triangles are the experimental data. Curves are the theoretical model results after the procedure of fitting (see the text). Experiment condition: 130 mg of $\beta$-galactosidase on 1 mL of the solution, T = 30°C, pH =7.2, $MgSO_4$ 6.7 mM, NaCl 10 mM.

allolactose and synthesis of trisaccharides. Which of these activities are dominating and how does the contribution of each activity depend on the concentration of the substrates and products? The answers to these questions enable us to understand better how the functioning of an enzyme is controlled by the substrate. To solve this problem we used the equations of the rates at fixed zero values of the concentration of allolactose, galactose, glucose and trisaccharides at changing concentration of lactose within the range of 0 to 5 mM (figure 6.21). It is possible to interpret this study as an attempt to forecast theoretically the experimental data on measuring initial rates of lactose utilization and synthesis of various products depending on the concentration of the main substrate.

As seen from figure 6.21 the dependence of the rate of lactose consumption on the concentration of the latter under the conditions employed shows a clear deviation from the classical law of Michaelis-Menten, namely, at lactose concentrations higher than 10 mM, a small decrease of the

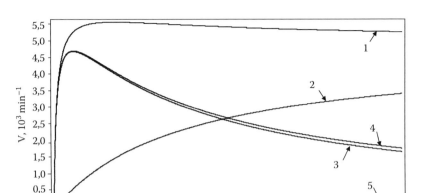

FIGURE 6.21  The model flux distribution towards every product of the reaction. 1, stationary rate of lactose consumption; 2, 3, 4, 5, stationary synthesis rates of allolactose, galactose, glucose and trisaccharides, correspondingly. The model conditions: alac=oligo = 0 mM, glc = from 0 to 1000 mM, gal = from 0 to 1000 mM.

consumption rate is seen. The rates of synthesis of glucose and galactose have marked optima; at lactose concentration of 25 mM the yield is maximal. The rate of allolactose synthesis under the same conditions increases monotonically within the whole range studied. At a lactose concentration of 25 mM, the rate of allolactose synthesis becomes equal to the rate of synthesis of monosaccharides. It should be noted that the situation in which the rate of allolactose synthesis can exceed the rate of monosaccharide formation *in vivo* is most probably impossible, since actual lactose concentration inside the bacterium is deliberately lower than the modeled one.

Within the whole range of lactose concentrations, the consumption of this substrate for the synthesis of by-products (trisaccharides) turns out to be very low and does not exceed 3 percent of the rate of lactose consumption. We can conclude that, under conditions close to the intracellular ones, the substrate consumption for the synthesis of trisaccharides can be ignored without any loss in exactness described.

Besides the data shown in figure 6.21, we studied how the permanent reaction rates depended on the concentrations of glucose and galactose (data are not available), since metabolites could be in higher concentrations in actual bacteria. It turned out that, at concentrations up to 1 mM of each of the monosaccharides, the values of the fixed rates remained unchanged.

It is possible to explain this by the fact that equilibrium of lactose hydrolysis is shifted to a great extent forward; hence, decomposition is, in fact, irreversible in the range studied resulting in weak sensitivity of the enzyme to monosaccharides.

## Kinetic Model of Imidazologlycerol-Phosphate Synthetase from *Escherichia coli*

Imidazologlycerol-phosphate synthetase (IGPS) belongs to the group of aminotransferases that catalyze the transfer of the amido group of glutamine onto various acceptor molecules—intermediates in biosynthesis of purine and pyrimidine nucleotides and amino acids (Zalkin 1993). IGPS from *Escherichia coli* is a key enzyme in the biosynthesis of histidine and structurally is a heterodimer encoded by *hisH* and *hisF* (Zalkin and Smith 1998). Similarly to other glutamine aminotransferases, imidazologlycerol-phosphate synthetase catalyzes two reactions (Zalkin 1993; Zalkin and Smith 1998):

$$PRFAR + Gln = AICAR + IGP + Glu \qquad (6.79)$$
$$Gln = Glu + NH_3 \qquad (6.80)$$

The essence of the first reaction is transfer of the amido group of glutamine accompanied by decomposition of 5′-phosphoribulosylformimino-5-aminoimidazolo-4-carboxamide-ribonucleotide (PRFAR) into 5-amino-imidazolo-4-carboxamido-1-β-d-ribofuranosyl 5′-monophosphate (AICAR) and imidazologlycerol phosphate (IGP) and synthesis of the imidazole ring. This reaction is catalyzed by the synthetase activity of IGPS. In the second reaction, glutamine (Gln) is decomposed to glutamate (Glu). The rate of this reaction is governed by the glutaminase activity of IGPS.

In this section we reconstruct the catalytic cycle of imidazologlycerol-phosphate synthetase, derive the rate equations for glutaminase and synthetase activities, and evaluate the unknown parameters characterizing kinetic properties of this enzyme. Based on the model, we predict how concentrations of the substrates and effectors determine the contributions of glutaminase and synthetase activities to the enzyme operation.

### Experimental Data

In this study, we used the following available data on structural and functional properties of IGPS: (a) the enzyme has two catalytic sites: one for binding to glutamine and another to PRFAR (Zalkin 1993; Zalkin and Smith 1998); (b) binding to the substrates and dissociation of the products occur accidentally (Zalkin 1993; Zalkin and Smith 1998); (c) the amido group cleaved from glutamine is transferred on the catalytic site binding PRFAR without emerging into solution; that is, an intramolecular transfer of the amido group occurs (Zalkin 1993; Zalkin and Smith 1998); (d) the glutaminase activity of imidazologlycerol-phosphate synthetase manifests itself essentially only with excess glutamine and depletion of PRFAR, the second substrate (Chittur et al. 2001; Klem and Davisson 1993); (e) whereas IGP, one of the products of the synthetase reaction catalyzed by imidazolo-glycerol-phosphate synthetase, catalyzes the glutaminase reaction, another product, AICAR, does not affect the latter (Chittur et al. 2001; Klem and Davisson 1993); (f) on the pathway of biosynthesis of histidine, the reaction catalyzed by IGPS is preceded by another reaction catalyzed by ProFAR isomerase, its substrate, 5′-phosphoribosylformimino-5-aminoimidazolo-4-carboxamido-ribonucleotide (ProFAR), being an activator of the gluta-minase activity of IGPS (Chittur et al. 2001; Klem and Davisson 1993).

Some kinetic properties of the enzyme functioning as synthetase, i.e., catalyzing only reaction (6.79), were investigated earlier (Klem and Davisson 1993). The Michaelis constants for Gln and PRFAR:

$$K^s_{m,Gln} = 240 \ \mu M, \quad K^s_{m,PRFAR} = 1.5 \ \mu M \quad (6.81)$$

and the turnover number of the enzyme:

$$k^s_{cat} = 8.5 \ s^{-1} \quad (6.82)$$

were determined at pH 8 and 25°C. In the same work, Klem and Davisson evaluated the Michaelis constant for Gln and the turnover number of the enzyme functioning as glutaminase; i.e., catalyzing only reaction (6.80):

$$K^g_{m,Gln} = 4800 \ \mu M, \quad k^g_{cat} = 0.07 \ s^{-1}. \quad (6.83)$$

To characterize quantitatively the effect of ProFAR and IGP on the glutaminase activity of IGPS, Klem and Davisson evaluated the apparent Michaelis constant for Gln and apparent turnover number of the enzyme functioning as glutaminase in the presence of the activators ProFAR and IGP:

$$[ProFAR]_0 = 2000 \ \mu M, \quad K^{g,ProFAR}_{m,app,Gln} = 2800 \ \mu M, \quad k^{g,ProFAR}_{cat,app} = 2{,}6 \ sec^{-1} \quad (6.84)$$

$$[IGP]_0 = 9000 \ \mu M, \quad K^{g,IGP}_{m,app,Gln} = 1900 \ \mu M, \quad k^{g,IGP}_{cat,app} = 2{,}7 \ sec^{-1} \quad (6.85)$$

Kinetic characteristics of the enzyme operation, that is, changes in concentrations of the substrates and products in the presence of 9 nM IGPS with time, were also monitored in Klem and Davisson (1993).

## Catalytic Cycle

The experimental data mentioned in items (a)–(f) of the previous section were accounted for by development of a kinetic model of the catalytic cycle of imidazologlycerol-phosphate synthetase, presented in figure 6.22. We suggest that the enzyme functions via the random bi ter mechanism according to the Cleland classification (Cleland 1963); this agrees with items (a) and (b) of the previous section. Transfer of the triple enzyme–substrate complex Gln·E·PRFAR into the enzyme–product complex Glu·E·IGP·AICAR corresponds with an intramolecular transfer of the amido group mentioned in item (c) of the previous section and subsequent transformation of PRFAR into AICAR and IGP. As a result of the accidental order of the substrate binding, in the absence of PRFAR a complex of the enzyme with glutamine Gln·E is formed and the glutaminase reaction (6.80) catalyzed by free enzyme proceeds (E → Gln·E → Glu·E* → E); this corresponds to item (d) of the previous section. Activation of the glutaminase reaction in the presence of ProFAR and IGP described in items (e) and (f) of the previous section is accounted for by inclusion of the additional reaction cycles (E → E·ProFAR → Gln·E·ProFAR → Glu·E*·ProFAR → E·ProFAR) and (E·IGP → Gln·IGP → Glu·E*·IGP → E·IGP) into the standard random bi ter mechanism. In the following description, we do not study how the rate of the glutaminase reaction depends on concentration of one of its products ($NH_3$); that is, why $NH_3$ is not represented directly in figure 6.22.

## Derivation of the Rate Equations

To derive the equations which describe the dependence of the rates of synthetase and glutaminase reactions on concentrations of the substrates, products and effectors, we apply the method described previously. Indeed, we suggest that the rates of all reactions of the substrate binding and dissociation of the products are significantly higher than the rates of catalytic reactions designated as 1, 2, 3, and 4 in figure 6.22.This means that each of these 'fast' reactions can be considered as a quasi-equilibrium one (the dissociation constants are given near the corresponding reactions in figure 6.22); thus, we can write the following relationships:

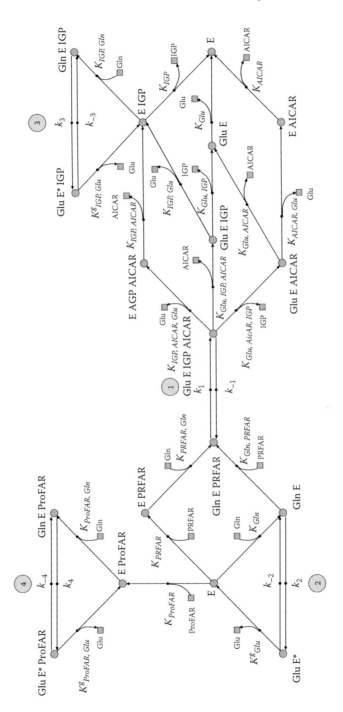

FIGURE 6.22 Catalytic cycle of imidazologlycerol-phosphate synthetase. The dissociation constants of the substrate–product complexes and the rate constants are presented near the corresponding stages of the catalytic cycle.

$$K_{Glu}^g = \frac{[Glu][E]}{[Glu \bullet E^*]} \qquad K_{Gln} = \frac{[Gln][E]}{[Gln \bullet E]} \qquad K_{PRFAR,Gln} = \frac{[Gln][E \bullet PRFAR]}{[Gln \bullet E \bullet PRFAR]}$$

$$K_{ProFAR,Glu}^g = \frac{[Glu][E \bullet ProFAR]}{[Glu \bullet E^* \bullet ProFAR]} \qquad K_{ProFAR,Gln} = \frac{[Gln][E \bullet ProFAR]}{[Gln \bullet E \bullet ProFAR]}$$

$$K_{ProFAR} = \frac{[ProFAR][E]}{[E \bullet ProFAR]}$$

$$K_{Glu} = \frac{[Glu][E]}{[Glu \bullet E]} \qquad K_{AICAR,Glu} = \frac{[Glu][E \bullet AICAR]}{[Glu \bullet E \bullet AICAR]} \qquad K_{AICAR} = \frac{[AICAR][E]}{[E \bullet AICAR]}$$

$$K_{IGP} = \frac{[IGP][E]}{[E \bullet IGP]} \qquad K_{Glu,IGP,AICAR} = \frac{[AICAR][Glu \bullet E \bullet IGP]}{[Glu \bullet E \bullet IGP \bullet AICAR]}$$

$$K_{Glu,IGP} = \frac{[IGP][Glu \bullet E]}{[Glu \bullet E \bullet IGP]}$$

$$K_{IGP,Glu}^g = \frac{[Glu][E \bullet IGP]}{[Glu \bullet E^* \bullet IGP]} \qquad K_{IGP,Gln} = \frac{[Gln][E \bullet IGP]}{[Gln \bullet E \bullet IGP]}$$

$$K_{IGP,AICAR} = \frac{[AICAR][E \bullet IGP]}{[E \bullet IGP \bullet AICAR]} \tag{6.86}$$

For concentrations of the enzyme states, the following conservation law is also fulfilled:

$$[E] + [E \bullet PRFAR] + [Gln \bullet E \bullet PRFAR] + [Gln \bullet E] + [Glu \bullet E^*] + [E \bullet ProFAR]$$
$$+ [Glu \bullet E^* \bullet ProFAR] + [Gln \bullet E \bullet ProFAR] + [E \bullet AICAR] + [Glu \bullet E \bullet AICAR]$$
$$+ [Glu \bullet E] + [Glu \bullet E \bullet IGP] + [E \bullet IGP] + [E \bullet IGP \bullet AICAR] + [Glu \bullet E^* \bullet IGP]$$
$$+ [Gln \bullet E \bullet IGP] + [Glu \bullet E \bullet IGP \bullet AICAR] = [HisHF] \tag{6.87}$$

where [HisHF] is the total concentration of imidazologlycerol-phosphate synthetase. Solving the system of linear (relative to concentrations of the enzyme states) algebraic equations (6.86) and (6.87), we obtain the following expressions for the enzyme forms:

$$[Gln \bullet E] = \frac{[Gln]}{K_{Gln}} \frac{[HisHF]}{\Delta}$$

$$[Gln \bullet E \bullet PRFAR] = \frac{[Gln]}{K_{PRFAR,Gln}} \frac{[PRFAR]}{K_{PRFAR}} \frac{[HisHF]}{\Delta}$$

$$[Gln \bullet E \bullet IGP] = \frac{[IGP]}{K_{IGP}} \frac{[Gln]}{K_{IGP,Gln}} \frac{[HisHF]}{\Delta} \qquad [Glu \bullet E^*] = \frac{[Glu]}{K_{Glu}^g} \frac{[HisHF]}{\Delta}$$

$$[\text{Gln} \bullet \text{E} \bullet \text{PRFAR}] = \frac{[\text{AICAR}]}{K_{\text{Glu,IGP,AICAR}}} \frac{[\text{IGP}]}{K_{\text{Glu,IGP}}} \frac{[\text{Glu}]}{K_{\text{Glu}}} \frac{[\text{HisHF}]}{\Delta}$$

(6.88)

$$[\text{Gln} \bullet \text{E} \bullet \text{ProFAR}] = \frac{[\text{ProFAR}]}{K_{\text{ProFAR}}} \frac{[\text{Gln}]}{K_{\text{ProFAR,Gln}}} \frac{[\text{HisHF}]}{\Delta}$$

$$[\text{Glu} \bullet \text{E}^* \bullet \text{IGP}] = \frac{[\text{IGP}]}{K_{\text{IGP}}} \frac{[\text{Glu}]}{K_{\text{IGP,Glu}}^{\text{g}}} \frac{[\text{HisHF}]}{\Delta}$$

$$[\text{Glu} \bullet \text{E}^* \bullet \text{IGP}] = \frac{[\text{ProFAR}]}{K_{\text{ProFAR}}} \frac{[\text{Glu}]}{K_{\text{ProFAR,Glu}}^{\text{g}}} \frac{[\text{HisHF}]}{\Delta}$$

$$\Delta = 1 + \frac{[\text{PRFAR}]}{K_{\text{PRFAR}}} + \frac{[\text{Gln}]}{K_{\text{PRFAR,Gln}}} \frac{[\text{PRFAR}]}{K_{\text{PRFAR}}} + \frac{[\text{Gln}]}{K_{\text{Gln}}} + \frac{[\text{Glu}]}{K_{\text{Glu}}^{\text{g}}} + \frac{[\text{ProFAR}]}{K_{\text{ProFAR}}} \cdot$$

$$\left(1 + \frac{[\text{Glu}]}{K_{\text{ProFAR,Glu}}^{\text{g}}} + \frac{[\text{Gln}]}{K_{\text{ProFAR,Gln}}}\right) + \frac{[\text{IGP}]}{K_{\text{IGP}}}\left(1 + \frac{[\text{Glu}]}{K_{\text{IGP,Glu}}^{\text{g}}} + \frac{[\text{Gln}]}{K_{\text{IGP,Gln}}}\right)$$

$$+ \frac{[\text{AICAR}]}{K_{\text{AICAR}}} + \frac{[\text{Glu}]}{K_{\text{AICAR,Glu}}} \frac{[\text{AICAR}]}{K_{\text{AICAR}}} + \frac{[\text{Glu}]}{K_{\text{Glu}}} + \frac{[\text{IGP}]}{K_{\text{Glu,IGP}}} \frac{[\text{Glu}]}{K_{\text{Glu}}}$$

$$+ \frac{[\text{AICAR}]}{K_{\text{IGP,AICAR}}} \frac{[\text{IGP}]}{K_{\text{IGP}}} + \frac{[\text{AICAR}]}{K_{\text{Glu,IGP,AICAR}}} \frac{[\text{IGP}]}{K_{\text{Glu,IGP}}} \frac{[\text{Glu}]}{K_{\text{Glu}}}$$

It should be noted that the equilibrium constant of the glutaminase reaction does not depend on which of the three forms of the enzyme (free enzyme, enzyme bound to ProFAR or to IGP) catalyzes the reaction; consequently, the following relationships are correct:

$$K_{\text{eq}}^{\text{g}} = \frac{K_{\text{Glu}}^{\text{g}}}{K_{\text{Gln}}} \frac{k_2}{k_{-2}} = \frac{K_{\text{ProFAR,Glu}}^{\text{g}}}{K_{\text{ProFAR,Gln}}} \frac{k_4}{k_{-4}} = \frac{K_{\text{IGP,Glu}}^{\text{g}}}{K_{\text{IGP,Gln}}} \frac{k_3}{k_{-3}}$$

(6.89)

which makes it possible to express $k_{-3}$ and $k_{-4}$ via $k_2$ and $k_{-2}$:

$$k_{-3} = k_3 \frac{K_{\text{IGP,Gln}}}{K_{\text{IGP,Glu}}^{\text{g}}} \frac{K_{\text{Glu}}^{\text{g}}}{K_{\text{Gln}}} \frac{k_2}{k_{-2}}, \quad k_{-4} = k_4 \frac{K_{\text{ProFAR,Gln}}}{K_{\text{ProFAR,Glu}}^{\text{g}}} \frac{K_{\text{Glu}}^{\text{g}}}{K_{\text{Gln}}} \frac{k_2}{k_{-2}}$$

(6.90)

According to the scheme of the catalytic cycle of IGPS presented in figure 6.22, the rate equations for synthetase and glutaminase reactions can be written as follows:

$$v_{\text{s}} = k_1[\text{Gln} \cdot \text{E} \cdot \text{PRFAR}] - k_{-1}[\text{Glu} \cdot \text{E} \cdot \text{IGP} \cdot \text{AICAR}]$$

(6.91)

$$v_g = (k_2[\text{Gln·E}] - k_{-2}[\text{Glu·E}^*\text{·IGP}]) + (k_3[\text{Gln·E·IGP}] - k_{-3}[\text{Glu·E}^*\text{·IGP}])$$
$$+ (k_4[\text{Gln·E·ProFAR}] - k_{-4}[\text{Glu·E}^*\text{·ProFAR}]) \tag{6.92}$$

Substitution of (6.88) and (6.90) into equations (6.91) and (6.92) gives the following equations for the rates of the synthetase and glutaminase reactions:

$$v_s = \frac{[\text{HisHF}]}{\Delta}\left( k_1 \frac{[\text{Gln}]}{K_{\text{PRFAR,Gln}}} \frac{[\text{PRFAR}]}{K_{\text{PRFAR}}} - k_{-1} \frac{[\text{AICAR}]}{K_{\text{Glu,IGP,AICAR}}} \frac{[\text{IGP}]}{K_{\text{Glu,IGP}}} \frac{[\text{Glu}]}{K_{\text{Glu}}} \right)$$

$$v_g = \frac{[\text{HisHF}]}{\Delta}\left( \frac{k_2}{K_{\text{Gln}}} + \frac{k_3}{K_{\text{IGP,Gln}}} \frac{[\text{IGP}]}{K_{\text{IGP}}} + \frac{k_4}{K_{\text{ProFAR,Gln}}} \frac{[\text{ProFAR}]}{K_{\text{ProFAR}}} \right)$$

$$\times \left( [\text{Gln}] - \frac{k_{-2}}{k_2} \frac{K_{\text{Gln}}}{K_{\text{Glu}}^g} [\text{Glu}] \right) \tag{6.93}$$

where $K_A$ and $K_A^g$ are the dissociation constants of the substrate (or product) A from free enzyme; $K_{BA}$ and $K_{BA}^g$ are the dissociation constants of the substrate (or product) A from the enzyme complex with the substrate (or product) B; $K_{CBA}$ are the dissociation constants of the substrate (or product) A from the enzyme complexes with the substrates (or products) B and C; $k_i$, $k_{-i}$ ($i = 1, 2, 3$ and 4) are the rate constants of catalytic stages of the enzyme cycle. The dissociation and rate constants are presented in figure 6.22 near the corresponding reactions.

In equations (6.92), there are twenty-two parameters—the rate and dissociation constants characterising kinetic properties of certain stages of the catalytic cycle presented in figure 6.22. However, in enzyme kinetics the rate equations are usually written using parameters which characterise kinetic properties of the enzyme as a whole. The Michaelis constants, the turnover number of the enzyme, and the equilibrium constant are used as such kinetic parameters (Cornish-Bowden 2001). To describe kinetic properties of imidazologlycerol-phosphate synthetase, we used seventeen kinetic parameters of this kind. The synthetase reaction was characterised by the turnover number of the enzyme catalyzing this reaction ($k_{\text{cat}}^s$), the equilibrium constant of the synthetase reaction ($K_{\text{eq}}^s$), and the Michaelis constants of the substrates and products of this reaction ($K_{m,\text{PRFAR}}^s$, $K_{m,\text{Gln}}^s$, $K_{m,\text{Glu}}^s$, $K_{m,\text{IGP}}^s$ and $K_{m,\text{AICAR}}^s$). The glutaminase reaction in the absence of activators was characterised by the turnover number of the enzyme catalyzing this reaction ($k_{\text{cat}}^g$), the equilibrium constant of the glutaminase reaction ($K^g_{eq}$), and the Michaelis constants of the substrates and products of this reaction ($K^g_{m,Gln}$ and $K^g_{m,Glu}$). To account for the effect of activators on the glutaminase reaction, we entered the following parameters: the turnover number of the enzyme at infinitely high concentration of IGP ($k^{g,IGP}_{cat}$) or ProFAR ($k^{g,ProFAR}_{cat}$) and the Michaelis constants of the substrates and products of the glutaminase reaction at infinitely high concentration of IGP ($K^{g,IGP}_{m,Gln}, K^{g,IGP}_{m,Glu}$) or Pro-FAR ($K^{g,IGP}_{m,Gln}, K^{g,IGP}_{m,Glu}$). Applying the method described earlier, we found functional interrelations between thus-defined kinetic parameters of the imidazologlycerol-phosphate synthetase and parameters characterizing kinetic properties of certain stages of its catalytic cycle. Using this interrelation, we expressed parameters of the catalytic cycle via kinetic parameters and wrote the rate equations (6.93) using traditional kinetic parameters:

$$v_s = \frac{[\text{HisHF}]}{\Delta} \frac{k^s_{cat}}{K^s_{m,PRFAR} K^g_{m,Gln}} \left([\text{PRFAR}][\text{Gln}] - [\text{AICAR}][\text{IGP}][\text{Glu}]/K^s_{eq}\right)$$

$$v_g = \frac{[\text{HisHF}]}{\Delta} \left( \frac{k^g_{cat}}{K^g_{m,Gln}} + \frac{k^{g,IGP}_{cat}}{K^{g,IGP}_{m,Gln}} \frac{[\text{IGP}]}{K_{IGP}} + \frac{k^{g,ProFAR}_{cat}}{K^{g,ProFAR}_{m,Gln}} \frac{[\text{ProFAR}]}{K_{ProFAR}} \right)$$

$$\times \left([\text{Gln}] - [\text{Glu}]/K^g_{eq}\right)$$

$$\Delta = 1 + \frac{Gln}{K^g_{m,Gln}} + \frac{PRFAR}{K^s_{m,PRFAR}} \cdot \left( \frac{K^s_{m,Gln}}{K^g_{m,Gln}} + \frac{Gln}{K^g_{m,Gln}} \right) + \frac{ProFAR}{K_{ProFAR}} \cdot \left( 1 + \frac{Glu}{K^{g,ProFAR}_{m,Glu}} + \frac{Gln}{K^{g,ProFAR}_{m,Glu}} \right)$$

$$+ \frac{IGP}{K_{IGP}} \cdot \left( 1 + \frac{Glu}{K^{g,IGP}_{m,Glu}} + \frac{Glu}{K^{g,IGP}_{m,Gln}} \right) + \frac{Glu}{K^g_{m,Glu}} + \frac{AICAR}{K_{AICAR}} + \frac{Glu}{K_{Glu}} + \frac{IGP}{K_{Glu,IGP}} + \left( 1 + \frac{K_{IGP} \cdot K^{g,IGP}_{m,Glu}}{K_{Glu,IGP} \cdot K_{Glu}} \right)$$

$$\times \left( \frac{AICAR}{K^s_{m,AICAR}} \cdot \frac{Glu}{K^{g,IGP}_{m,Glu}} \cdot \frac{K^s_{m,IGP}}{K_{IGP}} + \frac{AICAR}{K^s_{m,AICAR}} \cdot \frac{IGP}{K_{IGP}} \cdot \frac{K^s_{m,Glu}}{K^{g,IGP}_{m,Glu}} + \frac{AICAR}{K^s_{m,AICAR}} \cdot \frac{IGP}{K_{IGP}} \cdot \frac{Glu}{K^{g,IGP}_{m,Glu}} \right)$$

$$\text{(6.94)}$$

In equation (6.94), there are twenty-two parameters: seventeen are kinetic parameters and the other five are the dissociation constants of the products and effectors ($K_{AICAR}, K_{Glu}, K_{IGP}, K_{Glu,IGP}$ and $K_{ProFAR}$). From derivation of the expressions of parameters of the catalytic cycle in terms of kinetic parameters it follows that the dissociation constant of glutamate from the free enzyme must be higher than the Michaelis constant for glutamate in the glutaminase reaction:

$$K^g_{m,Glu} < K_{Glu} \tag{6.95}$$

*Evaluation of Parameters of the Rate Equations*

To find the values of parameters which appear in equations (6.94), we used experimental data from Klem and Davisson (1993) and Chittur et al. (2001). As mentioned previously, these authors evaluated Michaelis constants for Gln and PRFAR (6.81) and the turnover number (6.82) of the enzyme catalyzing only the synthetase reaction (6.79). The Michaelis constant for Gln and the turnover number (6.83) of the enzyme functioning as glutaminase; that is, catalyzing only reaction (6.80) was also evaluated in Klem and Davisson (1993) and Chittur et al. (2001). Besides this, Klem and Davisson (1993) found the values of apparent Michaelis constants for glutamine and apparent turnover number of the enzyme catalyzing only the glutaminase reaction in the presence of 2 mM ProFAR (6.84) or 9 mM IGP (6.85). We evaluated five kinetic parameters appearing in equations (6.94) using equations (6.81)–(6.83). The values of the apparent constants (6.84)–(6.85) were used to obtain four relationships between the remaining seventeen parameters; this allowed reduction of the number of unknown parameters to thirteen. These four relationships are algebraic expressions for the apparent constants, whose values are given in (6.84)–(6.85) via the parameters of equations (6.94).

To derive expressions for the apparent Michaelis constant for glutamine and the apparent turnover number of the enzyme catalyzing only the glutaminase reaction in the presence of 2 mM ProFAR, we rewrote the rate equation of the glutaminase reaction accounting for the conditions of the experiment, which gave evaluation of these apparent constants. In fact, in this experiment, the dependence of the initial rate of imidazologlycerol-phosphate synthetase on the glutamine concentration was measured at the following concentrations of the substrates, products and effectors: [PRFAR] = [Glu] = [IGP] = [AICAR] = 0, [ProFAR] = [ProFAR]$_0$ = 2 mM. Substitution of these concentration values into the rate equation for the glutaminase reaction (6.94) gives:

$$v_g = \frac{[\text{HisHF}] \left( \dfrac{k^g_{cat}}{K^g_{m,Gln}} + \dfrac{k^{g,\text{ProFAR}}_{cat}}{K^{g,\text{ProFAR}}_{m,Gln}} \dfrac{[\text{ProFAR}]_0}{K_{\text{ProFAR}}} \right)[\text{Gln}]}{1 + \dfrac{[\text{Gln}]}{K^g_{m,Gln}} + \dfrac{[\text{ProFAR}]_0}{K_{\text{ProFAR}}} \left( 1 + \dfrac{[\text{Gln}]}{K^{g,\text{ProFAR}}_{m,Gln}} \right)} \tag{6.96}$$

From equation (6.96) we obtained the following expressions for the apparent Michaelis constant for glutamine and apparent turnover number of the enzyme:

$$K_{m,app,Gln}^{g,ProFAR} = \frac{1 + \dfrac{[ProFAR]_0}{K_{ProFAR}}}{\dfrac{1}{K_{m,Gln}^{g}} + \dfrac{1}{K_{m,Gln}^{g,ProFAR}} \dfrac{[ProFAR]_0}{K_{ProFAR}}}$$

$$k_{cat,app}^{g,ProFAR} = \frac{\dfrac{k_{cat}^{g}}{K_{m,Gln}^{g}} + \dfrac{k_{cat}^{g,ProFAR}}{K_{m,Gln}^{g,ProFAR}} \dfrac{[ProFAR]_0}{K_{ProFAR}}}{\dfrac{1}{K_{m,Gln}^{g}} + \dfrac{1}{K_{m,Gln}^{g,ProFAR}} \dfrac{[ProFAR]_0}{K_{ProFAR}}} \qquad (6.97)$$

These two relationships allowed expression of $K_{ProFAR}$ and $k_{cat}^{g,ProFAR}$ as functions of the other kinetic parameters of equation (6.94), the apparent Michaelis constant, the apparent catalytic constant, and the $[ProFAR]_0$ value at which these constants were measured:

$$K_{ProFAR} = [ProFAR]_0 \frac{K_{m,Gln}^{g} \left( K_{m,app,Gln}^{g,ProFAR} - K_{m,Gln}^{g,ProFAR} \right)}{K_{m,Gln}^{g,ProFAR} \left( K_{m,Gln}^{g} - K_{m,app,Gln}^{g,ProFAR} \right)} \qquad (6.98)$$

$$k_{cat}^{g,ProFAR} = k_{cat,app}^{g,ProFAR} + \left( k_{cat,app}^{g,ProFAR} - k_{cat}^{g} \right) \frac{K_{m,app,Gln}^{g,ProFAR} - K_{m,Gln}^{g}}{K_{m,Gln}^{g} - K_{m,app,Gln}^{g,ProFAR}} \qquad (6.99)$$

Since $K_{ProFAR}$ and $k_{cat}^{g,ProFAR}$ are always positive, the values of parameters of the right-hand sides of equations (6.98) and (6.99) should satisfy the following inequalities:

$$k_{cat,app}^{g,ProFAR} > k_{cat}^{g}$$

$$K_{m,Gln}^{g} > K_{m,app,Gln}^{g,ProFAR} > K_{m,Gln}^{g,ProFAR} \qquad (6.100)$$

Using the data on the apparent kinetic parameters of the glutaminase reaction activated by IGP (6.85) and repeating all the reasoning and manipulations presented above, we obtained relationships which allowed us to express $K_{IGP}$ and $k_{cat}^{g,IGP}$ as functions of other kinetic parameters of equations (6.94), the apparent Michaelis constant, the apparent catalytic constant, and the $[IGP]_0$ value at which these constants were measured:

$$k_{cat}^{g,IGP} = k_{cat,app}^{g,IGP} + \left( k_{cat,app}^{g,IGP} - k_{cat}^{g} \right) \frac{K_{m,app,Gln}^{g,IGP} - K_{m,Gln}^{g,IGP}}{K_{m,Gln}^{g} - K_{m,app,Gln}^{g,IGP}}$$

$$K_{IGP} = [IGP]_0 \frac{K_{m,Gln}^{g} \left( K_{m,app,Gln}^{g,IGP} - K_{m,Gln}^{g,IGP} \right)}{K_{m,Gln}^{g,IGP} \left( K_{m,Gln}^{g} - K_{m,app,Gln}^{g,IGP} \right)} \qquad (6.101)$$

$$k_{\text{cat,app}}^{g,\text{IGP}} > k_{\text{cat}}^{g} \qquad K_{m,\text{Gln}}^{g} > K_{m,\text{app,Gln}}^{g,\text{IGP}} > K_{m,\text{Gln}}^{g,\text{IGP}} \qquad (6.102)$$

Thus, accounting for the values of kinetic parameters (6.81)–(6.83) and using equations (6.98), (6.99) and (6.101), we reduced the number of unknown parameters involved in the rate equation for IGPS from twenty-two to thirteen, and these thirteen parameters should satisfy inequalities (6.95), (6.100) and (6.102).

To obtain the values of the remaining thirteen parameters, we used kinetic data for imidazologlycerol-phosphate synthetase obtained at pH 8.0 and 25°C (Klem and Davisson 1993). Experimental conditions were as follows: to 90 µM PRFAR and 5 mM glutamine, imidazologlycerol-phosphate synthetase was added to a final concentration of 9 nM and changes in the PRFAR, AICAR and Glu concentrations were monitored with time. The kinetic model describing this experiment can be depicted as the kinetic scheme presented in figure 6.23; this model is given by a system of differential and algebraic equations:

$$d[\text{PRFAR}]/dt = -v_s,$$
$$d[\text{Glu}]/dt = v_s + v_g,$$
$$[\text{Gln}] + [\text{Glu}] = 5000 \ \mu M,$$
$$[\text{PRFAR}] + [\text{AICAR}] = 90 \ \mu M,$$
$$[\text{PRFAR}] + [\text{IGP}] = 90 \ \mu M \qquad (6.103)$$

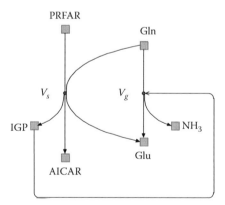

FIGURE 6.23 Kinetic scheme of the model corresponding with the experimental data on the kinetics of imidazologlycerol-phosphate synthetase. Solid arrows specify synthetase ($v_s$) and glutaminase ($v_g$) activities of the enzyme, and broken arrow shows that IGP activates the glutaminase reaction.

The initial values of variables in this model coincide with the initial experimental concentrations: [PRFAR] = 90 µM, [Gln] = 5000 µM, [Glu] = [IGP] = [AICAR] = 0. Eleven of thirteen kinetic parameters of the rate equations of the synthetase and glutaminase reactions were fitted so that time dependences of PRFAR, AICAR and Glu concentrations obtained as numerical solutions of the system (6.103) best coincided with corresponding experimental dependences (Klem and Davisson 1993). The values of the two parameters $K_{m,Gln}^{g,ProFAR}$ and $K_{m,Glu}^{g,ProFAR}$ characterizing kinetic properties of the enzyme in the presence of ProFAR (activator of the glutaminase reaction) could not be evaluated from the experimental data (Klem and Davisson 1993) because of zero concentration of this activator in the experiment considered. Figure 6.24 presents experimental data (squares; Klem and Davisson 1993) and model results (solid lines). Approximation of the experimental data (figure 6.24, curves 2, 4, and 6) appeared to be impossible, within a sufficient range of accuracy, as it involved changing values of eleven parameters, which remained nonevaluated (that is, free) after accounting for the experimental data (Chittur et al. 2001; Klem and Davisson 1993) by using the values of true and apparent catalytic and Michaelis constants in our model. However,

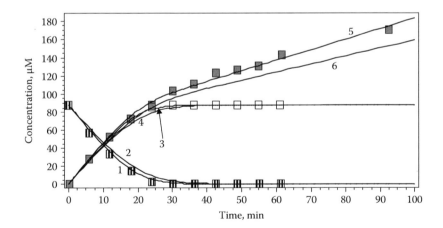

FIGURE 6.24 Time dependences of PRFAR (curves 1 and 2 and hatched squares), AICAR (curves 3 and 4 and open squares) and Glu (curves 5 and 6 and gray squares) concentrations, experimentally measured (hatched, open and gray squares) and obtained from the kinetic model based on this experiment (curves 1–6).

if we suppose that the values of these true and apparent catalytic and Michaelis constants from expressions (6.81)–(6.85) are measured with 30 percent accuracy and that we can therefore vary these values within this 30 percent interval together with the eleven nonevaluated parameters, the results of modelling will exactly coincide with experimental data (figure 6.24, curves 1, 3, and 5). The parameter values thus derived are given in table 6.5.

The remaining two parameters $K_{m,Gln}^{g,ProFAR}$ and $K_{m,Glu}^{g,ProFAR}$ characterizing the kinetic properties of imidazologlycerol-phosphate synthetase in the presence of ProFAR, an activator of the glutaminase reaction, were evaluated by the data of Jurgens et al. (2000); in this work, Jurgens and coauthors studied the properties of ProFAR isomerase, a precursor of IGPS in the pathway of the biosynthesis of histidine and a catalyst of the isomerization reaction of ProFAR into PRFAR. Experimental conditions were as follows: pH 7.5 and 25°C. To 0.25 µM ProFAR isomerase, 1 µM imidazologlycerol-phosphate synthetase and 5 mM glutamine, ProFAR was added to a final concentration of 20 µM, and the time dependence of its concentration was monitored. The kinetic model describing this experiment can be depicted as the kinetic scheme presented in figure 6.25; this model is given by a system of differential and algebraic equations:

$$\begin{cases} d[\text{ProFAR}]/dt = -v_{\text{ProFAR\_isomerase}} \\[2mm] d[PRFAR]/dt = v_{\text{ProFAR\_isomerase}} - v_s^{pH} \\[2mm] d[\text{Glu}]/dt = v_s^{pH} + v_g^{pH} \\[2mm] [\text{Gln}]+[\text{Glu}] = 5000\ \mu M \\[2mm] [\text{ProFAR}]+[\text{PRFAR}]+[\text{AICAR}] = 20\ \mu M \\[2mm] [\text{ProFAR}]+[\text{PRFAR}]+[\text{IGP}] = 20\ \mu M \end{cases} \qquad (6.104)$$

The initial values of variables in the model coincide with the initial experimental concentrations: [ProFAR] = 20 µM, [Gln] = 5000 µM, [Glu] = [PRFAR] = [IGP] = [AICAR] = 0. Since most of the imidazologlycerol-phosphate synthetase parameters were evaluated from experimental data obtained at pH 8.0 (Klem and Davisson 1993), for evaluation of remaining

TABLE 6.5  Values of Kinetic Parameters of Imidazol Glycerolphosphate Synthase and ProFAR Isomerase

| Parameter | Experimentally Measured Values, µM, min$^{-1}$ | Fitted Values, µM, min$^{-1}$ |
|---|---|---|
| $k_{cat}^s$ | 510 | 585 |
| $K_{eq}^s$ | | 4849 |
| $K_{m,PRFAR}^s$ | 1.5 | 1.86 |
| $K_{m,Gln}^s$ | 240 | 180 |
| $K_{m,Glu}^s$ | | 0.01 |
| $K_{m,IGP}^s$ | | 0.14 |
| $K_{m,AICAR}^s$ | | 20572 |
| $k_{cat}^g$ | 4.2 | 5.28 |
| $K_{eq}^g$ | | 1028 |
| $K_{m,Gln}^g$ | 4800 | 5960 |
| $K_{m,Glu}^g$ | | 51355 |
| $k_{cat}^{g,IGP}$ | | 210.3 |
| $k_{cat}^{g,ProFAR}$ | | 156.2 |
| $K_{m,Gln}^{g,IGP}$ | | 1323 |
| $K_{m,Glu}^{g,IGP}$ | | 10379 |
| $K_{m,Gln}^{g,ProFAR}$ | | 946.2 |
| $K_{m,Glu}^{g,ProFAR}$ | | 46225 |
| $K_{AICAR}$ | | 31690 |
| $K_{Glu}$ | | 51415 |
| $K_{IGP}$ | | 61.3 |
| $K_{Glu,IGP}$ | | 30907 |
| $K_{ProFAR}$ | | 4.658 |
| $k_{cat}^{HisA}$ | 42 | 42 |
| $K_{m,ProFAR}^{HisA}$ | 0.6 | 0.6 |
| $K_{m,PRFAR}^{HisA}$ | | 0.028 |
| $k_{cat,app}^{g,IGP}$ | 162 | 210 |
| $K_{m,app,Gln}^{g,IGP}$ | 1900 | 1330 |
| $k_{cat,app}^{g,ProFAR}$ | 156 | 156 |
| $K_{m,app,Gln}^{g,ProFAR}$ | 2800 | 2800 |
| $K_H^1$ | | 0.01 |
| $K_H^2$ | | 0.933 |

parameters from experimental data obtained at pH 7.5 (Jurgens et al. 2000), it is necessary to account for the pH dependence of the activity of this enzyme. Following the method for describing pH dependence of the rate of enzymatic reaction (Cornish-Bowden 2001), we suggest that a

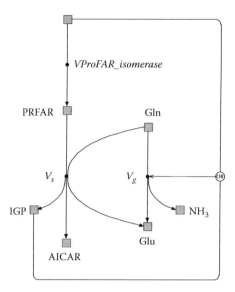

FIGURE 6.25 Kinetic scheme of a model corresponding to the kinetic study of imidazologlycerol-phosphate synthetase–ProFAR isomerase system. Solid arrows designate reactions catalyzed by ProFAR isomerase ($v_{\text{ProFAR\_isomerase}}$) and also the synthetase ($v_s$) and glutaminase ($v_g$) activities of IGPS. Broken arrows show that IGP and ProFAR stimulate the glutaminase activity of IGPS.

catalytic site (one or several amino acid residues directly participating in catalysis) can be deprotonated and protonated singly or doubly, the singly protonated form of the enzyme being catalytically active. We also suggest that the values of the proton dissociation constants do not depend on the enzyme state. These suggestions do not contradict experimental data on kinetics of this enzyme; as a result, the pH dependence of the rate of IGPS functioning is given as follows:

$$
v_g^{\text{pH}} = v_g \frac{1 + \dfrac{K_H^1}{10^{6-\text{pH}_1}} + \dfrac{10^{6-\text{pH}_1}}{K_H^2}}{1 + \dfrac{K_H^1}{10^{6-\text{pH}}} + \dfrac{10^{6-\text{pH}}}{K_H^2}}
\qquad
v_s^{\text{pH}} = v_s \frac{1 + \dfrac{K_H^1}{10^{6-\text{pH}_1}} + \dfrac{10^{6-\text{pH}_1}}{K_H^2}}{1 + \dfrac{K_H^1}{10^{6-\text{pH}}} + \dfrac{10^{6-\text{pH}}}{K_H^2}}
\qquad (6.105)
$$

where $v_s$ and $v_g$ are given by equations (6.94) describing the synthetase and glutaminase activities of imidazologlycerol-phosphate synthetase, $K_H^1, K_H^2$

are the proton dissociation constants from nonprotonated and singly protonated forms of the enzyme, respectively, and $pH_1 = 8.0$ coincides with pH in the experiment (Klem and Davisson 1993) used for evaluation of most of the enzyme constants. Since the data (Jurgens et al. 2000) were obtained at pH 7.5, in the rate equations for imidazologlycerol-phosphate synthetase involved in the kinetic model (6.104), which describes these data, the pH should also be 7.5.

The rate equation for ProFAR isomerase can be written as follows:

$$v_{ProFAR\_isomerase} = \frac{k_{cat}^{HisA}[HisA][ProFAR]}{K_{m,ProFAR}^{HisA} + [ProFAR] + \frac{K_{m,ProFAR}^{HisA}}{K_{m,PRFAR}^{HisA}} \cdot [PRFAR]} \quad (6.106)$$

The catalytic and Michaelis constants for ProFAR evaluated from the data obtained at pH 7.5 and 25°C were taken from Jurgens et al. (2000): $k_{cat}^{HisA} = 0,7\,sec^{-1}$, $K_{m,ProFAR}^{HisA} = 0,6\,\mu M$. The remaining unknown parameters of model (6.104) ( $K_{m,Gln}^{g,ProFAR}$, $K_{m,Glu}^{g,ProFAR}$ and $K_{m,PRFAR}^{HisA}$, $K_H^1, K_H^2$) were fitted so that the time dependence of ProFAR concentration obtained as a result of numerical solution of the system (6.104) best coincided with the corresponding experimentally obtained dependence (Jurgens et al. 2000). As shown in figure 6.26, the parameters thus found provide a precise coincidence of the results of modelling (solid line) with experimental data

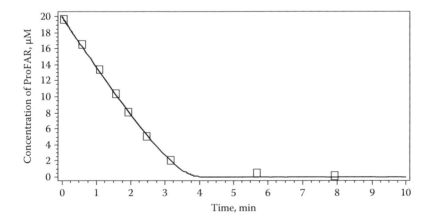

FIGURE 6.26 Time dependence of ProFAR concentrations: experimentally measured (open squares) and obtained from the kinetic model based on this experiment (solid line).

(squares; Jurgens et al. 2000). The values of fitted parameters are given in table 6.5.

### Application of the Model to Predict How the Synthetase and Glutaminase Activities of Imidazologlycerol-Phosphate Synthetase Depend on Concentrations of the Substrates and Effectors

As shown in Zalkin (1993) and Zalkin et al. (1998), imidazologlycerol-phosphate synthetase can catalyze two different processes: synthesis of imidazologlycerol phosphate (6.79) and decomposition of glutamine (6.80). To study what activity dominates and what is an accompanying one and how it depends on concentrations of the substrates and effectors, we plotted (figure 6.27) the rates of the glutaminase and synthetase reactions versus the substrate (PRFAR) concentration at various concentrations of effectors (IGP and ProFAR), zero concentrations of the products, pH 8.0, and a fixed concentration of glutamine ([Gln] = 1 mM), using the results of the previous section (the rate equations (6.11) and the values of parameters from the table). It appeared that in the absence of effectors the synthetase activity can be considered as dominating. In fact, as shown in figure 6.27a, at PRFAR concentrations exceeding 10 nM, the rate of the synthetase reaction is higher than the rate of the glutaminase reaction. Moreover, beginning from ~100 nM PRFAR, the rate of the synthetase reaction is more than ten times higher than that of the glutaminase reaction, and this difference grows with increase in PRFAR concentration, achieving 1000 times at [PRFAR] = 10 μM. In the presence of effector (ProFAR), the situation is not so clear-cut. As shown in figure 6.27b, at 10 μM ProFAR the rate of the synthetase reaction (curve 1) becomes equal to the rate of the glutaminase reaction (curve 2) at PRFAR concentration ~13 μM. At lower PRFAR concentrations, the rate of the glutaminase reaction is higher than the rate of the synthetase reaction, whereas at higher PRFAR concentrations the synthetase activity dominates. At higher Pro-FAR concentrations, the interval of PRFAR concentrations corresponding to domination of the glutaminase reaction expands. In fact, at 1 mM Pro-FAR the rate of the synthetase reaction (curve 3) becomes equal to the rate of the glutaminase reaction (curve 4) at PRFAR concentration ~1300 μM. Moreover, at PRFAR concentrations lower than 130 μM, the rate of the glutaminase reaction is more than ten times higher than the rate of the synthetase reaction; that is, the glutaminase activity becomes dominating. Analogous conclusions are true also for another effector—IGP (figure 6.27c). At 1 mM IGP, the rate of the synthetase reaction (curve 3)

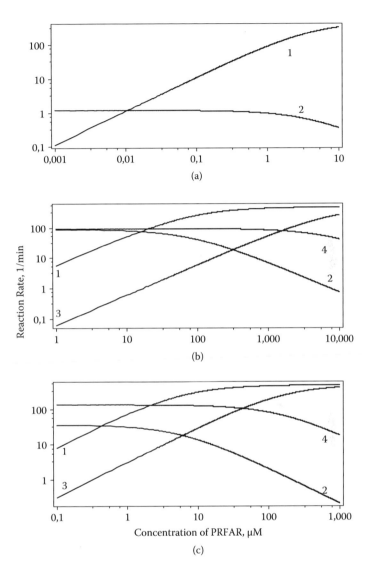

FIGURE 6.27 Rates of the synthetase (curves 1 and 3) and glutaminase (curves 2 and 4) reactions catalyzed by Imidazologlycerol-phosphate synthetase versus PRFAR concentration: a) In the absence of activators of the glutaminase reaction; b) At various ProFAR concentrations (10 μM Pro-FAR, curves 1 and 2; 1 mM ProFAR, curves 3 and 4); c) At various IGP concentrations (10 μM IGP, curves 1 and 2; 1 mM IGP, curves 3 and 4). All the curves were obtained at zero concentrations of the products and 1 mM Gln.

becomes equal to the rate of the glutaminase reaction (curve 4) at PRFAR concentration ~23 µM. As shown in figure 6.27c, increase in IGP concentration also results in expansion of the interval of PRFAR concentrations at which the glutaminase activity dominates.

As follows from figure 6.27, independently of the presence of the effectors—ProFAR and IGP—increase in the substrate (PRFAR) concentration results in decrease in the glutaminase and increase in the synthetase activities of imidazologlycerol-phosphate synthetase.

## ALLOSTERIC ENZYMES

Allosteric enzymes are enzymes that display nonhyperbolic sigmoidal dependences of reaction rate on concentration of substrates/products. An allosteric enzyme is an oligomer whose biological activity is affected by changing the conformation of its tertiary structure. In this section we describe an example of an allosteric enzyme operating in *E. coli*.

### Principles Used for Description of the Functioning of Allosteric Enzymes

Catalytic mechanisms of allosteric enzymes cannot be fully described in the framework of Michaelis-Menten kinetics. There are several mathematical models for allosteric enzymes available. The most well-known among them are the Hill model (Cornish-Bowden 2001), the Monod-Wyman-Changeux model (MWC; Monod, Wyman, and Changeux 1965), the Koshland-Nemethy-Filmer model (Koshland, Menethy, and Filmer 1966) and the Frieden-Kurganov model (Frieden and Colman 1967; Kurganov 1968). The most popular approaches to understanding the kinetics and regulation of individual allosteric enzymes are the first and the second one (Chassagnole et al. 2002; Shuster and Holzhutter 1995; Torres, Guixe, and Babul 1997).

These approaches take into account properties of the oligomeric enzymes; i.e., enzymes which consist of several identical subunits/monomers (in the case of phosphofructokinase-1 the number of monomers is four; Babul 1978; Berger and Evans 1991; Blangy, Buc, and Monod 1968; Deville-Bonne, Laine and Garel. 1991; Kotlarz and Buc 1982). The principal assumptions of the MWC approach are: (a) each subunit can exist in two conformational states, and each state is characterized by different dissociation constants of the substrates; (b) all subunits of the oligomeric

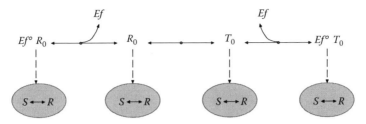

FIGURE 6.28 Generalized scheme of Monod-Wyman-Changeux model. $R_o$ and $T_o$ are free enzyme concentrations in R- and T-states, respectively; $Ef$ is the concentration of allosteric effector and $K_d{}^R$, $K_d{}^T$ its constant of dissociation with the enzyme states; $L_0$ is a constant of transition between enzyme states; $S \to P$ is subgraph representing interactions between enzyme and its substrates and products (in ellipse for R-state; in rectangle for T-state).

enzyme are expected to be only in the same conformational state, and for all the monomers transitions between different conformational states occur simultaneously (the concerted transition theory; Monod, Wyman, and Changeux 1965). This implies that the number of different conformational states of the oligomeric enzyme is equal to the number of conformational states of its single subunit. For example, if each monomer can be in two states, the catalytic cycle of the oligomeric enzyme should include only two conformational states, R (relax) and T (tense). Figure 6.28 shows the general scheme of this approach (Monod, Wyman, and Changeux 1965).

There is a generalization of the MWC approach proposed by Popova and Sel'kov (described in Popova and Sel'kov 1975, 1976, 1978). This approach allows us to take into account the following characteristics of enzyme operation:

(i)   reaction reversibility;

(ii)  catalytic cycles of the enzyme monomer including kinetic mechanisms with more than one substrate and product and competitive inhibition;

(iii) R (relax) and T (tense) states of the enzyme differ from each other not only in substrate affinity and Michaelis constants, but also in catalytic constants (Ewings and Doelle 1980).

Taking account of these properties of enzyme function enables us to develop a kinetic model of complex native enzymes.

A catalytic cycle of the enzyme represents a set of all the states of the enzyme and transitions between them. In the case of an oligomeric enzyme, a catalytic cycle is more complicated due to the presence of multiple subunits of the enzyme and, therefore, the higher number of substrates- and products-binding sites. For instance, the catalytic cycle of an enzyme consisting of a single subunit and operating in accordance with the ordered bi bi mechanism (Cleland 1963) includes four possible states, whereas the number of states of a dimer of the same enzyme increases up to ten (figure 6.29a). This multiplication of states is due to the fact that each monomer of oligomeric enzyme may run independently of other monomers. Thus, the catalytic cycle of the oligomer is a superposition of all the possible catalytic cycles of monomers of the enzyme. For the dimeric enzyme operating in accordance with the ordered bi bi mechanism, the catalytic cycle consists of four cycles of monomers (figure 6.29a). At this stage of a reconstruction of the catalytic cycle, competitive inhibition of the enzyme must be taken into account. Following the methods described in Cornish-Bowden (2001), it becomes possible to branch a catalytic cycle of the monomer of the enzyme (figure 6.29b), therefore increasing the number of enzyme states (extra complexes with metabolic regulators).

Allosteric regulation of the enzyme in the framework of this approach is taken into consideration via binding of effectors to regulatory sites of the enzyme and formation of additional complexes (figure 6.28). The ratio between R- and T-states influences the enzyme rate because of different catalytic properties of the R- and T-forms of the enzyme. An activator with more affinity to the R-form shifts the balance towards a more active R-state, leading to an increase in reaction rate; and an inhibitor, in turn, which binds more efficiently to the T-form, shifts the balance towards the T-state. This may essentially decrease the activity of the enzyme. The modulating properties of the effector (activation or inhibition) will depend only on the ratio of its constants of dissociation from different forms of the enzyme (Kurganov 1978).

## Kinetic Model of Phosphofructokinase-1 from *Escherichia coli*

Phosphofructokinase-1 (PfkA) belongs to the family of phosphotransferases (EN:2.7.1.11). PfkA catalyzes the transfer of $\gamma$-phosphate from ATP

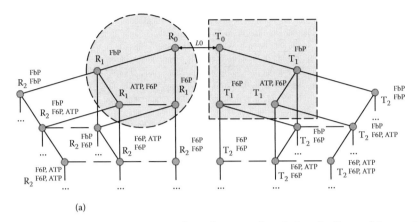

(a)

FIGURE 6.29 Reconstruction of catalytic cycle of phophofructokinase-1. (a) Binding of substrates (dissociation of products) with enzyme which operates in accordance with ordered bi bi mechanism. Solid black line corresponds to binding of the first substrate (F6P); dotted black line: binding of the second substrate (ATPMg$^{2-}$); dotted gray line: dissociation of the first product (ADPMg$^-$); solid gray line: dissociation of the second product (FbP). Catalytic cycles of single subunit of the enzyme at R- and T-states are marked out with gray ellipse and rectangle, correspondingly. (b) Detailed catalytic cycle of single subunit of phosphofructokinese-1 from *Escherichia coli*. In this scheme we summarized all information about functioning of the enzyme monomer, which we have taken into account. The kinetic mechanism is ordered bi bi in accordance with assumption (i) (see section "Reconstruction of a Catalytic Cycle of Phosphofructokinase-1"). Competitive inhibition with ATP$^{4-}$ leads to formation of the additional complex ($^{H}$E F6P ATP$^{4-}$) with constant of inhibition, $K_{i\_ATP}$. $K_{d\_ATPMg}$ is a constant of dissociation of ATP complex with magnesium ion, ATPMg$^{2-}$. Influence of pH on enzyme activity is shown as binding of two H$^+$ with enzyme. Inactive deprotonated and twice-protonated states of PfkA monomer are marked with gray color (see explanations in text).

to fructoso-6-phosphate (F6P). This reaction produces ADP and fructose-1,6-bisphosphate (FbP):

ATP + Fructoso-6-phosphate = ADP + Fructoso-1,6-bisphosphate.

PfkA is activated by ADP (Babul 1978; Blangy, Buc, and Monod 1968; Kotlarz and Buc 1982) and GDP (Ausat, Bras, and Garel 1997) and inhibited

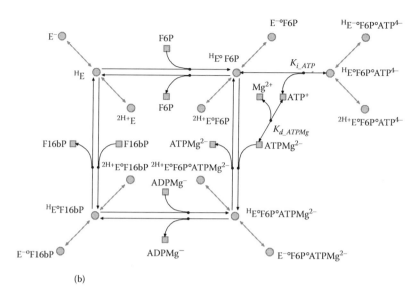

(b)

FIGURE 6.29 (Continued).

by phosphoenolpyruvate (PEP; Babul 1978; Blangy, Buc, and Monod 1968; Kotlarz and Buc 1982). It has been found (Reeves and Sols 1973) that binding sites for ADP, GDP and PEP differ from catalytic sites. This means that they are allosteric regulators of the enzyme [v].

In this section we develop a kinetic model of phosphofructokinase-1 from *E. coli* based on available structural and kinetic information on the one hand and taking into account catalytic, regulatory and allosteric properties of the enzyme as well as their dependence on pH and magnesium on the other hand. We apply the model to answer the following questions: (a) Is there a cooperativity of ATP binding to the enzyme; (b) What is the contribution of ADP inhibition to the enzyme function; (c) How does magnesium concentration affect operation of the enzyme?

*Available Experimental Data*

In this study, we used the following available data on structural and functional properties of phosphofructokinase-1 from *E. coli*:

1. PfkA has a quaternary structure, and the protein tetramer is the predominant form of enzyme in the cell (Babul 1978; Berger and Evans 1991; Blangy et al. 1968; Kotlarz and Buc 1982);

2. Sigmoid dependence of the initial reaction rate on the concentration of F6P has been demonstrated (Ausat, Bras, and Garel 1997; Berger and Evans 1991);

3. A complex of ATP with magnesium ions, $ATPMg^{2+}$ acts as the actual substrate (Babul 1978; Blangy, Buc, and Monod 1968; Kotlarz and Buc 1982);

4. PfkA is activated by purine nucleotide diphosphates, ADP (Babul 1978; Blangy, Buc, and Monod 1968; Kotlarz and Buc 1982) and GDP (Ausat, Bras, and Garel 1997) and inhibited by phosphoenolpyruvate (PEP; Babul 1978; Blangy, Buc, and Monod 1968; Kotlarz and Buc 1982);

5. Regulatory binding sites which are separate from catalytic ones have been described for the effectors of this enzyme, PEP and ADP (Reeves and Sols 1973);

6. Apparent Michaelis and binding constants depending on concentrations of ADP and PEP have been measured (Pham, Janiak-Spens and Reinhart 2001; Pham and Reinhart 2001);

7. The phosphofructokinase reaction runs slowly in the presence of other nucleotide phosphates (Blangy, Buc, and Monod 1968).

Most experimental data characterizing the kinetics of phosphofructokinase-1 from *E. coli* cells describe the dependences of the initial reaction rate on substrates, products, effectors and pH (Babul 1978; Blangy, Buc, and Monod 1968; Kotlarz and Buc 1982). Other quantitative data represent the dependence of fluorescence emission intensity on concentration of substrates and effectors of the enzyme (Reinhart 2001a, 2001b).

*Reconstruction of a Catalytic Cycle of Phosphofructokinase-1*

On the basis of the approach described previously, we have reconstructed a catalytic cycle of phosphofructokinase-1 from *E. coli*. For this purpose we used experimentally established findings on structural and functional properties of this enzyme described before and the following assumptions:

(i) The kinetic mechanism of Pfk-1 monomer is ordered bi bi. This assumption is based on the fact that Pfk-2 monomer operates in accordance with the same kinetic mechanism (Campos, Guixe, and Babul 1984; Guixe and Babul 1985).

(ii) Since the magnesium complex of ATP, $ATPMg^{2-}$ (1–3) is an actual substrate of the reaction, a free form of $ATP^{4-}$ can be considered as an inhibitor, which competes with $ATPMg^{2-}$ for catalytic site of the enzyme;

(iii) ADP as well as GDP can be bound to the allosteric site, which is specific for purine nucleotides.

In accordance with the assumption (i), the catalytic cycle of Pfk-1 monomer includes four key states which characterize interaction of the enzyme with substrates and products (figure 6.29). First F6P and then ATPMg bind to the enzyme and then phosphate is transferred; dissociation of products from the enzyme occurs in the reverse order; i.e., ADP first and then FbP (Campos, Guixe, and Babul 1984; Guixe and Babul 1985). In accordance with clause 3 (in the section 'Available Experimental Data') and assumption (ii), we have introduced into the catalytic cycle of the monomer a state of the enzyme bound to the free form of ATP ($ATP^{-4}$). This allows us to take into account competitive inhibition Pfk-1 by the free form of ATP under the conditions of an increased total concentration of ATP in the system with fixed concentration of $Mg^{2+}$. Based on clause 6 of the section 'Available Experimental Data', the ability of phosphofructokinase to catalyze reactions with other nucleotides (GTP, UTP, etc.) has been omitted (Blangy, Buc, and Monod 1968). The effect of pH on the enzyme activity has been taken into account by applying the method suggested by Cornish-Bowden (2001). This method is based on the assumption that there are two proton-binding sites in an active center of a protein. It implies that each state of the enzyme can be found in three forms: deprotonated, monoprotonated and diprotonated. Only the monoprotonated form is active. Summarizing the above, we have constructed a scheme of the catalytic cycle of a single subunit of phosphofructokinase-1 (figure 6.29b).

The existing experimental data (clauses 4 and 5 of 'Available Experimental Data') imply that the regulation of phosphofructokinase-1 is typical for most allosteric enzymes. On the basis of the data and also in accordance with the assumption (iii), allosteric regulation of PfkA can be represented in terms of the scheme depicted in figure 6.30. This scheme demonstrates that taking into account binding of allosteric regulators, PEP or purine nucleotides (ADP or GDP) adds five supplementary complexes to the catalytic cycle of PfkA for each state of the enzyme.

The complete catalytic cycle of Pfk-1 represents a superposition of the schemes depicted in figure 6.29a, figure 6.29b and figure 6.30. All details of the monomer catalytic cycle (figure 6.29b) should be incorporated into the catalytic cycle of the tetramer (figure 6.29a) to produce the complete catalytic cycle (figure 6.30). The subgraphs designated as S->P (specifying a set of reactions converting substrate, S, to product, P) of the complete catalytic cycle, enclosed in elliptical and rectangular panels, should be substituted for the R- and T-subgraphs, respectively, of the tetramer catalytic cycle.

## Derivation of a Rate Equation

In accordance with the method of generalization of the MWC model described in the section 'Allosteric Enzymes' (Ivanicky, Krinsky, and Sel'kov 1978), the rate equation for the oligomeric enzyme can be written in the following manner:

$$V = \frac{n^* f^* \left(1 + \left(\frac{f'}{f}\right) * Q\right)}{(1+Q)};$$

$$Q = L_o * \left(\frac{\left(1 + \frac{Effector}{K_{ef}^T}\right)}{\left(1 + \frac{Effector}{K_{ef}^R}\right)}\right)^n * \frac{E_R}{E_T}; \qquad (6.107)$$

where $f$ is a rate equation for the single subunit in the R-state; $f'$ is the same for the T-state; Q is a function that determines the ratio between the R and T forms of the enzyme; $E_R$ is a concentration of the free enzyme in the R-state; $E_T$ is a concentration of the free enzyme in the T-state; $L_o$ is a constant of transition between the R and T states of the enzyme; *Effector* is

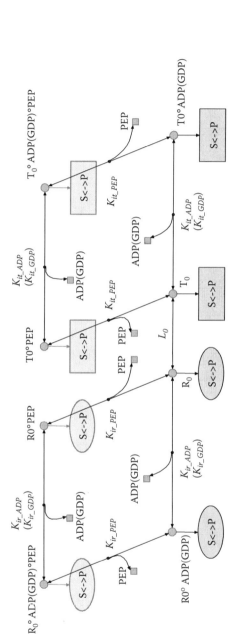

FIGURE 6.30 Scheme of allosteric regulation of phosphofructokinase-1. This corresponds to assumption about presence of two regulatory sites for PEP and purine nucleotides, ADP and GDP. $K_{ir\_PEP}$, $K_{ir\_ADP}$, $K_{ir\_GDP}$ are constants of effectors binding with R-state of phophofructokinase-1 and $K_{it\_PEP}$, $K_{it\_ADP}$, $K_{it\_GDP}$ with T-state. $L_0$ is a constant of transition between enzyme states; $S \to P$ is a subgraph representing interactions between enzyme and its substrates and products (in ellipse for R-state; in rectangle for T-state).

a concentration of the allosteric effector; $K_{ef}{}^R$ and $K_{ef}{}^T$ are constants of dissociation of the effector for the R and T forms of the enzyme, respectively; and $n$ is the number of subunits of the enzyme.

The rate equation of the reaction, catalyzed by the PfkA monomer in the R-state, has been derived on the basis of its catalytic cycle (figure 6.29b) using algorithms described in Cleland (1963):

$$f = \frac{V_{mr}^{forward} * \left( ATPMg^{2-} * F6P - \dfrac{ADPMg^- * F16bP}{Keq} \right)}{Z_{SP}^R * Z_{pH}};$$

$$Z_{pH} = 1 + \dfrac{H^+}{K_{d\_H\_1}} + \dfrac{K_{d\_H\_2}}{H^+};$$

$$Z_{SP}^R = K_{ir\_F6P} * K_{mr\_ATPMg^{2-}} + K_{mr\_ATPMg^{2-}} * F6P * \left( 1 + \dfrac{ATP^{4-}}{K_{ir\_ATP^{4-}}} \right)$$

$$+ K_{mr\_F6P} * ATPMg^{2-} + ATPMg^{2-} * F6P$$

$$+ K_{mr\_F6P} * ATPMg^{2-} * \dfrac{FbP}{K_{ir\_FbP}} + \dfrac{ATPMg^{2-} * F6P * ADP}{K_{ir\_ADP}}$$

$$+ K_{mr\_F6P} * \dfrac{ATPMg^{2-} * ADP * FbP}{Wr * K_{mr\_ADP} * K_{ir\_FbP}} + \dfrac{Wr}{Keq}$$

$$* \left( K_{mr\_FbP} * \dfrac{F6P * ADP}{K_{ir\_F6P}} + K_{mr\_ADP} * FbP + K_{mr\_FbP} * ADP + ADP * FbP \right);$$

$$(6.108)$$

Here, $K_{mr\_ATPMg2-}$, $K_{mr\_F6P}$, $K_{mr\_ADP}$, $K_{mr\_FbP}$ are Michaelis constants of substrates and products; $K_{ir\_F6P}$, $K_{ir\_ADP}$, $K_{ir\_F6P}$, $K_{ir\_ATP}{}^{4-}$ are constants of dissociation of substrates, competitive inhibitor $ATP^{4-}$ and products; $K_{d\_H\_1}$, $K_{d\_H\_2}$ are constants of proton binding in the active center of the enzyme; $V_{mr}^{forward}$, $V_{mr}^{reverse}$ are maximum rates of forward and reverse reactions, respectively; $Wr = V_{mr}^{forward}/V_{mr}^{reverse}$ represents the ratio of the maximum rates of forward and reverse reactions for the R-state of the monomer; $Wt$ is the similar value for the T-state of the monomer; and $K_{eq}$ is the equilibrium constant of the reaction. In accordance with the Holdein equation (Cleland 1963) we can express $K_{eq}$ in terms of kinetic constants:

$$Keq = Wr * \frac{K_{ir\_FbP} * K_{mr\_ADP}}{K_{ir\_F6P} * K_{mr\_ATPMg^{2-}}} = Wt * \frac{K_{it\_FbP} * K_{mt\_ADP}}{K_{it\_F6P} * K_{mt\_ATPMg^{2-}}}; \quad (6.109)$$

The rate equation of the reaction catalyzed by the monomer of the enzyme in the T-state ($f'$), can be derived in a similar way and differs from equation (6.108) only in parameters quantifying kinetic properties of the T-state. The expression for $f'$ can be obtained from equation (6.108) by the following substitutions:

$$K_{mr\_ATPMg} \to K_{mt\_ATPMg}; K_{mr\_F6P} \to K_{mt\_F6P}; K_{ir\_F6P} \to K_{it\_F6P}; K_{ir\_ATP} \to K_{it\_ATP};$$
$$K_{efr\_PEP} \to K_{eft\_PEP}; K_{efr\_ADP} \to K_{eft\_ADP}; K_{efr\_GDP} \to K_{eft\_GDP}; K_{ir\_FbP} \to K_{it\_FbP};$$
$$K_{mr\_FbP} \to K_{mt\_FbP}; K_{ir\_ADP} \to K_{it\_ADP}; K_{mr\_ADP} \to K_{mr\_ADP}; V_{mr} \to V_{mt}; W_{mr} \to W_{mt}.$$

Functions $Q$, $Er$ and $Et$ have been derived on the basis of the main regulatory features of PfkA depicted in figure 6.30 in such a way as to take into account the action of all the allosteric effectors. According to clauses 4 and 5 of 'Available Experimental Data', we assume that ADP and GDP bind in the same regulatory site, while PEP binds to another site. Thus, an expression for the functions $Q$, $Er$ and $Et$ of phosphofructokinase-1 can be written as follows:

$$Q = Lo * \left( \frac{\left(1 + \frac{ADP}{K_{eft\_ADP}} + \frac{GDP}{K_{eft\_GDP}}\right)\left(*(1 + \frac{PEP}{K_{eft\_PEP}})\right)}{\left(1 + \frac{ADP}{K_{efr\_ADP}} + \frac{GDP}{K_{efr\_GDP}}\right) * (1 + \frac{PEP}{K_{efr\_PEP}})} * \frac{E_r}{E_t} \right)^{,n}$$

$$Er = \frac{K_{ir\_F6P} * K_{mr\_ATPMg^{2-}} * Eo}{Z_{SP}^R};$$

$$Et = \frac{K_{it\_F6P} * K_{mt\_ATPMg^{2-}} * Eo}{Z_{SP}^T};$$

$$(6.110)$$

where $Eo$ is the total concentration of the enzyme.

Substituting equations (6.108), (6.109) and (6.110) into equation (6.107), we obtain the complete rate equation of the reaction catalyzed by phosphofructokinase-1.

### Verification of the Model against Experimental Data

The kinetic model of PfkA includes thirty parameters. Values of model parameters are given in table 6.6. The values of three of them have been taken from literature sources. The number of enzyme subunits, $n$, is equal to 4, since phosphofructokinase-1 is a tetramer (Babul 1978; Blangy, Buc, and Monod 1968; Kotlarz, Buc (1982); Berger, Evants 1991). The dissociation constant of ATP with magnesium ions, $K_{d\_ATPMg}$, is equal to 58.8 μM (Taqui Khan and Martell 1962). The ratio of the maximum rates

TABLE 6.6   Values of PfkA Model Parameters and Ranges of Sensitivity

| Parameter | Value (mM) | Range of Sensitivity (mM) |
|---|---|---|
| $K_{mr\_ATPMg}$ | 8.1e-5 | 6.5e-5 –1.14e-4 |
| $K_{mr\_F6P}$ | 2.1e-5 | 1.2e-5 –2.8-4 |
| $K_{ir\_F6P}$ | 1.84 | 0.53-7.5 |
| $K_{ir\_ATP}$ | 2.5e-05 | 7e-6 – 7e-5 |
| $K_{efr\_PEP}$ | 20 | 7 – 35 |
| $K_{efr\_ADP}$ | 0.074 | 0.05 – 0.11 |
| $K_{efr\_GDP}$ | 0.33 | 0.14 – 0.58 |
| $K_{ir\_FbP}$ | 0.046 | 0.04 – 0.055 |
| $K_{mr\_FbP}$ | 10 | 2.8 – 21 |
| $K_{ir\_ADP}$ | 55 | 12 – 95 |
| $K_{mr\_ADP}$ | 0.69 | 0.45 – 1.16 |
| $K_{mt\_ATPMg}$ | 3.35 | 1.86 – 11.5 |
| $K_{mt\_F6P}$ | 33.1 | 12 – 45 |
| $K_{it\_F6P}$ | 0.0086 | 0.0022 – 0.014 |
| $K_{it\_ATP}$ | 0.014 | 0.003 – 0.07 |
| $K_{eft\_PEP}$ | 0.26 | 0.22 – 0.3 |
| $K_{eft\_ADP}$ | 9 | 2 – 16 |
| $K_{eft\_GDP}$ | 9 | 3 – 15 |
| $K_{it\_FbP}$ | 50.5 | 41 – 68 |
| $K_{mt\_FbP}$ | 10 | 1 – 200 |
| $K_{it\_ADP}$ | 80 | 20 – 300 |
| $K_{mt\_ADP}$ | 2 | 0.1 – 10 |
| $L_o$ | 14.4 | 12 – 17 |
| $W_t$ | 0.75 | 0.1 – 10 |
| $W_r$ | 12.5 | ND[b] |
| w | $10^4$ | ND |
| $K_{d\_ATPMg}$ | 0.058 | ND |
| $K_{d\_H\_1}$ | 3.8e–12 | 1e-12 – 3e-8 |
| $K_{d\_H\_2}$ | 6.97e-5 | 5.7-5 – 8.9e-5 |
|  | 0.197 | 0.157 – 0.233 |
|  | 0.0011 | 0.0005 – 0.005 |
|  | 30 | 9.8 – 61 |
| $K_{efr\_GD}$ | 0.012 | 001 – 0.018 |
| $K_{ir\_GDP}$ |  |  |
| $K_{eft\_GDP}$ |  |  |
| $K_{it\_GDP}$ |  |  |

of forward and reverse reactions of the R-state of the monomer, $Wr$, has been calculated in the following manner. As has been found in previous work (Babul 1978), the rate of the reverse reaction, i.e., ADP phosphorylation with FbP, constitutes about 8 percent of the forward rate under physiological conditions. This means that $Wr$ is equal to 12.5. Two parameters of the model ($V_{mr}^{forward}$ and $V_{mt}^{forward}$) cannot be estimated on the basis of *in vitro* experimental data. The values of the parameters depend on experimental conditions and the amount of the enzyme in the experimental assay. However, based on experimental data, their ratio ω has been estimated as $10^4$. This means that the enzyme in the T-state is actually inactive, which is typical of most allosteric enzymes (Kurganov 1978).

Values of other model parameters have been estimated by fitting of the rate equation to corresponding experimental data to provide the best coincidence between experimental and model calculated curves (see figure 6.31 and figure 6.33a). Overall, fourteen experimental curves consisting of ninety-six experimentally measured points published in Ausat, Bras, and Garel (1997), Babul (1978), Deville-Bonne, Laine and Garel. (1991); Pham, Janiak-Spens and Reinhart 2001; Pham and Reinhart 2001 were used for estimation of twenty-six model parameters. These experimental data represent *in vitro* measured dependencies of the initial rate of a reaction catalyzed by phosphofructokinase-1 on the concentrations of substrates, products and effectors of the enzyme and on pH.

All model parameters have been examined for their sensitivity (see table 6.6). For two of the parameters ($K_{mt\_FbP}$, $K_{it\_ADP}$), a very low level of sensitivity has been found, so we have taken for them the same values as for the similar R-form parameters ($K_{mr\_FbP}$, $K_{ir\_ADP}$). This means that in spite of much experimental information about phosphofructokinase-1 there is a lack of data characterizing the kinetics of the reverse reaction and of product inhibition of the enzyme. Time dependences of concentration changes of products and substrates at different initial concentration could fill the gap and would be very helpful for better verification of kinetic models of phosphofructokinase-1.

*Predictions of the Model*

In the previous section we have shown that the model of phosphofructokinase-1 has been verified against experimental data; i.e., it simulates such experimentally established properties of PfkA as sigmoid dependencies of the initial rate of the reaction on F6P concentration, the effect of allosteric effectors and $Mg^{2+}$ ions on the reaction rate and the dependence of the maximum rate of the reaction on pH of the medium. Below we consider the properties of PfkA, which are predicted by our model. These properties of the enzyme have not yet been clearly supported by experimental data but are derived from the catalytic mechanism and values of kinetic parameters.

Phosphofructokinase-1 represents a pronounced cooperativity of binding both fructose-6-phosphate and $ATPMg^{2-}$. The difference between F6P and $ATPMg^{2-}$ Michaelis constants in the R- and T-states is several orders of magnitude (table 6.6). Generally, researchers focus their attention only on cooperativity of F6P binding since the inflection point of the dependence of the initial rate of the reaction on the concentration of this substrate can be observed within the range 0.2–0.4 mM (figure 6.31b and figure 6.31d). For the second substrate, $ATPMg^{2-}$, no significant cooperativity has been observed (Ausat, Bras, and Garel 1997; Babul 1978; Blangy, Buc, and Monod 1968). However, our model predicts that cooperativity of $ATPMg^{2-}$ is important but an inflection point can be observed at lower concentration that that for F6P (near 0.05 mM). To demonstrate this effect, it is necessary to plot the dependence of the initial rate of the reaction on ATP at lower concentrations of this substrate (figure 6.32). This has not yet been done in any experiments aimed at the study of the kinetic properties of phosphofructokinase-1 (Ausat, Bras, and Garel 1997; Babul 1978; Blangy, Buc, and Monod 1968).

One more interesting prediction of our model is the observation that at a total concentration of ATP of more than 10 mM, and a fixed concentration of $Mg^{2+}$ ions (10 mM), substrate inhibition occurs. At a total concentration of ATP of 20 mM, the rate of the phosphofructokinase reaction is only 20 percent of maximum rate (figure 6.33a). The mechanism of this inhibition is explained by the presence in the system of free-form $ATP^{4-}$ (figure 6.33b), which, as has been already mentioned, acts as an inhibitor. It competes with the actual substrate of the reaction, $ATPMg^{2-}$, for the binding site in the active center of the enzyme. Values of estimated parameters also point to the possibility of substrate inhibition (table 6.6). Thus, the free form of ATP exhibits a stronger affinity to the enzyme compared to the magnesium form. The physiological role of substrate inhibition of this type is as yet unknown. On the one hand, it seems possible that under *in vivo* conditions this inhibition would not play an important role since the total concentration of magnesium ions in *E. coli* cells (about 10 mM; Sundararaj et al. 2004) exceeds the total concentration of ATP (under normal conditions it ranges within 3–5 mM; Chassagnole et al. 2002; Hoque et al. 2005; curve 1 in figure 6.33c). This implies that the greater part of the ATP pool in the cell is complexed with magnesium ions and inhibition of phosphofruktokinase-1 by the

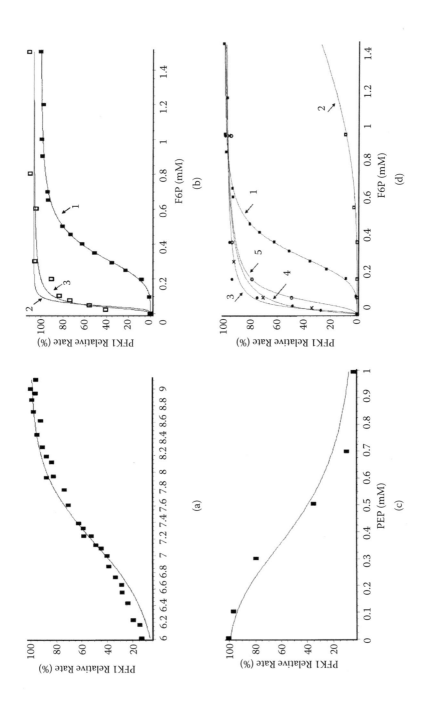

FIGURE 6.31 Coincidence between model curves and experimental data. (a) pH optimum of phosphofructo-kinase-1, experimental data (■). (b) Cooperativity of F6P binding and GDP activation of the phosphofruc-tokinase-1. The first set of experimental data (■) was measured at ATP = 1 mM, Mg$^{2+}$ = 10 mM, pH = 8.2 and absence of GDP; the second one (□) was measured under the same conditions with addition of 2 mM of GDP. Curve 1 corresponds to the first set of data; curves 2 and 3 correspond to the second one and show how two variants of the model (with competitive inhibition of GDP, curve 3 and without, curve 2) coincide with experimental data. (c) Allosteric character of PEP inhibition of phosphofructokinase -1, experimental data (■). Experimental conditions are same as for the first set of data in figure 6.32b with concentration of the second substrate F6P equal to 1 mM. (d) Cooperativity of ADP activation and product inhibition. Curve 1 and set of the data (■) correspond to absences of any allosteric regulators; curve 2 and set of the data (□) correspond to presence of 1 mM of PEP and absence of ADP; curve 3 and set of the data (●) correspond to absence of PEP and presence 0.5 mM of ADP; curve 4 and set of the data (×) correspond to presence of 1 mM PEP and 0.5 mM of ADP; curve 5 and set of the data (○) correspond to presence of 1 mM PEP and 1 mM of ADP; other conditions are the same as for the first set of data in figure 6.32b.

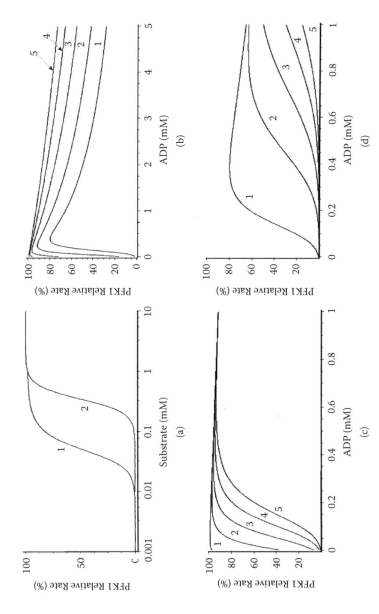

FIGURE 6.32 Cooperativity of phosphofructokinase-1 substrates binding. Curve 1 corresponds to ATPMg$^{2-}$ and curve 2 to F6P.

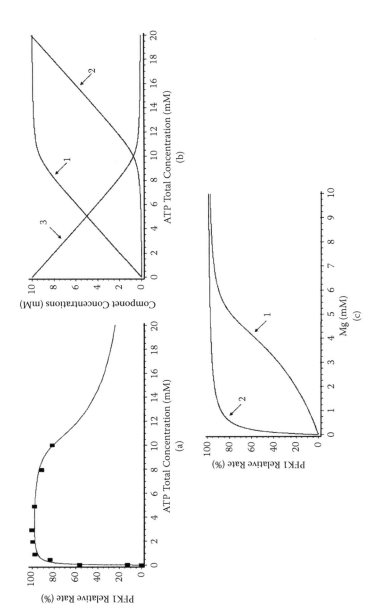

FIGURE 6.33  Effect of competitive inhibition by ATP⁴⁻. (a) Coincidence between model curves and experimental data (■) published in Babul (1978). Experimental conditions are F6P = 1 mM, Mg²⁺ = 10 mM, pH = 8.2. (b) Appearance of ATP free form (curve 2) under increase of total ATP concentration in system with constant concentration of magnesium ions; curve 1, magnesium complex of ATP and curve 3, free Mg²⁺. (c) Influence of competitive inhibition of ATP⁴⁻ on activity of phosphofructokinase-1 at physiological substarte concentrations (F6P = 1mM, ATP = 4 mM). Curve 1 is a dependence of activity of the enzyme from total concentration of magnesium ions in the system; curve 2 is a dependence of activity of the enzyme from concentration of free magnesium ions.

free form of ATP should be neglible (only 3 percent). On the other hand, the total concentration of magnesium ions in the cell cannot be a realistic indicator of this kind of inhibition, because most magnesium ions are bound with different intracellular substrates. So, a more useful characteristic for this is the intracellular concentration of free magnesium ions. There have been several studies aimed at estimating this parameter in eukaryotic cells, where it varies in range between 0.4 and 1.2 mM (Romani and Scarpa 2000). These values of concentration of free magnesium ions correspond to $ATP^{4-}$ inhibition of phosphofructokinase-1 in the range between 11 and 17 percent (curve 2 in figure 6.33c). Measurements of the concentration of free $Mg^{2+}$ in bacterial cells have been made only once using *Salmonella enterica* (Froschauer et al. 2004). In this work, the authors showed that the concentration of free $Mg^{2+}$ varied in the range between 0.95 and 1.4 mM depending on the external concentration of magnesium ions in the medium. These values correspond to $ATP^{4-}$ inhibition of phosphofructokinase-1 in the range between 9 and 13 percent (curve 2 in figure 6.33c). Summing all these facts, we conclude that, under normal conditions, inhibition of phosphofructokinase-1 in *E. coli* by $ATP^{4-}$ is about 10 percent. However, it is possible that, under some specific conditions, this kind of inhibition increases (up to 15–20 percent) and becomes essential for enzyme function.

One further property of phosphofructokinase-1 which has been predicted by our model is the following: in addition to allosteric activation of ADP, a noticeable product inhibition of enzyme activity by this metabolite has been found. In other words, ADP at high concentrations could compete with $ATPMg^{2-}$ for binding to the catalytic site. At small concentrations of ADP (less than 2 mM), product inhibition has been found not to have an obvious effect on the rate of the forward reaction. However, this inhibition has a perceptible contribution under the joint action of allosteric effectors (curves 3 and 4 in figure 6.31d). Indeed, to fit the PfkA model to the experimental data depicted in figure 6.31d properly, it was necessary to introduce ADP product inhibition into the catalytic cycle (and the corresponding rate equation). The value of the parameter responsible for ADP product inhibition, $K_{mr\_ADP}$, is equal to 0.69 mM. This evidences quite a high affinity of ADP molecule to the catalytic site of phosphofructokinase-1 and supports our hypothesis that ADP product inhibition can have a perceptible contribution to enzyme operation under physiological conditions. It should also be noted that the ability of ADP to inhibit PfkA activity at a concentration of more than 0.5 mM has been shown previously (Blangy, Buc, and Monod 1968).

Our model predicts that in addition to allosteric activation by GDP, a noticeable product inhibition of enzyme activity by this metabolite has been found. Indeed, the model of PfkA which does not take into account GDP product inhibition demonstrates perceptible discrepancy with experimental data (curve 2 in figure 6.31b). This discrepancy can be explained in terms of the fact that activation of phosphfructokinase-1 with the GDP, as in the case with ADP, is modulated by competitive inhibition, when GDP is binding at the active center of the enzyme instead of $ATPMg^{2-}$. To fit the experimental data depicted in figure 6.31b properly we should introduce into the catalytic cycle (and the corresponding rate equation) GDP product inhibition (see curve 3 in figure 6.31b). This property of phosphofructokinase-1 can be related to the possibility of this enzyme to catalyze phosphorylation of F6P with GTP as another phosphate donor. The value of the apparent Michaelis constant of PfkA for $GTPMg^{2-}$, which is equal to 1.2 mM (Blangy, Buc, and Monod 1968), is indirect evidence for a rather high affinity of GDP for phosphofructokinase-1. Hence, there is the possibility of product inhibition by GDP. This situation is observed for various kinases able to utilize different nucleotide phosphates as substrates, for instance, for pyruvatekinase-1 from *E. coli* cells (Waygood and Sanwal 1974).

## TRANSPORTERS

Transporters are integral membrane proteins that catalyze reactions of transfer of substances from one compartment to another without chemical transformation of the substances. Due to intramembrane localisation, the function of the transporters can depend on electric potential across the membrane and gradients of different substances in different compartments. All these peculiarities which distinguish the function of transporters from that of ordinary enzymes are taken into account in the rate equations.

### Kinetic Model of Mitochondrial Adenine Nucleotide Translocase

Adenine nucleotide translocase (ANT) catalyzes electrogenic ATP and ADP exchange across the inner mitochondrial membrane. The transporter provides ATP efflux into the cytosol in exchange for ADP entering the matrix. It maintains a high concentration of ADP in mitochondria. It has been shown that ANT limited flux via oxidative phosphorylation under many different sets of conditions (Davis and Davies-Thienen 1984; Kholodenko 1984a, 1984b; Kholodenko 1988).

In this section we develop a valid kinetic model of mitochondrial adenine nucleotide translocase on the basis of all the available structural and functional information. We describe the dependence of the exchange rate on membrane potential and adenylate distributions on each side of the mitochondrial membrane.

Using the model and estimated parameters we have predicted how the rate of the ATP/ADP translocation depends on concentrations of adenine nucleotides, pH and electric potential difference under conditions close to the physiological. We have predicted how the effectiveness of ANT operation is controlled by adenylate concentrations and potential.

### Experimental Data for Model Verification

To develop and verify a kinetic model of ANT quantitatively describing its functioning, we used experimental data obtained by Kraemer and Klingenberg (1982). To characterize exchange kinetics they used a translocator built in liposomes. This experimental technique enabled them to avoid influences of intramitochondrial components and inhibitory effects of bivalent cations and allowed them to measure dependences of the initial rate of influx of labelled adenylates on their bulk phase concentration at different concentrations of internal adenylate, pH and membrane potential. We used these experimental data to verify our model and estimate its parameters. It is worth mentioning that in all experiments ATP and ADP concentrations inside liposomes were much higher than the dissociation constants of the internal ANT-binding site. This allowed us to estimate only the ratio of dissociation constants for ATP and ADP but not the values themselves.

### Antiporter Functioning Mechanism

Based on X-ray structural analysis (Pebay-Peyroula et al. 2003), we assumed that the adenylate nucleotide translocase functional unit was a dimer operating as two mutually concerted channels transporting two molecules of adenylates in opposite directions. As the properties of the subunits are

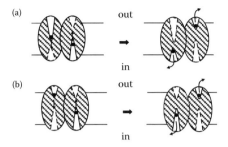

FIGURE 6.34    Possible positions of the subunits in the functional unit of ANT. (a) Isotopic position of the *trans*-type: on both sides of the membrane there are identities for binding and transfer. (b) Anisotropic position of the *cis*-type: there is only one type of site on each side of the membrane.

anisotropic, i.e., there are different binding sites of the protein facing into the mitochondria matrix and intramembrane space (Pebay-Peyroula et al. 2003), two possible mechanisms could be proposed. Figure 6.34 illustrates the possible inter-arrangement of the subunits in the dimer. The trans-scheme (figure 6.34a) refers to the situation in which adenylate binding sites for both monomers on each side of the membrane are the same in either state of the antiporter functional unit. If the trans-scheme were realized, the transporter's dimer in any state would have the same properties on each side of the membrane, contradicting the ANT anisotropy properties (Brandolin et al. 1980). For this reason, the cis-scheme of the location (figure 6.34b) is more likely, because the transporter in any state has only one type of binding site on each side of the membrane.

### Kinetic Scheme

In accordance with the functioning mechanism (cis-scheme, figure 6.34b) of adenine nucleotide translocase chosen in the previous section as the most probable one we propose an appropriate kinetic scheme as depicted in figure 6.35. This scheme corresponds directly to the bi-bi random mechanism according to the Cleland's (1963) classification. Two stages are involved in the formation of the ternary complexes between the antiporter and adenylates (*TET, TED, DET, DED*). As shown in figure 6.35, ATP or ADP binds to binding sites of ANT facing each side of the membrane

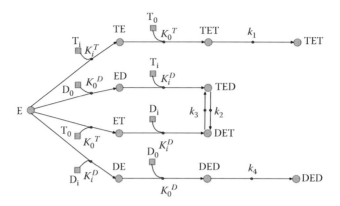

FIGURE 6.35 The kinetic scheme of the catalytic cycle of the mitochondrial adenine nucleotide translocase. $T$, $D$ are understood as ATP, ADP, respectively; the lower index means the location: $i$, inside the matrix; $o$, within the intramembrane space. $E$ is free form of enzyme. $XEY$ is ternary complex of an ANT dimer with molecules $X$ from matrix and $Y$ from intramembrane space. Lines show processes of the complex formation. Arrows show transfer processes. The scheme shows constants of the transfer rates and dissociation constants for the corresponding stages.

(dissociation constants are indicated near the corresponding reactions). Formation of the complexes induces transfer of adenylates in opposite directions with their dissociation afterwards to leave binding sites for the next cycle. Rate constants of each elementary transfer stage are shown in the scheme. Rate constants $k_2$, $k_3$ describe rates of direct and indirect exchange of ATP for ADP. Rate constants $k_1$, $k_4$ correspond to exchange of identical adenylates (ATP for ATP or ADP for ADP) as studied by Kraemer and Klingenberg (1982).

*Derivation of Rate Equation*

To simplify the process of model construction and analysis we assume that:

1. Association/dissociation stages of the ANT catalytic cycle (designated as lines in figure 6.35) are in quasi-equilibrium. This directly corresponds to the conclusion of Kraemer and Klingenberg (1982) that transmembrane adenylate transfer (designated by arrows) limits ANT functioning.

2. ANT is able to carry only deprotonated and magnesium-free forms of adenylates. Indeed, it was previously shown that adenylates bound to magnesium are not involved in exchange catalyzed by ANT (Klingenberg 1980; Pfaff, Heldt, and Klingenberg 1969). Furthermore, it has been demonstrated that only deprotonated ADP and ATP molecules carrying the charges 3- and 4-, respectively, were able to be transferred across the membrane (Klingenberg and Rottenberg 1977; Vignais et al. 1985).

3. The electrostatic field of the charged membrane influences both the stage of transfer and the stage of ternary complex formation.

According to the scheme of the catalytic cycle, the full turnover number per unit of time may be expressed as follows:

$$v = k_1 \cdot TET + k_2 \cdot TED + k_3 \cdot DET + k_4 \cdot DED, \qquad (6.111)$$

where $TET, TED, DET, DED$ are concentrations of corresponding antiporter states with respect to the internal volume of liposomes; i.e., the ratio of the number of molecules in each state to the volume of liposomes. The concentrations of different states may be expressed in terms of adenylate concentrations using equilibrium relationships corresponding to association/dissociation stages:

$$TE = E \frac{T_i}{K_i^T}, \quad ED = E \frac{D_o}{K_o^D}, \quad ET = E \frac{T_o}{K_o^T}, \quad DE = E \frac{D_i}{K_i^D},$$

$$TET = E \frac{T_i \cdot T_o}{K_i^T K_o^T}, \quad DET = E \frac{D_i \cdot T_o}{K_i^D K_o^T}, \quad TED = E \frac{T_i \cdot D_o}{K_i^T K_o^D}, \quad DED = E \frac{D_i \cdot D_o}{K_i^D K_o^D},$$

$$(6.112)$$

where $T$, $D$ refer to concentrations of deprotonated adenylates (ATP and ADP, respectively). The lower indices $o$ and $i$ imply their location: $o$, outside the liposome or $i$, inside the liposome. The concentration of the free form of transporter $E$ may be derived from the equation expressing the total ANT concentration as the sum of concentrations of its different states:

$$E + TE + ET + DE + TET + DET + TED + DED = E_0. \qquad (6.113)$$

By rearranging equations (6.111), (6.112) and (6.113) we obtain

$$E = E_0/\Delta$$

where

$$\Delta = 1 + \frac{T_i}{K_i^T} + \frac{T_o}{K_o^T} + \frac{D_i}{K_i^D} + \frac{D_o}{K_o^D} + \frac{T_i \cdot T_o}{K_i^T K_o^T} + \frac{D_i \cdot T_o}{K_i^D K_o^T} + \frac{T_i \cdot D_o}{K_i^T K_o^D} + \frac{D_i \cdot D_o}{K_i^D K_o^D}$$

$$= \left(1 + \frac{T_o}{K_o^T} + \frac{D_o}{K_o^D}\right)\left(1 + \frac{T_i}{K_i^T} + \frac{D_i}{K_i^D}\right),$$

$$v' = \frac{v}{E_0} = \frac{1}{\Delta}\left(k_1 \frac{T_i \cdot T_o}{K_i^T K_o^T} + k_2 \frac{T_i \cdot D_o}{K_i^T K_o^D} + k_3 \frac{D_i \cdot T_o}{K_i^D K_o^T} + k_4 \frac{D_i \cdot D_o}{K_i^D K_o^D}\right), \qquad (6.114)$$

where $v'$ is the turnover number for a single functional unit of ANT. To estimate parameters of the equation we fitted it against experimentally measured dependencies of initial exchange rate on concentration of external adenylates (Kraemer and Klingenberg 1982). Since these dependencies have been measured under the condition that concentration of inner adenylates is much higher than their affinity to the ANT-binding site facing the internal space of the liposome, the following inequality is true:

$$T_i/K_i^T + D_i/K_i^D \gg 1.$$

By using this inequality, we can simplify the denominator $\Delta$ of equation (6.114):

$$\Delta \cong \left(1 + \frac{T_o}{K_o^T} + \frac{D_o}{K_o^D}\right)\left(\frac{T_i}{K_i^T} + \frac{D_i}{K_i^D}\right).$$

Multiplying the numerator and denominator of equation (6.114) by $K_i^D$ we can reduce the number of parameters by one.

$$v' = \frac{1}{\Delta'}\left(k_1 q \frac{T_i \cdot T_o}{K_o^T} + k_2 q \frac{T_i \cdot D_o}{K_o^D} + k_3 \frac{D_i \cdot T_o}{K_o^T} + k_4 \frac{D_i \cdot D_o}{K_o^D}\right),$$

$$\qquad (6.115)$$

$$\Delta' = \Delta \cdot K_i^D = \left(1 + \frac{T_o}{K_o^T} + \frac{D_o}{K_o^D}\right)(D_i + q T_i).$$

where $q = K_i^D/K_i^T$ shows the affinity ratio of the internal ANT-binding site for ATP and ADP.

Equation (6.115) represents the turnover rate of the antiporter. It describes both the 'productive cycles' (exchange of ATP for ADP) and 'futile cycles' (exchange of ATP for ATP or ADP for ADP). Since each turnover results in transport of an adenylate molecule across the membrane, equation (6.115) describes the dependence of the influx rate of the labelled adenylates on their external and internal concentrations. These dependencies have been measured by Kraemer and Klingenberg (1982) and we will utilize them to identify the unknown parameters of the model.

To describe ANT functioning in mitochondria we have to derive an equation expressing the rate of changes in adenylate concentrations (not turnover rate!) as a function of their matrix and intermembrane space concentrations. This means that only productive cycles of ANT should be taken into account. Since the stoichiometry of ANT is 1:1, ADP concentration change is equal to that of ATP but with the opposite sign:

$$\frac{dD_i}{dt} = -\frac{dT_i}{dt} = v_{exchange}$$

where $v_{exchange}$ stands for the exchange rate; i.e., the rate of changes in adenylate concentrations.

According to the scheme of the ANT catalytic cycle (figure 6.35) the exchange rate may be expressed as follows:

$$v_{exchange} = k_2 \cdot TED - k_3 \cdot DET.$$

This equation takes into account only productive cycles; i.e., those resulting in exchange of ATP for ADP but not ATP for ATP or ADP for ADP. Similarly to equations (6.111)–(6.115), we can derive the exchange rate per dimer of ANT as a function of concentrations of adenylates and the kinetic parameters.

$$v'_{exchange} = \frac{1}{\Delta'}\left(k_2 q \frac{T_i D_o}{K_o^D} - k_3 \frac{D_i T_o}{K_o^T}\right) \qquad (6.116)$$

Using this equation we can express the equilibrium constant of exchange.

$$\begin{cases} K_{eq} = \left(\frac{T_o \cdot D_i}{D_o \cdot T_i}\right)_{eq}, \\ \left(\frac{dD_i}{dt}\right)_{eq} = \left(\frac{dT_i}{dt}\right)_{eq} = \left(\frac{dD_o}{dt}\right)_{eq} = \left(\frac{dT_o}{dt}\right)_{eq} = 0. \end{cases}$$

From this it follows that

$$v'_{exchange} = \left( \frac{1}{\Delta'} \left( k_2 q \frac{T_i D_o}{K_o^D} - k_3 \frac{D_i T_o}{K_o^T} \right) \right)_{eq} = 0,$$

$$K_{eq} = \left( \frac{T_o \cdot D_i}{D_o \cdot T_i} \right)_{eq} = q \cdot \frac{k_2}{k_3} \cdot \frac{K_o^T}{K_o^D}; \qquad (6.117)$$

Using this equation we can express the parameter $q$ in terms of other parameters:

$$q = \frac{k_3}{k_2} \cdot \frac{K_o^D}{K_o^T} K_{eq}. \qquad (6.118)$$

and then exclude it from equation (6.115).

Equations (6.115) and (6.116) include concentrations of the adenylates in free deprotonated forms. They could be expressed in terms of total adenylate concentrations. By applying the equilibrium relationships for complex formation of adenine nucleotides with magnesium and proton, and relationships setting the total concentration of adenylates as the sum of concentrations of their different forms, we can express concentrations of free adenylates in terms of their total concentrations:

$$T = T_{total} \left( 1 + \frac{Mg}{K_{Mg}^T} + \frac{H}{K_H^T} + \frac{Mg \cdot H}{K_{Mg}^{TH} \cdot K_H^T} \right)^{-1},$$

$$D = D_{total} \left( 1 + \frac{Mg}{K_{Mg}^D} + \frac{H}{K_H^D} + \frac{Mg \cdot H}{K_{Mg}^{DH} \cdot K_H^D} \right)^{-1},$$

Here, $Mg$, $H$ stand for concentrations of free magnesium and proton ions, respectively. $K_{Mg}^T, K_H^T, K_{Mg}^{TH}, K_{Mg}^D, K_H^D, K_{Mg}^{DH}$ are dissociation constants of proton/magnesium and adenylates. The values of these constants have been taken from Alberty (1992).

*Dependence of Kinetic Constants on Membrane Potential*

The transmembrane transfer of the adenylates depends on the electrostatic field of the charged membrane. To take this dependence into account we assume that kinetic parameters characterising ANT functioning (rate and dissociation constants) depend on membrane potential which, in its turn, unambiguously determines the electrostatic field of the charged membrane. We develop a method enabling us to derive the dependences of the parameters on membrane potential. In the framework of the method we assume that the total value of the membrane potential is the sum of 'local' electric potentials and each of the local potentials influences a corresponding stage of the ANT catalytic cycle. To express these influences in terms of dependencies of kinetic constants of the stages of the catalytic cycle on the corresponding local electric potentials we use well-known laws of thermodynamics and electrostatics as well as the superposition rule.

Every position of the adenylate can be characterized by the electric potential value. In accordance with the superposition rule, the sum of potential differences between the consecutive adenylate positions is equal to the total potential difference across the membrane. We have assumed that the difference in potentials between the adjacent positions of adenylate is proportional to the total potential difference across the membrane. The applied approach divides the drop in potentials into elementary stages. We have considered the influence of the electric field on them in terms of the drop in potential at each stage. The scheme depicted in figure 6.36

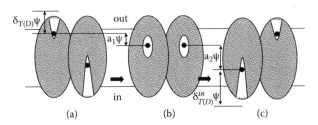

FIGURE 6.36 The scheme of adenine nucleotide translocase used to derive the dependence of the transfer rate on the electrostatic membrane potential. A, B, C are consecutive states of ANT functional unit in the process of the adenylates translocation. $\delta_{T(D)}$ is the ratio of the potential difference between the adenilate bound at the site of ANT faced to the external side of the membrane and adenilate in the bulk phase to the total membrane potential. $\delta_{T(D)}^{in}$ is the ratio of the potential difference between the adenilate bound at the site of ANT faced to the internal side of the membrane and adenilate in the bulk phase to the total membrane potential. $a_1$ is the displacement of the external adenilate to the coordinate of the maximum of the potential barrier, and $a_2$ is the displacement of the internal adenilate to the coordinate of the maximum of the potential barrier.

illustrates the potential influence on the rate of the antiporter operation. Values of the potential drop are marked for all stages of the scheme. We have applied the Nernst equation to derive the dependence of the equilibrium constant of ANT-catalyzed ADP/ATP exchange on potential. Since ATP/ADP exchange has 1:1 stoichiometry and, consequently, results in transfer of only one elementary charge across the membrane, we obtain:

$$K_{eq} = \exp(\phi),\ \phi = F\varphi/(RT),$$

$$K_{eq}(0) = 1,$$

where $\varphi$ is the potential difference across the membrane (further we use the dimensionless quantity of potential $\phi$ ).

The dependence of the dissociation constant of binding of the external ATP to the free form of ANT upon the potential can be evaluated using the Nernst equation for the equilibrium:

$$\Delta\bar{\mu} = \Delta\bar{\mu}_0 - 4F\delta\varphi + RT\ln\{T_o \cdot E/(TE)\} = 0,$$

and then:

$$K_o^T = (T_o \cdot E/(ET))_{eq} = K_o^{T,0}\exp(4F\delta_T\varphi/(RT)) = K_o^{T,0}\exp(4\delta_T\phi),$$

$$K_o^{T,0} = K_o^T(\phi = 0),$$

(6.119)

where $\delta_T = \Delta\varphi/\varphi$ is the ratio of the potential difference between the ATP-binding site and the space outside the membrane to the total membrane potential; −4 is ATP charge in terms of elementary charges. We have assumed that $\delta_T$ is a constant value. Similarly, we have derived the potential dependence of the dissociation constant of binding of the external ADP to the free form of ANT.

$$K_o^D = K_o^{D,0}\exp(3\delta_D\phi),$$

$$K_o^{D,0} = K_o^D(\phi = 0).$$

(6.120)

The influence of potential on the transfer constants can be accounted for using Eyring's theory of absolute reaction rates. We have assumed that the energy profile of the transfer across the membrane (limiting stage) is a single barrier (figure 6.37), and the transfer is a jump over the barrier from one potential well to another. We have defined the reaction coordinate

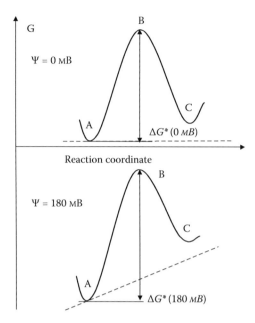

FIGURE 6.37 The potential barrier profile to derive the rate dependence upon the membrane potential along the reaction coordinate. The dashed line shows the profile of the potential imposed by the field of the charged membrane. A, B, C correspond to the ANT states depicted in figure 6.37.

as the coordinate of adenylate bound to the inner side along its transfer direction across the membrane. In this case, the transport rate is determined by the probability of the system transferring to this or the other stage and depends on the barrier's height. The A and C states (figure 6.36) refer to ANT bound to adenylates and ready to proceed with turnover. The stage B refers to the maximum of the potential barrier for the transfer process. In the general case, where the free energy has different profiles for different types of exchange (ATP for ATP, ADP for ADP and ATP for ADP) we take into consideration different values of rate constants and charges of the transported molecules.

   If the potential increases, the parameters of the energy profile change as well (figure 6.37), which induces the change of the transfer constants:

$$k \sim \exp\{-\Delta G^{*}/(RT)\}, \qquad (6.121)$$

where $\Delta G^{*}$ is the value of the energy barrier.

If the membrane potential is not zero, the height of the energy barrier is determined by the interactions of the membrane electrostatic field with the charges of the enzyme and bound adenylates; i.e., it depends on their displacement in the process of transfer from stage A to stage C. Assuming the additivity of the energy of interaction between these charges, the free energy, and the field of the charged membrane, the activation energy of charges $Z_j$ displaced by the distance $\delta_j$ in the process of transfer from stage A into stage C can be expressed as:

$$\Delta G^* = \Delta G_o^* + F\varphi \sum_j Z_j \delta_j.$$

In our case,

$$\Delta G^* = \Delta G_o^* + F\varphi(Z_1 a_1 + Z_2 a_2 + a_3), \tag{6.122}$$

where $a_1$ is the displacement of the external adenylate to the maximum coordinate, $a_2$ is the displacement of the internal adenylate, and $a_3 = \Sigma Z_k \delta_k$ is the effective parameter characterizing the displacement of all charges of the enzyme.

By using equations (6.121) and (6.122) we obtain:

$$k_i = k_i^0 \exp(Z_1 a_1 \phi + Z_2 a_2 \phi + a_3 \phi),$$
$$k_i^0 = k_i \big|_{\phi=0}, \quad i = 1\ldots 4. \tag{6.123}$$

The potential fall $\delta_{T(D)}^{in}$ resulted from the adenylate dissociation/binding from/with the matrix site of ANT is not independent and is expressed through other potential falls:

$$\delta_{T(D)}^{in} = 1 - \delta_{T(D)} - a_1 - a_2,$$

This parameter has been already taken into account in the equation expressing the dependence of the equilibrium constant of ATP/ADP exchange on electric potential difference and, consequently, there is no need to use it again.

*Estimation of Parameters*

As was mentioned above, to verify a quantitative model it is necessary to identify its kinetic parameters in accordance with some experimental

TABLE 6.7    The Values of the Apparent kinetic Parameters
Calculated on the Basis of Two Sets of the Experimental
Data Measured at $\Psi = 0 mV$ and $\Psi = 180 mV$

| Parameter | $\Psi = 0\,mV$ | $\Psi = 180\,mV$ |
|---|---|---|
| $k_1$ | 37 min$^{-1}$ | 21min$^{-1}$ |
| $k_2$ | 8.8 min$^{-1}$ | 48 min$^{-1}$ |
| $k_3$ | 17 mim$^{-1}$ | 1.7 min$^{-1}$ |
| $k_4$ | 31 min$^{-1}$ | 30 min$^{-1}$ |
| $K_o^D$ | 57$\mu$M | 57$\mu$M |
| $K_o^T$ | 68 $\mu$M | 402$\mu$M |
| $K_i^D / K_i^T$ * | 1.6 | 4.9 |

*Only ratio of the parameters can be estimated on the basis of
used experimental data.

study. As a criterion of fitness of the model to experimental data the following function was used:

$$f(k_j, K_j) = \sum_i^n \left( \frac{v_i - \hat{v}_i}{\hat{v}_i} \right)^2. \tag{6.124}$$

Here $n$ is the total number of the experimental points, $\hat{v}_i$ is the experimentally measured value of the influx rate (Kraemer and Klingenberg
1982), and $v_i$ is the value of the influx rate calculated on the basis of the
model at points corresponding to experimental ones. To estimate values
of unknown parameters, the relative error of the model ($\sqrt{f/n}$) has been
minimized. In this way we have identified two sets of kinetic parameters
for two series of experiments (in the presence or absence of the membrane
potential). Parameter values are shown in table 6.7.

Using equations (6.119), (6.120) and (6.123), we derive a system of algebraic equations giving parameters characterizing the potential dependence of the ANT operation as functions of parameters whose values were
estimated against experimental data and summarized in table 6.7.

$$\begin{cases} \exp\{((-4)a_1 + (-4)a_2 + a_3)6,9\} = k_1^{6,9}/k_1^0, \\ \exp\{((-3)a_1 + (-4)a_2 + a_3)6,9\} = k_2^{6,9}/k_2^0, \\ \exp\{((-4)a_1 + (-3)a_2 + a_3)6,9\} = k_3^{6,9}/k_3^0, \\ \exp\{4\delta_T 6,9\} = K_o^{T,6,9}/K_o^{T,0}, \\ \exp\{3\delta_D 6,9\} = K_o^{D,6,9}/K_o^{D,0}. \end{cases}$$

Solving the system we have obtained the values of the unknown parameters.

$$a_1 = -0.32, \quad a_2 = 0.25, \quad a_3 = 0.21, \quad \delta_D = 0, \quad \delta_T = 0.06.$$

It is noteworthy that these values are approximate. Their accuracy is affected by the tolerance of the numerical evaluation and of precision experimental techniques.

## Model Verification

In the previous section we have found two sets of kinetic parameters that correspond to two sets of experimental data (in the presence and absence of membrane potential). Using equations (6.119), (6.120) and (6.123) we can proceed to a single set of parameters (table 6.8) that fits all the experimental points (Kraemer and Klingenberg 1982). Both the theoretical simulations and the experimental points are depicted in figure 6.38.

Since rate equation (equation (6.115)) is a nonlinear function, it has different sensitivity with respect to different parameters. This means that the accuracies of parameter estimation differ as well. To study sensitivity of the model solution we have varied each parameter individually, calculated the relative error of the model ($\sqrt{f/n}$) and found a range of parameter values providing no more than 10 percent deviation of the model solution from corresponding experimental data (see table 6.8).

Values of the model parameters shown in table 6.1 and table 6.2 enable us to conclude that about 6 percent of the electric potential falls on the

TABLE 6.8 The Values of Parameters of the Model and Ranges of Sensitivity

| Parameter | Value | Range of 10% Sensitivity |
|---|---|---|
| $k_1^o$ | 37 min$^{-1}$ | (31-39) min$^{-1}$ |
| $k_2^o$ | 8.8 min$^{-1}$ | (10.1-11.6) min$^{-1}$ |
| $k_3^o$ | 17 min$^{-1}$ | (13-43) min$^{-1}$ |
| $k_4^o$ | 31 min$^{-1}$ | (26-32) min$^{-1}$ |
| $K_o^{D,0}$ | 57 μM | (45-57) μM |
| $K_o^{T,0}$ | 68 μM | (45-71) μM |
| $a_1$ | −0.32 | −(0.265-0.271) |
| $a_2$ | 0.25 | (0.203-0.207) |
| $a_3$ | 0.31 | (0.179-0.196) |
| $\delta_T$ | 0.06 | (0.06-0.09) |
| $\delta_D$ | 0.00 | (0-0.006) |

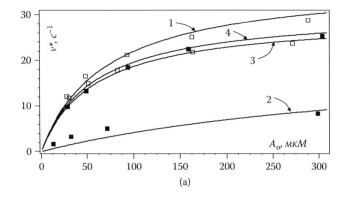

(a)

FIGURE 6.38 The results of parameter estimation. The experimental values (symbols) and model generated curves (solid lines) at $pH$ 8.0, $Mg^{2+}$ =0. (a) The dependence of influx rate of the labeled adenilate ($v^*$) on: (1) concentration of the external labeled ATP ($T_i$ = $10mM$, $D_i$ = $0mM$, $D_o$ = $mM$ $\Psi$ = $0mV$); (2) concentration of the external labeled ATP ($T_i$ = $10mM$, $D_i$ = $0mM$, $D_o$ = $0\mu M$, $\Psi$ = $0mV$);(3) concentration of the external labeled ADP ($T_i$ = $0mM$, $D_i$ = $10mM$, $T_o$ = $0\mu M$, $\Psi$ = $180mV$); (4) concentration of the external labeled (ADP $T_i$ = $0mM$, $D_i$ = $10mM$, $T_o$ = $0mM$, $\Psi$ = $180mV$). (b) The dependence of influx rate of the labeled adenilate ($v^*$) on concentration of the external labeled ADP.

$$T_i = 5mM, D_i = 5mM, T_o = 0\mu M, \Psi = 180mV; \tag{1}$$

$$T_i = 5mM, D_i = 5mM, T_o = 100\mu M, \Psi = 180mV; \tag{2}$$

$$T_i = 5mM, D_i = 5mM, T_o = 400\mu M, \Psi = 180mV; \tag{3}$$

$$T_i = 5mM, D_i = 5mM, T_o = 0\mu M, \Psi = 0mV; \tag{4}$$

$$T_i = 5mM, D_i = 5mM, T_o = 20\mu M, \Psi = 0mV; \tag{5}$$

$$T_i = 5mM, D_i = 5mM, T_o = 100\mu M, \Psi = 0mV. \tag{6}$$

(c) The dependence of inf-lux rate of the labeled adenilate ($v^*$) on: (1) concentration of the external labeled ADP ($T_i$ = $5mM$, $D_i$ = $mM$, $T_o^* = D_o^*, \Psi = 180mV$); (2) concentration of the external labeled ADP ($T_i = 5mM, D_i = 5mM, T_o^* = D_o^*, \Psi = 0mV$); (3) concentration of the external labeled ATP ($T_i$ = 5mM, $D_i$ = 5mM, $D_o^* = T_o^*, \Psi = 180mV$); (4) concentration of the external labeled ATP($T_i = 5mM, D_i = 5mM, D_o^* = T_o^*, \Psi = 0mV$).

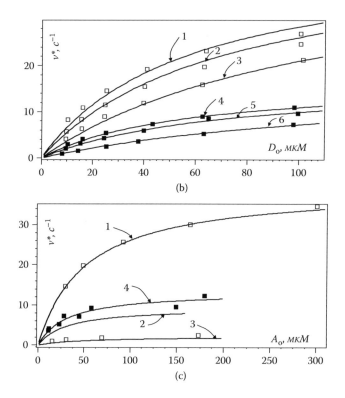

FIGURE 6.38 (Continued).

ATP binding stage and the binding constant for the external ATP significantly depends upon the potential. Indeed, variation of the electric potential from 0 to 180 mV results in about 600 percent increase in the value of $K_o^T$. In contrast, we have found that about 0 percent of the electric potential falls on the ADP binding stage and, consequently, $K_o^D$ does not depend on electric potential.

Essential deviations of the values of the parameters $a_1$ and $a_2$ from zero imply that changes in the conformation of the dimer of ANT result in significant changes in the position of adenylates. The value of the parameter $a_3$ equal to 0.31 indicates the presence of charged groups on the surface of the ANT molecule that displace during adenylate transfer.

*Model Predictions*

The model was applied to study the antiporter behaviour under conditions close to the physiological ones. Theoretical dependences depicted in figure 6.39 have been calculated by applying concentration values of the metabolites equal to those in mitochondria.

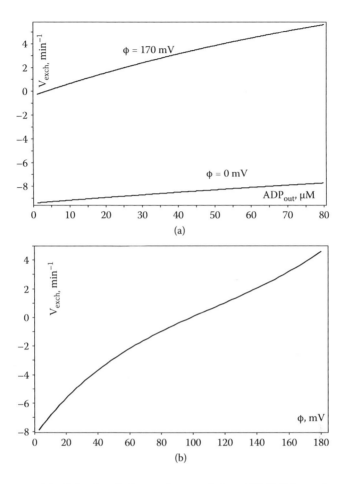

FIGURE 6.39 (a) Calculated dependence of the ATP/ADP exchange rate upon concentration of the external ADP at various fixed values of the membrane potential. $pH_i 7.8, pH_o 7.2, Mg^{2+} = 1mM, D_i = T_i = 5mM, T_o = 2mM$. (b) Calculated dependence of the ATP/ADP exchange rate upon the membrane potential. Parameter values: $pH_i 7.8, pH_o 7.2, Mg^{2+} = 1mM$, $D_i = T_i = 5mM, T_o = 2mM, D_o = 50\mu M$. (c) Calculated dependence of the ATP/ADP exchange effectiveness (ratio of exchange rate to the total number of turnovers) upon the membrane potential. Parameter values: $pH_i 7.8$, $pH_o 7.2$, $Mg^{2+} = 1mM$, $D_i = T_i = 5mM, T_o = 2mM, D_o = 50\mu M$. (d) Calculated dependence of the ATP/ADP exchange rate upon pH inside and outside mitochondria. Each curve involves the change of the pH values in one of the compartments at the fixed pH in the other one. Parameter values: $pH_i 7.8, pH_o 7.2, Mg^{2+} = 1mM, D_i = T_i = 5mM, T_o = 2mM, D_o = 50\mu M$, $j = 170mV$.

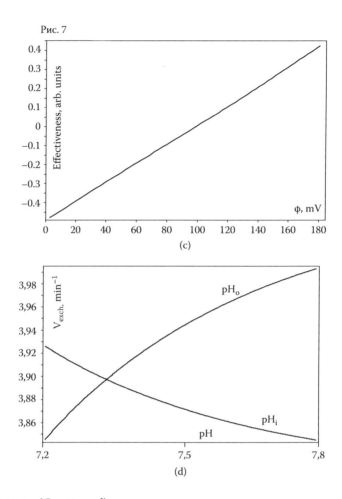

FIGURE 6.39 (Continued).

In the uncoupled state, the model predicts ADP efflux out of mitochondria (negative values of the exchange velocities in figure 6.39a). This is mainly due to dependence of the equilibrium constant of the ADP/ATP exchange upon the potential. During the energization of mitochondria, the positive direction of the transfer resulting in the accumulation of ADP in the mitochondria matrix is maintained even under very low concentrations of external ADP. This stimulates the rate increase of ATP synthesis in mitochondria. Figure 6.39b shows the dependence of the exchange rate upon the potential. It has a positive direction only at a potential of not less than 100 mV.

To characterise the effectiveness of the antiporter operation we have introduced a new function, the ratio of the exchange rate to the total turnover per time unit.

$$effectiveness = v'_{exchange} / v'.$$

This function relates changes in ADP concentration in the mitochondrial matrix to the total number of ANT transfer cycles. The dependence of the effectiveness upon the membrane potential is shown in figure 6.39c. Negative values of effectiveness represent the opposite transfer direction. The effectiveness of the transfer is apparently not more than 0.5 within the given range of parameters even for energized mitochondria. This means that no more than half of the transfer cycles result in outward flux of ATP. This can be explained by low concentration of the external ADP because the probability of complex formation with ADP from the intramembrane space is lower than with ATP.

To describe the processes of adenylate protonation/deprotonation we have taken account of the fact that only free forms of adenylates can be transferred by ANT across the membrane and the affinity of ADP to ANT binding sites differs significantly from that of ATP. From this it follows that variation of internal and external pH changes the ratio between free forms of ADP and ATP and, consequently, the exchange rate. Figure 6.39D shows that the exchange rate increases with the increase of external pH and decreases with the increase of internal pH. It is worth mentioning that to describe ANT functioning properly the model has to take into account the pH dependence of both adenylate distribution between the free and protonated forms and the ability of ANT to transfer bound adenylates across the membrane.

## CHAPTER 7

# Kinetic Models of Biochemical Pathways

The subject of systems biology encompasses, on the one hand, any biological systems, the characteristics of which can be quantitatively measured, and, on the other hand, the mathematical models that mimic the behavior of these biological systems. Systems biology primarily intends to construct a model which is the most closely related to the biological system under study and to reveal thereby new regulatory and dynamic properties as well as structure-mechanism relationships of this system. The criterion of adequacy of the mathematical model to the relevant biological system is the quality of simulation of all possible experimental data obtained for this biological object.

Any model represents a caricature of an actual biochemical system. First of all, this is due to lack of information about the mechanism of functioning of separate segments of the biological system and about the interactions between these segments. Without such information, the construction of the model is substantiated with some assumptions, which are derived from common sense or from comparison with functioning of analogous systems and allows one to eliminate these 'blank spots'. By paraphrasing the statement made above, we can conclude that one of the main objectives of systems biology is finding such a caricature of real biological system that will resemble as much as possible the original prototype. This chapter presents several examples of mechanistic models of biochemical pathways such as the Krebs cycle of mitochondria in hepatocytes

(Mogilevskaya, Demin, and Goryanin 2006) and the biosynthesis of branched-chain amino acids in *Escherichia coli* (Demin et al. 2005). These models are represented by systems of algebraic and ordinary differential equations, have a high predictive power and allow analysis and interpretation of multilevel experimental data.

## MODELLING OF THE MITOCHONDRIAL KREBS CYCLE

The citric acid cycle (also known as the tricarboxylic acid cycle, the TCA cycle or the Krebs cycle) is a series of chemical reactions of central importance in all living cells that use oxygen as part of cellular respiration. In aerobic organisms, the citric acid cycle is part of a metabolic pathway involved in the chemical conversion of carbohydrates, fats, carbon dioxide and water to generate a form of usable energy. Together with oxidative phosphorylation, TCA is involved in fuel molecule catabolism and ATP production.

In this section, we develop a kinetic model of the TCA cycle operating in mitochondria oxidizing glutamate as a substrate. We derive rate equations of the individual reactions from catalytic cycles based on protein structural data of the corresponding enzyme and estimate the kinetic parameters of the rate equations by fitting them to *in vitro* data. Furthermore, we validate our model using experimental data measured on suspensions of mitochondria.

### Model Development

Traditionally, the Krebs cycle is described as a sequence of nine reactions resulting in the formation of oxaloacetate from citrate through cis-aconitate, isocitrate, α-ketoglutarate, succinyl-coenzyme A, succinate, fumarate and malate. The cycle is closed because of the condensation of oxaloacetate with acetyl-coenzyme A and the formation of citrate. Acetyl-coenzyme A is formed from pyruvate, fatty acids or amino acid oxidation. It has been found (Kondrashova 1989) that, under conditions of high energy demand, the Krebs cycle may not work in its full form, but a shunt exists via transamination of glutamate and oxaloacetate with the subsequent formation of α-ketoglutarate (figure 7.1). Such a truncated Krebs cycle is also found in Morris hepatoma 3924A mitochondria (Parlo and Coleman 1984). In this case, glutamate (Glu) is the carbon atom source, and Glu is transported to the mitochondrial matrix by the aspartate-glutamate carrier (AGC, figure 7.1), which exchanges external glutamate $Glu_{out}$ with an internal aspartate ($Asp_{in}$) carrying a proton from the intermembrane space of the mitochondria to the matrix. Glutamate donates its amino group to

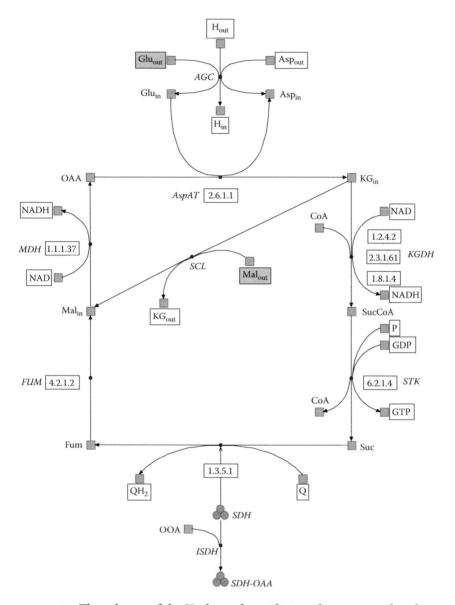

FIGURE 7.1 The scheme of the Krebs cycle oxidizing glutamate and malate as substrates.

oxaloacetate (OAA) to form α-ketoglutarate ($KG_{in}$). This reaction (AspAT, figure 7.1) is catalyzed by aspartate aminotransferase. The next reaction, an oxidative decarboxylation of $KG_{in}$ to succinyl-coenzyme A (SucCoA), is catalyzed by α-ketoglutarate dehydrogenase (KGDH). $KG_{in}$ may be also exchanged with external malate ($Mal_{out}$) (KMC). This reaction is catalyzed by the dicarboxylate carrier. SucCoA is broken up to form succinate (Suc). This reaction is associated with phosphorylation of GDP and catalyzed by succinate thiokinase (STK). Oxidation of Suc to fumarate (Fum) is accompanied by reduction of ubiquinone (Q) to ubiquinol ($QH_2$) and catalyzed by succinate dehydrogenase (SDH). Fum is hydrated to Mal in the reaction catalyzed by fumarase (FUM). Oxidation of Mal to OAA catalyzed by malate dehydrogenase (MDH) closes the Krebs cycle. The model also includes a reaction which corresponds to the binding of OAA to the catalytically active state of succinate dehydrogenase (SDH) and results in formation of a catalytically inactive state of this enzyme, SDH-OAA. $Glu_{in}$, $Asp_{in}$, OAA, $KG_{in}$, SucCoA, CoA, Suc, Fum, $Mal_{in}$, SDH, SDH-OAA are the model variables, and the concentrations denoted $Glu_{out}$, $Asp_{out}$, $H_{out}$, $H_{in}$, $Mal_{out}$, $KG_{out}$, P, ATP, ADP, $Ca^{2+}$, GTP, GDP, Q, $QH_2$, NAD, NADH are model parameters. The subscripts *in* and *out* denote intra- and extra-mitochondrial concentrations, respectively. The constant metabolites appear in figure 7.1 in square boxes. Values of the parameters are listed in table 7.1. The model is described by the following system of differential equations:

$$\frac{dGlu_{in}}{dt} = V_{AGC} - V_{AspAT}; \qquad \frac{dAsp_{in}}{dt} = V_{AspAT} - V_{AGC};$$

$$\frac{dOAA}{dt} = V_{MDH} - V_{AspAT} - V_{ISDH}; \qquad \frac{dKG_{in}}{dt} = V_{AspAT} - V_{KGDH} - V_{KMC};$$

$$\frac{dSucCoA}{dt} = V_{KGDH} - V_{STK}; \qquad \frac{dSuc}{dt} = V_{STK} - V_{SDH}; \qquad \frac{dFum}{dt} = V_{SDH} - V_{FUM};$$

$$\frac{dMal_{in}}{dt} = V_{FUM} - V_{MDH} + V_{KMC}; \qquad \frac{dCoA}{dt} = V_{STK} - V_{KGDH}; \qquad \frac{dSDH}{dt} = -V_{ISDH};$$

$$\frac{d(SDH-OAA)}{dt} = V_{ISDH} \qquad\qquad\qquad\qquad (7.1)$$

There are four conservation laws in the system: $N_{tot}$ (conservation of amino groups), $CoA_{tot}$ (conservation of CoA), $C_{tot}$ (conservation of four-carbon skeleton) and $SDH_{tot}$ (conservation of succinate dehydrogenase). Pool values are listed in table 7.1.

## Description of Individual Enzymes of the Krebs Cycle

To describe the kinetics of the individual enzymes we use *in vitro* data available from the literature. Where experimental data for hepatocytes

TABLE 7.1  Krebs Cycle Metabolite Concentrations Obtained from Literature and from the Model

| Substrates | Concentration (mM) | | Substrates | Concentration (mM) | |
|---|---|---|---|---|---|
| | Literature Data | Model Data | | Literature Data | Model Data |
| $KG_{in}$ | 0.15 (Hoek 1971); 1.13–1.6 (Siess, Kientsch-Engel, and Wieland 1984) | 0.018 | Suc | — | 0.007 |
| $KG_{out}$ | 0.54 (Hoek 1971) | 0.54 | $Asp_{out}$ | — | 0 |
| $Glu_{in}$ | 7.3 (Hoek 1971) | 7.3 | Fum | — | 1.94 |
| $Glu_{out}$ | 0.86 (Hoek 1971) | 20 | NAD | 2 (Panov and Scaduto 1995) | 2 |
| OAA | 0.002–0.006 (Siess, Kientsch-Engel, and Wieland 1984) | 0.0002 | NADH | 1 (Panov and Scaduto 1995) | 1 |
| $Mal_{in}$ | 0.324 (Garber and Hanson 1971); 0.5–2.5 (Williamson, Lund, and Krebs 1967) | 1.16 | ATP + ADP | 12 (Wilson, Nelson, and Erecinska 1982) | 12 |
| $Mal_{out}$ | 0.495 (Garber and Hanson 1971) | 0 | ATP/ADP | 9.4 (Wilson, Nelson, and Erecinska 1982) | 9.4 |
| $Asp_{in}$ | 0.192–0.297 (Siess, Kientsch-Engel, and Wieland 1984) | 0.3 | P | — | 5 |
| SucCoA | 0.36–0.91 (Hansford RG, Johnson RN.. 1975) | 0.63 | GDP | — | 0.2 |
| CoA | 0.16–0.79 (Hansford et al. 1975) | 0.37 | GTP | — | 1.8 |
| $N_{tot}$ | — | 7.6 | Q | — | 19 |
| $SDH_{tot}$ | 0.05 (Vinogradov 1986) | 0.05 | $QH_2$ | — | 1 |
| $Ca^{2+}$ | 0.001 (McCormack and Denton 1979) | 0.001 | $H_{in}$ | 5.2e-6 (Haas et al. 1985) | 5.2e-6 |
| $\Delta\psi$ | 139 mV (Haas et al. 1985) | 139 mV | $H_{out}$ | — | 3.98e-5 |
| $C_{tot}$ | — | 3.801 | Gly | — | 1 |
| $CoA_{tot}$ | — | 1 | SDH-OAA | — | 0.0458 |
| | | | SDH | — | 0.0042 |

are missing, we use kinetic and structural data for the corresponding heart enzymes. Derivation of the rate equations and estimation of kinetic parameters is accomplished in accordance with the principles formulated in chapters 4, 5 and 6.

### α-Ketoglutarate Dehydrogenase

Catalyzes the irreversible reaction of oxidative decarboxylation of α-ketoglutarate with the formation of succinyl-coenzyme A and the reduction of NAD. The enzyme is described according to the ping pong ter-ter mechanism (McCormack and Denton 1979). It is assumed that $CO_2$ concentration does not influence the reaction rate. The influence of activators (ADP and $Ca^{2+}$), inhibitors (ATP) and products (NADH, SucCoA) is taken into account. We assume that the enzyme exists in both active and inactive forms. ADP or $Ca^{2+}$ transforms the enzyme to its active form, whereas ATP transforms the enzyme to its inactive form (see catalytic cycle depicted in figure 7.2 for details). The rate equation is derived in the following form:

$$V_{KGDH} = \frac{KGDH \cdot \left(1 + \frac{ADP}{K_i^{ADP}}\right) \cdot k_f \frac{KG_{in}}{K_m^{KG_{in}}} \frac{CoA}{K_m^{CoA}} \frac{NAD}{K_m^{NAD}}}{\frac{CoA}{K_m^{CoA}} \frac{NAD}{K_m^{NAD}} \left(\frac{KG_{in}}{K_m^{KG_{in}}} + \frac{1 + \frac{ATP}{K_i^{ATP}}}{1 + \frac{Ca}{K_i^{Ca}}}\right) + \frac{KG_{in}}{K_m^{KG_{in}}} \left(\frac{CoA}{K_m^{CoA}} + \frac{NAD}{K_m^{NADH}}\right)\left(1 + \frac{NADH}{K_i^{NADH}} + \frac{SucCoA}{K_i^{SucCoA}}\right)}$$

(7.2)

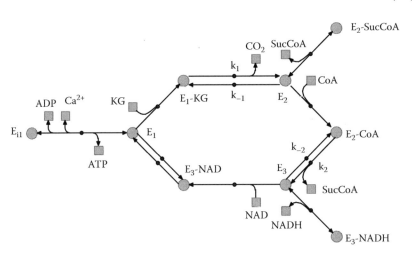

FIGURE 7.2 The scheme of the α-ketoglutarate dehydrogenase catalytic cycle.

Here, KGDH is the concentration of α-ketoglutarate dehydrogenase; $K_m^{CoA}, K_m^{NAD}, K_m^{KG_{in}}$ are Michaelis constants for substrates; $K_i^{ADP}, K_i^{ATP}, K_i^{Ca}$, $K_i^{NADH}, K_i^{SucCoA}$ are inhibition constants for effectors; $k_f$ is the catalytic constant. Parameters known from the literature are KGDH (Fahien and Teller 1992), $k_f$ (Massey 1960), $K_m^{CoA}$ (Smith, Bryla, and Williamson 1974), $K_m^{NAD}$ (Hamada 1975), $K_i^{ADP}, K_i^{ATP}, K_i^{Ca}, K_m^{KG_{in}}$ (McCormack and Denton 1979). To estimate inhibition constants for NADH and SucCoA and make values of known parameters, more precise experimental data (McCormack and Denton 1979) is used in which reaction was started by addition of α-ketoglutarate dehydrogenase to a solution of the substrates α-ketoglutarate, CoA and NAD and time dependences of NADH accumulation were monitored with and without addition of effectors ADP and ATP. To quantitatively describe these experiments the following system of differential equations has been developed (figure 7.3):

$$\frac{dKG_{in}}{dt} = -V_{KGDH}; \qquad \frac{dCoA}{dt} = -V_{KGDH}; \qquad \frac{dNAD}{dt} = -V_{KGDH};$$

$$\frac{dNADH}{dt} = V_{KGDH}; \qquad \frac{dSucCoA}{dt} = V_{KGDH} \qquad\qquad (7.3)$$

Here, $V_{KGDH}$ is the rate equation for α-ketoglutarate dehydrogenase given by equation (7.2); substrates and products of reaction catalyzed by α-ketoglutarate dehydrogenase are variables of the model (7.3). Initial conditions for the system of differential equations (7.3) are set in accordance with initial concentrations of substrates used in the experiment (McCormack and Denton 1979): $KG_{in} = 0.1$ mM; NAD = 1 mM; CoA = 0.25 mM; NADH = 0; SucCoA = 0.

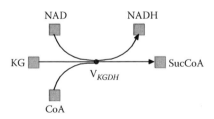

FIGURE 7.3 The scheme of α-ketoglutarate dehydrogenase reaction described by the system (7.2) to describe time dependences of NADH production catalyzed by the enzyme.

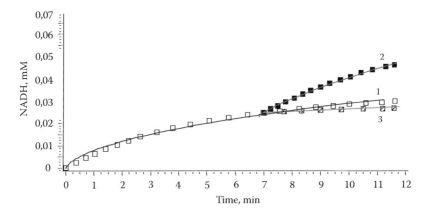

FIGURE 7.4 Time dependence of NADH production by $\alpha$-ketoglutarate dehydrogenase reaction presented by experimental points and described by the curves according to the system (7.3) with the following initial values (enzyme concentration was equal to 0.4 nM): 1) $KG_{in} = 0.1$ mM; NAD = 1 mM; CoA = 0.25 mM; NADH = 0; SucCoA = 0 (white squares); 2) 1.5 mM ADP was added on the seventh minute (black squares); 3) 1.5 mM ATP was added on the seventh minute (squares with oblique hatching).

The system of differential equations (7.3) has four conservation laws:

$$KG_{in} + SucCoA = C_{tot}^{KGDH} \text{ (conservation of four-carbon skeleton)}$$

$$CoA + SucCoA = CoA_{tot}^{KGDH} \text{ (conservation of CoA)}$$

$$KG_{in} + NADH = e_{tot}^{KGDH} \text{ (conservation of electrons)}$$

$$NADH + NAD = NAD_{tot}^{KGDH} \text{ (conservation of pyridine nucleotides)}$$

Values of $NAD_{tot}^{KGDH}$, $e_{tot}^{KGDH}$, $CoA_{tot}^{KGDH}$, $C_{tot}^{KGDH}$ can be calculated from the initial conditions. To estimate parameter values of equation (7.2) we fit the solution of the system of differential equations (7.3) to experimentally measured time dependencies of NADH accumulation. Figure 7.4 demonstrates a good fit of experimental data from McCormack and Denton (1979; symbols) to theoretical curves generated by the system of differential equations (7.3). Values of kinetic parameters are listed in table 7.2.

## Aspartate-Glutamate Carrier (AGC)

The inner mitochondrial membrane is not permeable to glutamate. Only a membrane carrier in exchange for an aspartate can transfer protonated

TABLE 7.2 Kinetic Parameters Values of the Krebs Cycle Enzymes Known from Literature and Estimated via Fitting of Rate Equations to Literature Experimental Data (Michaelis Constants, Dissociation Constants and Enzymes Concentrations Are in mM, Rate Constants Are in 1/min)

| Enzyme Designation | Literature Values of Kinetic Parameters (Ref.) | Kinetic Parameters Values Estimated via Fitting to Experimental Data (Ref.) |
|---|---|---|
| AGC | $K_m^{Glu_{out}} = 0.25$; $K_m^{Glu_{in}} = 3$; $K_m^{Asp_{out}} = 0.12$ (Dierks and Kramer 1988) $K_m^{Asp_{in}} = 0.0435$ (La Noue et al. 1979) | *$AGC = 2$; $k_{1,0} = 99,800$; $k_{-1,0} = 9940$; $k_2 = 100,000$ $k_{-2} = 9940$; $K_m^{H_{in}} = 0.00004$; $K_m^{H_{out}} = 0.01$; *$K_m^{H_{out}} = 0.1$; $K_m^{Glu_{out}}* = 9.3$; (La Noue et al. 1979) |
| AspAT | $AspAT = 0.14(38)$; $k_r = 51,870$; $K_m^{KG_{in}} = 6.9$; $K_m^{Asp_{in}} = 1.9$ (Kuramitsu 1985); $K_m^{OAA} = 0.088$ (Garber and Hanson 1971); $K_{eq} = 6.6$ (Siess, Kientsch-Engel, and Wieland 1984) | *$AspAT = 1.5$; $k_1 = 5e7$; $k_f = 10,000$ (La Noue et al. 1979); $K_m^{Glu_{in}} = 0.55$; $k_{-1} = 51,999$ (Kuramitsu 1985) |
| KGDH | $KGDH = 0.002$ (Fahien and Teller 1992); $k_f = 83,110$ (Massey 1960) $K_m^{CoA} = 0.0027$ (Smith, Bryla, and Williamson 1974); $K_m^{NAD} = 0.005$ (Hamada 1975); $K_m^{KG_{in}} = 0.2$; $K_i^{ATP} = 0.1$; $K_i^{ADP} = 0.1$; $K_i^{Ca} = 0.0012$ (McCormack and Denton 1979) | *$KGDH = 1$; *$K_m^{ADP} = 0.005$ (La Noue et al. 1979) $K_m^{KG_{in}} = 0.03$; $K_m^{CoA} = 0.002$; $K_m^{NAD} = 0.93$; $K_i^{SucCoA} = 0.011$; $K_i^{NADH} = 0.0018$; $K_i^{ATP} = 0.01$; $K_i^{ADP} = 0.56(35)$; $K_i^{Sal} = 0.001$ (Kaplan, Kennedy, and Davis 1954) |
| STK | $K_m^{Suc} = 0.4 - 0.8$ $K_m^{CoA} = 0.005 - 0.02$; $K_m^{SucCoA} = 0.01 - 0.06$; $K_m^{GDP} = 0.002 - 0.008$; $K_m^{GTP} = 0.05 - 0.01$; $K_m^P = 0.2 - 0.7$ (Cha and Parks 1964a); $k_f = 10,780$ (Cha and Parks 1964b); $K_{eq} = 3.7$ (Kaufman and Alivisatos 1955) | $K_m^{Suc} = 0.81$; $K_m^{CoA} = 0.017$; $K_m^{SucCoA} = 0.024$ $K_m^{GDP} = 0.007$; $K_m^{GTP} = 0.000068$; $K_m^P = 1.5$ $k_1 = 1,700,000$; $k_{-1} = 1149$; $k_2 = 10,000$; $k_{-2} = 1,990,000$; $K_d^{CoA} = 0.029$; $K_{d,E-GDP-CoA}^{CoA} = 0.00038$; $K_d^{GDP} = 0.14$; $K_{d,E-GTP-SucCoA}^{SucCoA} = 0.49$ (Cha and Parks 1964a); *$STK = 1$ (La Noue et al. 1979) |

(Continued)

TABLE 7.2 Kinetic Parameters Values of the Krebs Cycle Enzymes Known from Literature and Estimated via Fitting of Rate Equations to Literature Experimental Data (Michaelis Constants, Dissociation Constants and Enzymes Concentrations Are in mM, Rate Constants Are in 1/min)(Continued)

| Enzyme Designation | Literature Values of Kinetic Parameters (Ref.) | Kinetic Parameters Values Estimated via Fitting to Experimental Data (Ref.) |
|---|---|---|
| SDH | SDH = 0.05; $k_f$ = 10,000; $k_r$ = 102 (Vinogradov 1986) $K_m^{Suc}$ = 0.13; $K_m^Q$ = 0.0003; $K_m^{QH_2}$ = 0.0015; $K_m^{Fum}$ = 0.025 (Grivennikova et al. 1993) $K_{d,E-Suc}^{Suc}$ = 0.01; $K_{d,E-Fum}^{Fum}$ = 0.29 (Kotlyar and Vinogradov 1984) | $K_m^{Suc}$ = 0.084; $K_{d,E-Suc}^{Suc}$ = 0.29 (Kotlyar and Vinogradov 1984) $k_f$ = 1e6 (La Noue et al. 1979) $K_i^{Sal}$ = 7e-5 (Kaplan, Kennedy, and Davis 1954) |
| FUM | FUM = 2.27e-4; $K_m^{Fum}$ = 0.047; $K_m^{Mal_{in}}$ = 0.017; $k_f$ = 90721; $k_r$ = 71,342 (Alberty 1961) | *FUM = 0.5; *$K_m^{Fum}$ = 0.01 (La Noue et al. 1979) $K_m^{Fum}$ = 0.036; $k_f$ = 90,722 (Greenhut, Umezawa, and Rudolph 1985) |
| MDH | $k_f$ = 5.4e5; $k_r$ = 8.6e3; MDH = 9.03e-4; $K_m^{OAA}$ = 0.0795; $K_m^{Mal_{in}}$ = 0.386; $K_m^{NAD}$ = 0.0599; $K_m^{NADH}$ = 0.26; $K_i^{OAA}$ = 0.0055; $K_i^{Mal_{in}}$ = 1.1; $K_i^{NAD}$ = 0.36 (Heyde and Ainsworth 1968) $K_i^{NADH}$ = 0.0136; $K_{eq}$ = 8000 | *MDH = 1 (La Noue et al. 1979) |
| KMC | $k_f$ = 325; $k_r$ = 309; $K_m^{Mal_{out}}$ = 1.36; $K_m^{Mal_{in}}$ = 0.71; $K_m^{KG_{in}}$ = 0.17; $K_m^{KG_{out}}$ = 0.31; $K_{eq}$ = 1 (Indiveri et al. 1991) | *KMC = 2 (La Noue et al. 1979) $k_1$ = 858; $K_i^{KG_{out}}$ = 4.2e-3 (Indiveri et al. 1991) |
| Oxaloacetate binding to SDH (ISDH) | $k_i$ = 1200 1/min*mM; $k_{-i}$ = 0.02 1/min (Vinogradov 1986) | |
| SCL | $V_f$ = 0.75 mM/min [Forman, Davidson, Webster, 1971]; $K_m^{Sal}$ = 2 [Vessey, Hu, Kelly, 1996]; $K_m^{CoA}$ = 0.63 [Ricks, Cook, 1981] | $K_i^{CoA}$ = 0.63 |
| SGT | $V_f$ = 900 mM/min; $K_m^{SalCoA}$ = 0.008; $K_m^{Gly}$ = 20 [Forman, Davidson, Webster, 1971] | $K_i^{Gly}$ = 20 |

* Parameters values estimated from verification of the whole model.

glutamate to the matrix. The membrane electrochemical potential is consumed in the reaction. The rate equation for AGC was derived assuming that it functions according to a random ter-ter mechanism (Dierks and Kramer 1988). By the Cleland classification (Cleland 1963) this means that binding of substrates and release of products occur in an arbitrary order (see figure 7.5). We assume that the affinity of the enzyme to substrate/product does not depend on the enzyme state. In this case, dissociation constants for substrates and products are equal to the Michaelis constants. The mechanism includes two slow stages (with rate constants $k_1, k_{-1}, k_2, k_{-2}$) corresponding to reorientation of transporter with respect to the inner mitochondrial membrane. Using these assumptions and applying rapid equilibrium and quasi-steady-state approaches we derive the following rate equations for AGC:

$$V_{AGC} = AGC \cdot \cfrac{k_1 k_2 \frac{Glu_{out}}{K_m^{Glu_{out}}} \frac{Asp_{in}}{K_m^{Asp_{in}}} \frac{H_{out}}{K_m^{H_{out}}} - k_{-1} k_{-2} \frac{Glu_{in}}{K_m^{Glu_{in}}} \frac{ASP_{out}}{K_m^{Asp_{out}}} \frac{H_{in}}{K_m^{H_{in}}}}{\left( k_1 \frac{Glu_{out}}{K_m^{Glu_{out}}} \frac{Asp_{in}}{K_m^{Asp_{in}}} \frac{H_{out}}{K_m^{H_{out}}} + k_{-2} \right)\left( 1 + \frac{G_{in}}{K_m^{H_{in}}} \right) \cdot \left( 1 + \frac{Glu_{in}}{K_m^{Glu_{in}}} \right)\left( 1 + \frac{Asp_{out}}{K_m^{Asp_{out}}} \right)}$$

$$\left( k_1 \frac{Glu_{in}}{k_m^{Glu_{in}}} \frac{Asp_{out}}{K_m^{Asp_{out}}} \frac{H_{in}}{K_m^{H_{in}}} + k_2 \right)\left( 1 + \frac{H_{out}}{K_m^{H_{out}}} \right) \cdot \left( 1 + \frac{Glu_{out}}{K_m^{Glu_{out}}} \right)\left( 1 + \frac{Asp_{in}}{K_m^{Asp_{in}}} \right)$$

(7.4)

Here, AGC is the concentration of the aspartate-glutamate carrier, and $k_m^{Glu_{out}}, K_m^{Asp_{in}}, K_m^{H_{out}}, K_m^{Glu_{in}}, K_m^{Asp_{out}}, K_m^{H_{in}}$ are the Michaelis constants for intra- and extra-mitochondrial glutamate, aspartate and protons. Since transport of aspartate and glutamate is coupled with proton transport through the membrane the reaction rate of AGC depends on the transmembrane potential. We assume that all stages of the catalytic cycle associated with charge transfer across the membrane depend on potential. These are the stages of carrier reorientation corresponding to glutamate and aspartate transfer across the membrane (reaction 1 of catalytic cycle characterised by rate constants $k_1$ and $k_{-1}$, see figure 7.5) and the stages of proton binding and release characterised by the corresponding Michaelis constants. These parameters depend on electric potential (Boork and Wennerstrom 1984; Reynolds, Johnson, and Tanford 1985):

$$K_m^{H_{in}} = K_{m,0}^{H_{in}} * e^{\delta_1 \frac{\Delta \psi}{RT/F}}; \quad K_m^{H_{out}} = K_{m,0}^{H_{out}} * e^{-\delta_3 \frac{\Delta \psi}{RT/F}}; \quad k_1 = k_1^0 * e^{(1-\alpha)\delta_2 \frac{\Delta \psi}{RT/F}};$$

$$k_{-1} = k_{-1}^0 * e^{-\alpha \delta_2 \frac{\Delta \psi}{RT/F}}$$

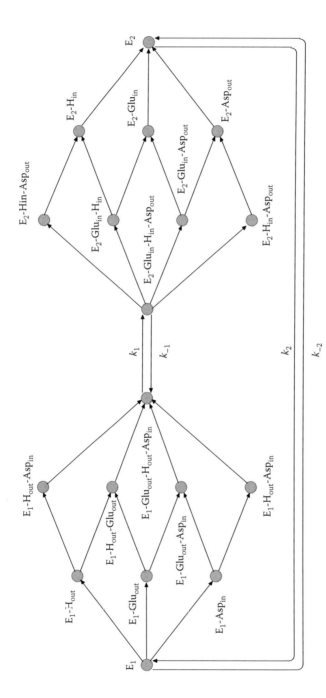

FIGURE 7.5 The scheme of the aspartate-glutamate carrier catalytic cycle.

where $\Delta\psi > 0$ is a transmembrane potential; $T$ is the absolute temperature; $R$ is the universal gas constant; $F$ is Faraday's constant; $\delta_i$ is a part of the potential, consumed by the ith stage; and $\alpha$ is a part of the potential that influences reverse reaction. We have assumed that $\delta_1 = 1, \delta_2 = 0.8, \delta_3 = 0.1, \alpha = 0.5$. Values of the Michaelis constants for substrates and products have been taken from the literature (Dierks and Kramer 1988). The remaining parameters were unknown and were estimated by fitting the model to experimental data (La Noue et al. 1979; parameters values are listed in table 7.2). Dependences of initial rates of aspartate influx into submitochondrial particles have been measured at different pH values and glutamate concentrations. Figure 7.6 demonstrates that experimental data from La Noue et al. (1979; symbols) and theoretical curves generated by rate equation (7.4) closely coincide. Two parameters could not be estimated from *in vitro* data: the concentration of AGC in mitochondria and the Michaelis constant for external proton.

## Aspartate Aminotransferase (AspAT)

Aspartate aminotransferase catalyzes transamination of glutamate and oxaloacetate with formation of α-ketoglutarate and aspartate. AspAT kinetics were

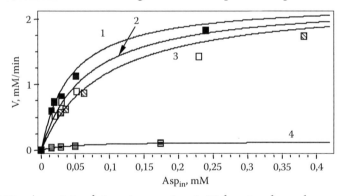

FIGURE 7.6 Aspartate-glutamate carrier initial rate dependence on the aspartate concentration presented by experimental points (obtained on submitochondrial particles with transmembrane potential of 180 mV) and by rate equation (7.4) in the following conditions (enzyme concentration in submitochondrial particles was defined to be 0.4 $\mu$M): 1) $Glu_{in} = 0$; $Glu_{out} = 96.25$ mM; $pH_{out} = 7.2$; $pH_{in} = 7.5$ (black squares). 2) $Glu_{in} = 0$; $Glu_{out} = 96.25$ mM; $pH_{out} = 7.2$; $pH_{in} = 6$ (white squares). 3) $Glu_{in} = 1$ mM; $Glu_{out} = 95.75$ mM; $pH_{out} = 7.2$; $pH_{in} = 7.4$ (squares with oblique hatching). 4) $Glu_{in} = 0$; $Glu_{out} = 61.25$ mM; $pH_{out} = 7.2$; $pH_{in} = 6$ (squares with horizontal hatching).

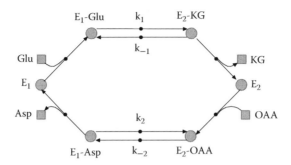

FIGURE 7.7   Scheme of catalytic cycle of aspartate aminotransferase.

described using a ping pong bi-bi mechanism (Cascante and Cortes 1988; see scheme of catalytic cycle in figure 7.7).

It is assumed that the mechanism includes two slow stages: the first corresponds to the transformation of glutamate to α-ketoglutarate (reaction 1 in figure 7.7) and the second corresponds to formation of aspartate from oxaloacetate (reaction 2 in figure 7.7). Using these assumptions and applying rapid equilibrium and quasi-steady-state approaches we derive the following rate equation for AspAT:

$$
V_{AspAT} = AspAT * \frac{k_f^2 \dfrac{OAA}{K_m^{OAA}} \dfrac{Glu_{in}}{K_m^{Glu_{in}}} - k_r^2 \dfrac{Asp_{in}}{K_m^{Asp_{in}}} \dfrac{KG_{in}}{K_m^{KG_{in}}}}{\left( k_f \dfrac{Glu_{in}}{K_m^{Glu_{in}}} + k_r \dfrac{Asp_{in}}{K_m^{Asp_{in}}} \right)\left( 1 + \dfrac{k_r}{k_{-1}} \dfrac{KG_{in}}{K_m^{KG_{in}}} + \dfrac{k_1 - k_f}{k_1} \dfrac{OAA}{K_m^{OAA}} \right)}
$$

$$
+ \left( k_r \dfrac{KG_{in}}{K_m^{KG_{in}}} + k_f \dfrac{OAA}{K_m^{OAA}} \right) * \left( 1 + \dfrac{k_f}{k_1} \dfrac{Glu_{in}}{K_m^{Glu_{in}}} + \dfrac{k_{-1} - k_r}{k_{-1}} \dfrac{Asp_{in}}{K_m^{Asp_{in}}} \right)
\tag{7.5}
$$

We have obtained the following constraints from this equation:

$$
k_1 > k_f; \quad k_{-1} > k_r
$$

Here, AspAT is the concentration of aspartate aminotransferase; $K_m^{OAA}$, $K_m^{ASP_{in}}, K_m^{Glu_{in}}, K_m^{KG_{in}}$ are the Michaelis constants for substrates and products; $k_f, k_r$ are turnover numbers in the forward and reverse directions; and $k_1, k_{-1}$ are rate constants for individual reaction steps (see catalytic cycle in figure 7.7). Parameters known from the literature are AspAT (Fahien and Teller 1992), $K_m^{ASP_{in}}, K_m^{KG_{in}}, k_r$ (Kuramitsu 1985), $K_m^{OAA}$ (Garber and Hanson 1971) and $K_{eq}$ (Siess, Kientsch-Engel, and Wieland 1984). Parameter $k_{-1}$ value was estimated from experimental data (Kuramitsu 1985) in which the dependence of the initial rate of the reverse reaction on α-ketoglutarate

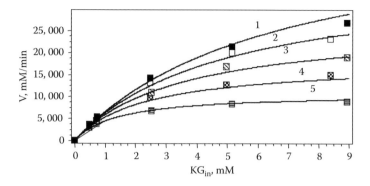

FIGURE 7.8 Aspartate aminotransferase initial rate dependence on the concentration of $\alpha$-ketoglutarate presented by experimental points and described by the curves according to the rate equation (7.5) with the following concentrations of aspartate (enzyme concentration was defined to be 1 mM): 1–50 mM (black squares); 2–5 mM (white squares); 3–2 mM (squares with oblique hatching); 4–1 mM (dotted squares); 5–0.5 mM (squares with horizontal hatching).

concentration was measured at different concentrations of aspartate. All parameter values are listed in table 7.2. Figure 7.8 demonstrates good fitting of experimental data from Kuramitsu (1985; symbols) to theoretical curves generated by equation (7.5). Other parameters ($k_m^{Glu_{in}}$, $k_f$ and $k_1$) could not be estimated from available *in vitro* data.

*Succinate Thiokinase (STK)*

Succinate thiokinase catalyzes the reaction of succinyl-CoA decomposition coupled with GDP phosphorylation. The catalytic mechanism of the STK reaction has been published in Cha and Parks (1964a). However, we have found that the rate equation derived in accordance with this mechanism could not describe reciprocal plots of initial velocities versus reciprocal concentrations of GDP and CoA measured with four fixed concentrations of phosphate. To avoid these discrepancies we suggest that the STK kinetics should be described in accordance with a random bi uni uni bi mechanism (see scheme of the catalytic cycle of STK in figure 7.9). According to this mechanism, substrates GDP and SucCoA bind to the enzyme in a random order. The decomposition of SucCoA occurs at the catalytic site followed by succinate release. The next step of the catalytic cycle is binding of phosphate followed by GDP phosphorylation and release of CoA and GTP in random order. The catalytic cycle depicted in figure 7.9 has two dead-end complexes

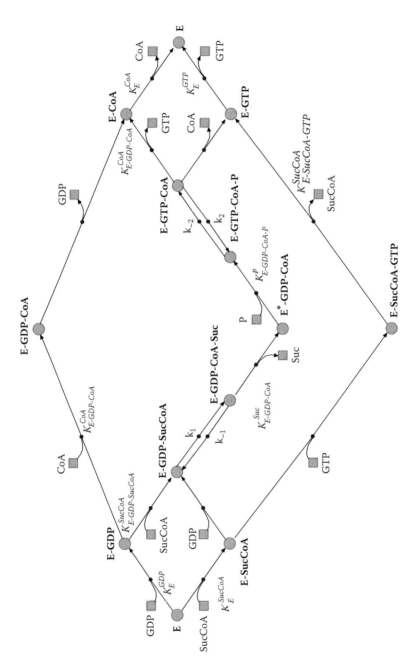

FIGURE 7.9   The scheme of the succinate thiokinase catalytic cycle.

E-GDP-CoA and E-SucCoA-GTP. The rate equation derived in accordance with the new mechanism allows us to describe almost all reciprocal plots of initial velocities versus substrates from Cha and Parks (1964b):

$$
V_{STK} = \frac{STK \cdot \left( k_1 k_2 \dfrac{SucCoA}{K_{E-GDP-SucCoA}^{SucCoA}} \dfrac{GDP}{K_{E-GDP}^{GDP}} \dfrac{P}{K_{E-P}^{P}} - k_{-1} k_{-2} \dfrac{Suc}{K_{E-Suc}^{Suc}} \dfrac{GTP}{K_{E-GTP-CoA}^{GTP}} \dfrac{CoA}{K_{E-CoA}^{CoA}} \right)}{\left( k_1 \dfrac{SucCoA}{K_{E-GDP-SucCoA}^{SucCoA}} \dfrac{GDP}{k_{E-GDP}^{GDP}} + k_{-2} \dfrac{GTP}{K_{E-GTP-CoA}^{GTP}} \dfrac{CoA}{K_{E-CoA}^{CoA}} \right) \left( 1 + \dfrac{Suc}{K_{E-Suc}^{Suc}} + \dfrac{P}{K_{E-P}^{P}} \right)}
$$

$$
\begin{aligned}
+ &\left( 1 + \frac{GDP}{K_{E-GDP}^{GDP}} + \frac{SucCoA}{K_{E-SucCoA}^{SucCoA}} + \frac{SucCoA}{K_{E-GDP-SucCoA}^{SucCoA}} \frac{GDP}{K_{E-GDP}^{GDP}} + \frac{GDP}{K_{E-GDP}^{GDP}} \frac{CoA}{K_{E-GDP-CoA}^{CoA}} + \frac{CoA}{K_{E-CoA}^{CoA}} \right. \\
&\left. + \frac{GTP}{K_{E-GTP}^{GTP}} + \frac{GTP}{K_{E-GTP-CoA}^{GTP}} \frac{CoA}{K_{E-CoA}^{CoA}} + \frac{SucCoA}{K_{E-CoA}^{GTP}} \frac{GTP}{K_{E-GTP-SucCoA}^{SucCoA}} \right) \left( k_2 \frac{P}{K_{E-P}^{P}} + k_{-1} \frac{Suc}{K_{E-Suc}^{Suc}} \right)
\end{aligned}
$$

(7.6)

Here, STK is the concentration of succinate thiokinase; $k_1, k_{-1}, k_2, k_{-2}$ are rate constants; and $K_E^S$ is the dissociation constant for compound S from enzyme form E. Parameters known from the literature are $K_m^{CoA}, K_m^{Suc}, K_m^{SucCoA}, K_m^{GDP}, K_m^{GTP}, K_m^{P}$, Michaelis constants for CoA, Suc, SucCoA, GDP, GTP and P (Cha and Parks 1964a); the catalytic constant $k_f$ (Cha and Parks 1964b); and the equilibrium constant $K_{eq}$ (Kaufman and Alivisatos 1955). Using the approach suggested in Demin et al. (2004) we express a number of parameters from equation (7.6) in terms of kinetic parameters known from the literature. Rate constants $k_2$ and $k_{-1}$ have been expressed from $k_f$ and $K_{eq}$ values:

$$
k_2 = \frac{k_f k_1}{k_1 - k_f}; \quad k_{-1} = \frac{k_{-2} \sqrt[3]{W}}{1 - \sqrt[3]{W}} \quad \text{where} \quad W = \frac{k_f^3 K_m^{Suc} K_m^{GTP} K_m^{CoA}}{K_{eq} k_1 k_{-2}^2 K_m^{SucCoA} K_m^{GDP} K_m^{P}}
$$

These expressions have given us the following constraints for the parameters: $k_1 > k_f, W > 1$.

Six dissociation constants are expressed through Michaelis constants:

$$
K_{E-GDP-SucCoA}^{SucCoA} = K_m^{SucCoA} \frac{k_1 + k_2}{k_2}; \quad K_E^{GDP} = K_m^{GDP} \frac{k_1 + k_2}{k_2};
$$

$$
K_{E-GDP-CoA-P}^{P} = K_m^{P} \frac{k_1 + k_2}{k_2}; \quad K_E^{Suc} = K_m^{Suc} \frac{k_{-1} + k_{-2}}{k_{-2}};
$$

$$
K_{E-GTP-CoA}^{GTP} = K_m^{GTP} \frac{k_{-1} + k_{-2}}{k_{-2}}; \quad K_E^{CoA} = K_m^{CoA} \frac{k_{-1} + k_{-2}}{k_{-2}}.
$$

Taking into account these relationships we have decreased the number of unknown parameters of equation (7.6) from fourteen to six. Remaining undetermined parameters are $k_1$, $k_{-2}$, $K_{E-SucCoA}^{SucCoA}$, $K_{E-GDP-CoA}^{CoA}$, $K_{E-GTP}^{GTP}$ and $K_{E-GTP-SucCoA}^{SucCoA}$. We estimate their values from experimental data (Cha et al. 1964) in which dependences of the initial rate of succinate thiokinase on concentrations of substrates and products were measured. Moreover, the fitting of the rate equation (7.6) to experimental data allows us to identify Michaelis constants for substrates and products more precisely (see table 7.2). Figure 7.10 demonstrates that experimental data from (Cha and

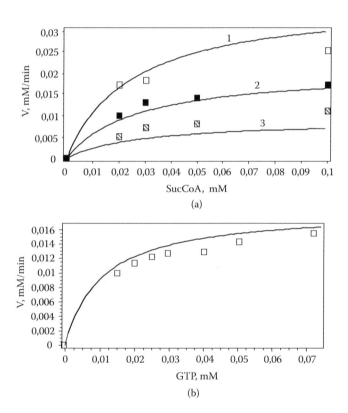

FIGURE 7.10 Succinate thiokinase initial rate dependence on the concentration of substrates and products presented by experimental points and described by the curves according to the rate equation (7.6) in the following conditions (enzyme concentration was defined to be 0.05 $\mu$M): a) GDP = 0.05 mM;SucCoA, mM: 1) 0.05 (white squares); 2) 0.03 (black squares); 3) 0.02 (squares with oblique hatching); b) Suc = 1 mM; CoA = 0.1 mM (white squares).

Parks 1964b; symbols) and theoretical curves generated by equation (7.6) closely coincide.

## Succinate Dehydrogenase

Succinate Dehydrogenase catalyzes the reaction of succinate oxidation to fumarate coupled with reduction of ubiquinone to ubiquinol. The enzyme is described according to a random bi bi mechanism. Applying the rapid equilibrium approach the following rate equation has been derived:

$$V_{SDH} = SDH \frac{k_f \frac{Suc}{K_{E-Suc}^{Suc}} \frac{Q}{K_m^Q} - k_r \frac{Fum}{K_{E-Fum}^{Fum}} \frac{QH_2}{K_m^{QH_2}}}{1 + \frac{Suc}{K_{E-Suc}^{Suc}} + \frac{Q}{K_m^Q} \frac{K_m^{Suc}}{K_{E-Suc}^{Suc}} + \frac{Suc}{K_{E-Suc}^{Suc}} \frac{Q}{K_m^Q} + \frac{Fum}{K_{E-Fum}^{Fum}} + \frac{QH_2}{K_m^{QH_2}} \frac{K_m^{Fum}}{K_{E-Fum}^{Fum}} + \frac{Fum}{K_{E-Fum}^{Fum}} \frac{QH_2}{K_m^{QH_2}}}$$

$$(7.7)$$

Here, SDH is the concentration of succinate dehydrogenase; $k_f, k_r$ are turnover numbers in the forward and reverse directions; $K_E^S$ is the dissociation constant for compound S from enzyme form E; and $K_m^Q, K_m^{QH2}, K_m^{Suc}, K_m^{Fum}$ are Michaelis constants for ubiquinone, ubiquinol, succinate and fumarate. The values of all parameters of equation (7.7) are available from the literature (see table 7.2). However, having fitted the rate equation (7.7) to experimental data published in Kotlyar and Vinogradov (1984) in which dependences of SDH initial rate on succinate concentration were measured at different concentrations of phenasine methosulfate (an electron acceptor analogous to ubiquinone) and malonate (an inhibitor competitive with succinate) we found that the values of $K_m^{Suc}$ and $K_{E-Suc}^{Suc}$ had to be changed substantially (see table 7.2) to provide the best coincidence between experimental data and theoretical curves as shown in figure 7.11.

It has been found (Grivennikova et al. 1993) that the mitochondrial succinate dehydrogenase is strongly inhibited by oxaloacetate. To take this fact into account we have assumed that catalytically active succinate dehydrogenase, SDH, could bind oxaloacetate to form a catalytically inactive complex, SDH-OAA. The process of SDH inactivation is described in accordance with the mass action law:

$$V_{ISDH} = k_i \cdot SDH \cdot OAA - k_{-i} \cdot (SDH - OAA) \qquad (7.8)$$

The values of rate constants of equation (7.8) were taken from Vinogradov (1986; see table 7.2). Taking into account both rate equations (7.7) and (7.8) we have reproduced experimental time dependences of SDH catalyzed succinate consumption (Grivennikova et al. 1993) on different

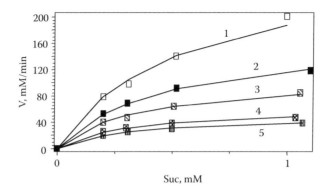

FIGURE 7.11 Succinate dehydrogenase initial rate dependence on the concentration of substrate succinate presented by experimental points and described by the curves according to the rate equation (7.7) with 20 μM malonate ($K_i$ = 1.2 $\mu$M) and the following concentrations of the second substrate phenasinemethosulphate (enzyme concentration in submitochondrial particles was defined to be 0.1 mM): 1) 250 $\mu$M (white squares); 2) 160 $\mu$M (black squares); 3) 50 $\mu$M (squares with oblique hatching); 4) 20 $\mu$M (dotted squares); 5) 10 $\mu$M (squares with vertical hatching).

concentrations of oxaloacetate in which reaction was started by addition of succinate dehydrogenase to a solution of the substrates succinate and ubiquinone and time dependences of succinate consumption were monitored at different concentrations of added oxaloacetate. To describe qualitatively these experiments the following system of differential equations has been developed (figure 7.12):

$$\frac{dSuc}{dt}=-V_{SDH};\quad \frac{dQ}{dt}=-V_{SDH};\quad \frac{dFum}{dt}=V_{SDH};\quad \frac{dQH_2}{dt}=V_{SDH};$$

$$\frac{dSDH}{dt}=-V_{ISDH};\quad \frac{dOAA}{dt}=-V_{ISDH};\quad \frac{d(SDH-OAA)}{dt}=V_{ISDH} \qquad (7.9)$$

Here, $V_{SDH}$, $V_{ISDH}$ are rate equations for succinate dehydrogenase and for the process of its inactivation given by equations (7.7) and (7.8); substrates, products and two forms of SDH (free and bound with OAA) are variables of the minimodel (7.9). Initial conditions for the system of differential equations (7.9) have been set in accordance with initial concentrations of substrates used in experiment (Grivennikova et al. 1993): Suc = 1.33 mM; Q = 10 μM; Fum = 0 mM; $QH_2$ = 0 mM; SDH = 1.5 nM; SDH-OAA = 0.

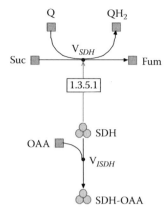

FIGURE 7.12 The scheme of SDH reaction described by the system (7.9) constructed to reproduce time dependences of the succinate concentration in succinate dehydrogenase reaction with added oxaloacetate.

The system of differential equations (7.9) has five conservation laws:

$$Suc + Fum = C_{tot}^{SDH} \text{ (conservation of four-carbon skeleton)}$$

$$Suc + QH_2 = e_{tot}^{SDH} \text{ (conservation of electrons)}$$

$$Q + QH_2 = Q_{tot}^{SDH} \text{ (conservation of ubiquinone)}$$

$$SDH + SDH\_OAA = SDH_{tot}^{SDH} \text{ (conservation of succinate dehydrogenase)}$$

$$OAA + SDH\_OAA = OAA_{tot}^{SDH} \text{ (conservation of oxaloacetate)}$$

Values of $C_{tot}^{SDH}$, $e_{tot}^{SDH}$, $Q_{tot}^{SDH}$, $SDH_{tot}^{SDH}$, $OAA_{tot}^{SDH}$ can be calculated from the initial conditions. Figure 7.13 demonstrates the reproduction of experimental time dependences from Grivennikova et al. (1993).

*Fumarase (FUM)*

The reaction of fumarate transformation to malate catalyzed by fumarase was described according to the uni-uni mechanism (Alberty 1961):

$$V_{FUM} = FUM \frac{k_f \frac{Fum}{K_m^{Fum}} k_r - \frac{Mal_{in}}{K_m^{Mal_{in}}}}{1 + \frac{Fum}{K_m^{Fum}} + \frac{Mal_{in}}{K_m^{Mal_{in}}}} \tag{7.10}$$

Here, FUM is the concentration of fumarase; $k_f, k_r$ are catalytic constants of the forward and reverse reactions; and $K_m^{Fum}, K_m^{Mal}$ are Michaelis

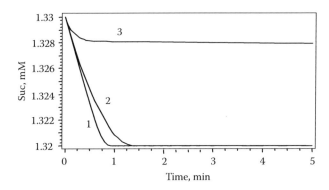

FIGURE 7.13   Succinate concentration time dependence in succinate dehydrogenase reaction in the following conditions: 1) Suc = 1.33 mM, $Q_2$ = 1 $\mu$M (ubiquinone homolog), OAA = 0; 2) Suc = 1.33 mM, $Q_2$ = 10 $\mu$M, OAA = 0.6 $\mu$M; 3) Suc = 1.33 mM, $Q_2$ = 10 $\mu$M, OAA = 6 $\mu$M.

constants for fumarate and malate. Parameter values of equation (7.10) are available from the literature (Alberty 1961; see table 7.2). However, having fitted the rate equation (7.10) to experimental data published in Greenhut, Umezawa, and Rudolph (1985) we have found that values of $K_m^{Fum}$ and $k_f$ had to be changed (see table 7.2) to provide a good fit (figure 7.14).

*Malate Dehydrogenase (MDH)*

The mechanism of the enzyme has been described according to the ordered bi-bi mechanism with enzyme-NAD complex isomerization

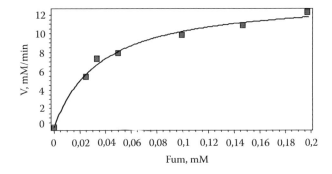

FIGURE 7.14   Fumarase initial rate dependence on the fumarate concentration described by the rate equation (7.10) and experimental points (enzyme concentration was defined to be 0.15 $\mu$M).

(Heyde and Ainsworth 1968):

$$V_{MDH} = \frac{MDH \cdot k_f k_r (Mal_{in} \cdot NAD / K_{eq} - NADH \cdot OAA)}{K_i^{NADH} K_m^{OAA} k_r + K_m^{OAA} k_r \cdot NADH + K_m^{NADH} k_r \cdot OAA + k_r \cdot NADH \cdot OAH}$$

$$+ K_m^{NAD} k_f \cdot Mal_{in}/K_{eq} + K_m^{Mal} k_f \cdot NAD/K_{eq} + k_f \cdot Mal_{in} \cdot NAD/K_{eq}$$

$$+ K_m^{NAD} k_f \cdot NADH \cdot Mal_{in}/K_i^{NADH} K_{eq} + K_m^{NADH} k_r \cdot OAA \cdot NAD/K_i^{NAD}$$

$$+ k_r \cdot NADH \cdot OAA \cdot Mal_{in}/K_i^{Mal} + k_f \cdot OAA \cdot Mal_{in} \cdot NAD/K_i^{OAA} K_{eq}$$

$$(7.11)$$

Here, MDH is the concentration of malate dehydrogenase; $k_f$, $k_r$ are catalytic constants in the forward and reverse directions; $K_{eq}$ is the equilibrium constant; $K_m^{OAA}$, $K_m^{NADH}$, $K_m^{NAD}$, $K_m^{Mal_{in}}$ are Michaelis constants for substrates and products; and $K_i^{OAA}$, $K_i^{NADH}$, $K_i^{NAD}$, $K_i^{Mal_{in}}$ are inhibition constants for substrates and products. All parameter values from equation (7.11) are available from the literature (Heyde and Ainsworth 1968; see table 7.2).

### α–Ketoglutarate-Malate Carrier (KMC)

α–Ketoglutarate-malate carrier catalyzes the exchange of external malate for intra-mitochondrial α-ketoglutarate. The rate equation for KMC was derived by assuming that it functions according to the random bi-bi mechanism (Indiveri et al. 1991; see figure 7.15). The mechanism includes two slow

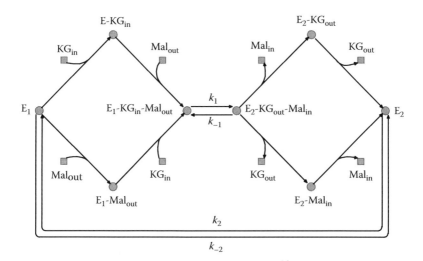

FIGURE 7.15   The scheme of $\alpha$-ketoglutarate/malate carrier catalytic cycle.

stages (with rate constants $k_1, k_{-1}, k_2, k_{-2}$) corresponding to the reorientation of transporter with respect to the inner mitochondrial membrane. Using these assumptions and applying rapid equilibrium and quasi-steady-state approaches we derive the following rate equation for KMC:

$$
V_{KMC} = KMC \frac{k_f^2 \frac{k_2}{k_1} \frac{KG_{in}}{K_m^{KG_{in}}} \frac{Mal_{out}}{K_m^{Mal_{out}}} - k_r^2 \frac{k_{-2}}{k_{-1}} \frac{KG_{out}}{K_m^{KG_{utt}}} \frac{Mal_{in}}{K_m^{Mal_{in}}}}{\left( \frac{k_f^2}{k_1} \frac{KG_{in}}{K_m^{KG_{in}}} \frac{Mal_{out}}{K_m^{Mal_{out}}} + k_{-2} \right) \left( 1 + \frac{k_r}{k_{-1}} \frac{KG_{out}}{K_m^{KG_{utt}}} \left( 1 + \frac{KG_{out}}{K_i^{KG_{out}}} \right) \right) \left( 1 + \frac{k_r}{k_{-1}} \frac{Mal_{in}}{K_m^{Mal_{in}}} \right)}
$$

$$
+ \left( k_2 + \frac{k_r^2}{k_{-1}} \frac{KG_{out}}{K_m^{KG_{out}}} \left( 1 + \frac{KG_{out}}{K_i^{KG_{out}}} \right) \frac{Mal_{in}}{K_m^{Mal_{in}}} \right) \left( 1 + \frac{k_f}{k_1} \frac{KG_{in}}{K_m^{KG_{in}}} \right) \left( 1 + \frac{k_f}{k_1} \frac{Mal_{out}}{K_m^{Mal_{out}}} \right)
$$

(7.12)

Here, KMC is the concentration of α-ketoglutarate-malate carrier; $k_f, k_r$ are catalytic constants in the forward and reverse directions; $K_m^{KG_{in}}$, $K_m^{Mal_{out}}$, $K_m^{KG_{out}}$, $K_m^{Mal_{in}}$ are Michaelis constants for substrates and products; and $k_1$, $k_{-1}, k_2, k_{-2}$ are rate constants of individual stages of the catalytic cycle of the carrier. Using the approach described in Demin et al. (2004) we can express some of the parameters of catalytic cycle α-ketoglutarate-malate carrier in terms of kinetic parameters known from the literature:

$$
k_2 = \frac{k_1 k_f}{k_1 - k_f}, \qquad k_{-2} = \frac{k_{-1} k_r}{k_{-1} - k_r},
$$

$$
k_{-1} = \frac{K_{eq}(k_1 - k_f) + k_f^3 K_m^{KG_{out}} K_m^{Mal_{in}} \big/ k_r^2 K_m^{KG_{in}} K_m^{Mal_{out}}}{k_f^3 K_m^{KG_{out}} K_m^{Mal_{in}} \big/ k_r^3 K_m^{KG_{in}} K_m^{Mal_{out}}}
$$

So we obtain the following constraints from these equations:

$$
k_1 > k_f; \quad k_{-1} > k_r
$$

Parameters known from the literature (Indiveri et al. 1991) are $k_f, k_r, K_m^{KG_{in}}, K_m^{Mal_{out}}, K_m^{KG_{out}}, K_m^{mal_{in}}$. The values of parameters $k_1$ and $K_i^{KG_{out}}$ have been estimated by fitting of the rate equation (7.12) to experimental dependences of the initial rate of KMC-catalyzed malate efflux on the concentrations of either internal malate or external α-ketoglutarate measured in Indiveri et al. (1991) at different concentrations of the second substrate (figure 7.16). The values of parameters are listed in table 7.2.

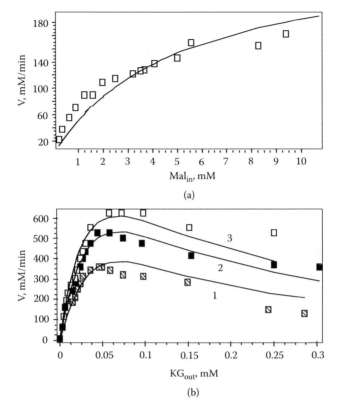

FIGURE 7.16 $\alpha$-Ketoglutarate-malate carrier initial rate dependence on substrates concentrations described by rate equation (7.12) and experimental points with following concentrations of the second substrate (enzyme concentration was defined to be 2 mM): a) $KG_{out}$ = 4.84 µM; b) $Mal_{in}$: 1–2 (squares with oblique hatching); 2–4 (black squares); 3–6 mM (white squares).

## Estimation of Model Parameters from *in Vivo* Data

As discussed in the previous section, some parameters could not be estimated from *in vitro* experimental data. These parameters were intramitochondrial concentrations of the enzymes—AGC, AspAT, KGDH, STK, FUM, MDH, KMC—and two kinetic parameters associated with the AGC rate equation—the Michaelis constants for external glutamate and proton. To estimate the values of these parameters we have adjusted the whole model to experimental data (La Noue et al. 1979) in which the following experiment was applied: a suspension of mitochondria was

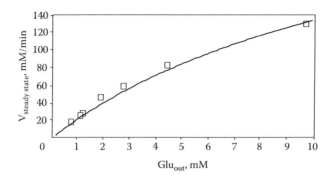

FIGURE 7.17 Dependence of stationary glutamate consumption rate by mitochondrial suspension on glutamate concentration. Simulation results and experimental points under the following conditions: $Glu_{out} = 0$–$20$ mM, $Mal_{out} = 3.7$ mM, $pH_{out} = 7.4$.

incubated in media with glutamate and malate and respiration and glutamate consumption rates have been measured. The steady-state flux was calculated from the dependence of aspartate-glutamate carrier concentration on the concentration of extra-mitochondrial glutamate. To estimate values of the unknown parameters we fitted the observed variation to experimentally measured dependence of glutamate consumption. Figure 7.17 demonstrates that experimental data from (La Noue et al. 1979) (symbols) and the theoretical curve generated by the system of differential equations (7.1) closely coincide. Values of intra-mitochondrial enzyme concentrations and kinetic parameters obtained are listed in table 7.2 (all parameters obtained by fitting to experimental data (La Noue et al. 1979) are marked by an asterisk).

Combining all of these approaches allows us to find all parameter values in the kinetic model (7.1). To test how the model (7.1) describes the kinetic and regulatory properties of the Krebs cycle we have compared calculated values of steady-state metabolite concentrations with those available in the literature measured on mitochondria oxidizing glutamate (see table 7.1). Electric potential difference, intra-mitochondrial concentrations of ubiquinone, ubiquinol, GTP, GDP, ATP, ADP, phosphate and protons were assigned values available from the literature and steady-state concentrations of intermediates of the Krebs cycle were calculated. As demonstrated in table 7.1, the calculated values of the steady-state concentrations of aspartate, glutamate, malate, coenzyme A and succinyl-CoA are quite close to those published in the literature.

The concentration of α-ketoglutarate differs from its literature value. This could be due to the different experimental conditions in Hoek (1971) under which α-ketoglutarate concentration was determined.

## MODELING OF THE *ESCHERICHIA COLI* BRANCHED-CHAIN AMINO ACID BIOSYNTHESIS

Valine and isoleucine are branched-chain amino acids that are widely used in biomedicine and biotechnology. Indeed, valine is used in the food industry for flavour additive production. Moreover, this amino acid is one of the sources of cephamycin antibiotic biosynthesis. In this section we develop a kinetic model of the biosynthesis of valine and isoleucine in *E. coli* and validate it against *in vitro* experimental data as described in chapters 4 and 6.

### Model Development

The initial substrates of isoleucine and valine biosynthesis pathways shown in figure 7.18 are pyruvate (*Pyr*), which is required for the synthesis of both amino acids and l-threonine (*Thr*), which is a precursor of l-isoleucine (*Ile*). To connect our model with the whole-cell metabolism we consider the influx to threonine (*Thr*) from oxaloacetate (*OA*) (step 1 in figure 7.18) and influx to Pyruvate (*Pyr*) from phosphoenolpyruvate (*PEP*) (step 8).

The reaction of deamination (step 2) catalyzed by threonine dehydratase (TDH) transforms l-threonine to α-ketobutyrate (*Kb*). Then decarboxylation of pyruvate catalyzed by acetolactate synthase (AHAS) results in α-aceto-α-hydroxybutyrate (*Ahb*) in the isoleucine pathway (step 3) and α-acetolactate (*Al*) in the valine pathway (step 9). Both acetohydroxybutyrate and acetolactate are converted to dihydroxy acids by isomeroreductase (IR), respectively, to dihydroxy-methylvalerate (*Dhv*) in the isoleucine pathway (step 4) and to dihydroxy-isovalerate (*Dhi*) in the valine pathway (step 10). These dihydroxy acids are converted to α-keto acids by dihydroxy-acid dehydratase (DHAD), respectively, to keto-methylvalerate (*Ktv*) in the isoleucine pathway (step 5) and to keto-isovalerate (*Kti*) in the valine pathway (step 11 in figure 7.18). The final step in the biosynthesis of both amino acids is a transamination reaction catalyzed by branched-chain amino acid transaminase (BCAT) with glutamate as an amino donor. Keto-methylvalerate is converted to isoleucine (*Ile*; step 6) and keto-isovalerate is converted to valine (*Val*; step 12). To couple our model to intracellular energy metabolism and to processes

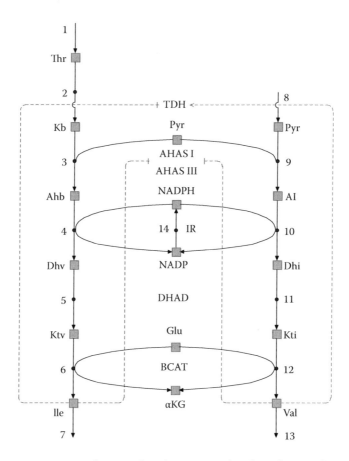

FIGURE7.18  Kinetic scheme of isoleucine and valine biosynthesis pathways. Metabolite notations are in text. Minus sign (−) corresponds to inhibition, plus sign (+) corresponds to activation.

consuming isoleucine and valine we take into account the recycling of NADPH (step 14) and the effluxes of isoleucine (step 7) and valine (step 13 in figure 7.18).

The kinetic model of branched-chain amino acid biosynthesis consists of ordinary differential equations describing the time behaviour of intracellular metabolites. Table 7.3 lists the complete set of these equations.

The pathway of isoleucine and valine biosynthesis has several features that are worthy of notice:

The pathway includes four enzymes (AHAS, IR, DHAD and BCAT), each of which is able to catalyze two different reactions: one belonging to

TABLE 7.3 The System of Differential Equations Describing Time Behaviour of Kinetic Model. Indices 1 Refer to Isoleucine Pathway, Indices 2 Refer to Isoleucine Pathway

| Isoleucine Pathway | Valine Pathway |
|---|---|
| $\dfrac{dThr}{dt} = v_{influx1} - v_{TDH}$ | $\dfrac{dPyr}{dt} = v_{influx2} - 2(v_{AHASI,2} + v_{AHASIII,1})$ |
| | $\quad - (v_{AHASI,1} + v_{AHAS,1})$ |
| $\dfrac{dKb}{dt} = v_{TDH} - (V_{AHASI,1} + v_{AHAS,1})$ | — |
| $\dfrac{dAhb}{dt} = v_{AHASI,1} + v_{AHASIII,1} - v_{IR1}$ | $\dfrac{dAl}{dt} = v_{AHASI,2} + v_{AHASIII,2} - v_{IR2}$ |
| $\dfrac{dDhv}{dt} = v_{IR1} - v_{DHAD1}$ | $\dfrac{dDhi}{dt} = v_{IR} - v_{DHAD2}$ |
| $\dfrac{dKtv}{dt} = v_{DHAD1} - v_{BCAT1}$ | $\dfrac{dKti}{dt} = v_{DHAD2} - v_{BCAT2}$ |
| $\dfrac{dIle}{dt} = v_{BCAT1} - v_{efflux1}$ | $\dfrac{dVal}{dt} = v_{BCAT2} - v_{efflux2}$ |
| $\dfrac{dNADP}{dt} = v_{IR1} + v_{IR2} - v_{NADPH\ recycling}$ | |
| $\dfrac{dNADPH}{dt} = v_{NADPH\ recycling} - v_{IR1} - v_{IR2}$ | |
| $NADP + NADPH = Const$ | |

the isoleucine biosynthesis pathway, the other belonging to the pathway of valine biosynthesis. From this it follows that intermediates of isoleucine biosynthesis pathway compete with intermediates of valine biosynthesis pathway for these four enzymes.

The reactions of deamination and decarboxylation (steps 2 and 3 in figure 7.18) are regulated by isoleucine and valine.

There are several isoenzymes of acetolactate synthase that give rise to differences in regulation by the end products of the pathways.

## Derivation of the Rate Equations

Since most of the enzymes involved in branched-chain amino acid formation are able to catalyze two reactions (one belonging to the isoleucine biosynthesis pathway and the other belonging to the pathway of valine

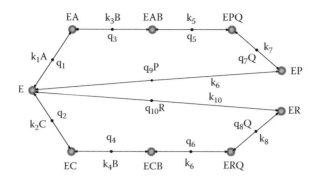

FIGURE 7.19 Catalytic cycle of the enzyme catalyzing two competing reactions. A, B and C are substrates, P, Q and R are products. E is enzyme in free form. $k_i$ and $q_i$ are the constants of elementary stages.

biosynthesis), they should be described so as to take into account competition between substrates of the reactions for the catalytic site of the enzyme. As a first step we construct a catalytic cycle describing this feature. Let us consider such a catalytic cycle for the reaction with two substrates and two products depicted in figure 7.19. Substrate A interacts with substrate B, giving products Q and P. At the same time, substrate C can interact with substrate B, giving products Q and R. Substrates A and C compete with each other for the binding site. Applying the quasi-steady-state approach we derive rate equations corresponding to both reactions, which in simplified form have the following appearance:

$$v_1 = \frac{\frac{V_{m1}}{F_1 K_{m1}^B K_m^A}\left(A \cdot B - \frac{P \cdot Q}{K_{eq}}\right)}{1+\left(\frac{P}{K_{m1}^P}+\frac{Q}{K_m^Q}-\right)\left(1+\frac{A}{F_1 K_m^A K_{m1}^B}\right)+\left(\frac{P}{K_{m2}^P}+\frac{R}{K_m^R}\right)\left(1+\frac{C}{F_2 K_m^c K_{m2}^B}\right)} \tag{7.13}$$

$$v_2 = \frac{\frac{V_{m2}}{F_2 K_{m2}^B K_m^C}\left(C \cdot B - \frac{P \cdot R}{K_{eq}}\right)}{1+\left(\frac{P}{K_{m1}^P}+\frac{Q}{K_m^Q}\right)\left(1+\frac{A}{F_1 K_m^A}\frac{B}{K_{m1}^B}\right)+\left(\frac{P}{K_{m2}^P}+\frac{R}{K_m^R}\right)\left(1+\frac{C}{F_2 K_m^C}\frac{B}{K_{m2}^B}\right)} \tag{7.14}$$

$$F_1 = 1+\frac{A}{K_m^A}+\frac{B}{K_{m1}^B}, \quad F_2 = 1+\frac{C}{K_m^C}+\frac{B}{K_{m2}^B}$$

Each of these corresponds to an ordered bi-bi mechanism (in Cleland terms), where $V_m$, $K_m$ and $K_{eq}$ are the maximal velocity, the Michaelis constants and the equilibrium constant respectively. The same approach can be applied to derive rate equations for competing reactions with two substrates and one product (ordered bi-uni according to the Cleland classification):

$$v_1 = \frac{\dfrac{V_{m1}}{F_1 K_{m1}^B K_m^A}\left(A \cdot B - \dfrac{Q}{K_{eq}}\right)}{1 + \dfrac{Q}{K_m^Q}\left(1 + \dfrac{A}{F_1 K_m^A}\dfrac{B}{K_{m1}^B}\right) + \dfrac{R}{K_m^R}\left(1 + \dfrac{C}{F_2 K_m^C}\dfrac{B}{K_{m2}^B}\right)},$$ 
(7.15)

$$v_2 = \frac{\dfrac{V_{m2}}{F_2 K_{m2}^B K_m^C}\left(C \cdot B - \dfrac{Q}{K_{eq}}\right)}{1 + \dfrac{Q}{K_m^Q}\left(1 + \dfrac{A}{F_1 K_m^A}\dfrac{B}{K_{m1}^B}\right) + \dfrac{R}{K_m^R}\left(1 + \dfrac{C}{F_2 K_m^C}\dfrac{B}{K_{m2}^B}\right)},$$ 
(7.16)

as well as for competing reactions with one substrate and one product (uni-uni according to the Cleland classification):

$$v_1 = \frac{\dfrac{V_{m1}}{K_m^A}\left(A - \dfrac{Q}{K_{eq}}\right)}{1 + \dfrac{Q}{K_m^Q} + \dfrac{A}{K_m^A} + \dfrac{R}{K_m^R} + \dfrac{C}{K_m^C}},$$ 
(7.17)

$$v_2 = \frac{\dfrac{V_{m2}}{K_m^C}\left(C - \dfrac{R}{K_{eq}}\right)}{1 + \dfrac{Q}{K_m^Q} + \dfrac{A}{K_m^A} + \dfrac{R}{K_m^R} + \dfrac{C}{K_m^C}},$$ 
(7.18)

### Detailed Description of Pathway Steps

For estimation of kinetic parameters of rate equations describing individual enzymes of the pathway we fitted rate equations to *in vitro* experimental data taken from different literature sources. When experimental curves were not available we used values of kinetic parameters available from electronic databases BRENDA and EMP and other published information on the individual enzyme. In the cases of lack of any experimental data we used 'free' values (notation f.p in table 7.4) of parameters.

### *Influxes*

We describe the reactions supplying initial substrates to the branched-chain amino acid biosynthesis in the simplest form according to the mass action law. Threonine, the precursor of isoleucine, is derived from

TABLE 7.4   Values of Kinetic Parameters of Isoleucine and Valine Biosynthesis Pathway in *Escherichia coli*

| Enzyme | Parameter Values | Source or Estimation |
|---|---|---|
| TDH | $K_r^{Thr} = 654.99$ <br> $K_t^{Thr} = 2.62$ <br> $K_r^{Kb} = 1.0 \text{ f.p}^*$ <br> $K_t^{Kb} = 1.0 \text{ f.p}$ <br> $K_r^{Ile} = 0.0058$ <br> $K_t^{Ile} = 7.98$ <br> $K_r^{Val} = 180.15$ <br> $K_t^{Val} = 0.05 \ L = 0.09$ <br> $n = 4$ | Fitting of literature data (Wessel et al. 2000) |
| AHAS | $K_m^{Kb} = 0.053$ <br> $K_{m1}^{Pyr} = 0.145$ <br> $K_m^{Ahb} = 1.0 \text{ f.p}$ <br> $K_{m2}^{Pyr} = 0.00098$ <br> $K_m^{Al} = 1.0 \text{ f.p}$ <br> $K_i^{Ile} = 1.416$ <br> $K_i^{Vol} = 0.033$ | Fitting of literature data (Engel et al. 2000; Vyazmensky et al. 1996; Bar-Ilan 2001; Hill, Duggleby 1998; Hill, Pang, Duggleby 1997; Eoyang L, Silverman PM. 1984; Barak, Chipman, and Gollop 1987) |
| IR | $K_{m1}^{NADPH} = 0.073$ <br> $K_m^{Ahb} = 0.002$ <br> $K_{m1}^{NADP} = 0.0042$ <br> $K_m^{Dhv} = 1 \text{ f.p}$ <br> $K_{m2}^{NADPH} = 0.073$ <br> $K_m^{Al} = 0.014$ <br> $K_{m2}^{NADP} = 0.206$ <br> $K_m^{Dhi} = 1 \text{ f.p}$ | Brenda and literature data (Madhavi et al. 1997; Aulabaugh and Schloss 1990; Chunduru, Mrachko, and Calvo 1998) |
| DHAD | $K_m^{Dhv} = 0.13$ <br> $K_m^{Dhi} = 0.1$ <br> $K_m^{Ktv} = 0.1 \text{ f.p}$ <br> $K_m^{Kti} = 0.1 \text{ f.p}$ | Fitting of literature data (Limberg, Klaffke, and Thiem 1995; Myers 1961) |
| BCAT | $K_m^{Ktv} = 0.2$ <br> $K_{m1}^{Glt} = 1 \text{ f.p}$ <br> $K_m^{Ile} = 1.1$ <br> $K_{m1}^{Ktg} = 0.6$ <br> $K_m^{Kti} = 0.2$ <br> $K_{m2}^{Glt} = 1 \text{ f.p}$ <br> $K_m^{Val} = 5$ <br> $K_{m2}^{Ktg} = 3$ | Brenda and literature data (Hall et al. 1993; Lee-Peng, Hermodson, and Kohlhaw 1979; Inoue et al. 1988) |

*\* F.p. Means Free Parameter.*

oxaloacetate (OA). Pyruvate, which takes part in both pathways, is derived from phosphoenolpyruvate:

$$V_{influx1} = V_{influx1}\left(OA - \frac{Thr}{K_{eq}^{in1}}\right), \quad V_{influx2} = V_{influx2}\left(PED - \frac{Pyr}{K_{eq}^{in2}}\right)$$

Parameters of $V_m$ and $K_{eq}$ were evaluated by the procedure described below.

### Threonine Dehydratase (TDH)

Threonine dehydratase catalyzes the conversion of threonine to ketobutyrate; it is inhibited by isoleucine and is activated by valine. According to results published in (Umbarger 1996; Wessel et al. 2000) isoleucine inhibits TDH by enhancing the cooperativity of substrate binding while valine abolishes cooperativity and allows the enzyme to bind threonine with normal Michaelis-Menten kinetics. To describe the reversible kinetics of TDH and to take into account allosteric properties of the enzyme we used the following rate equation taken from Ivanitzky, Krinsky, and Selkov (1978):

$$v_{TDH} = \frac{\frac{V_m^{TDH}}{K_m^{Thr}}\left(Thr - \frac{Kb}{K_{eq}^{TDH}}\right)}{\left(1 + \frac{Thr}{K_m^{Thr}} + \frac{Kb}{K_m^{Kb}}\right)} \frac{1 + LF^{n-1}H^n}{1 + LF^n H^n}$$

where

$$F = \frac{1 + \frac{Thr}{K_m^{Thr}}\frac{K_r^{Thr}}{K_t^{Thr}} + \frac{Kb}{K_m^{Kb}}\frac{K_r^{Kb}}{K_t^{Kb}}}{1 + \frac{Thr}{K_m^{Thr}} + \frac{Kb}{K_m^{Kb}}}, \quad H = \frac{\left(1 + \frac{Ile}{K_t^{Ile}}\right)\left(1 + \frac{Val}{K_t^{Val}}\right)}{\left(1 + \frac{Ile}{K_r^{Ile}}\right)\left(1 + \frac{Val}{K_r^{Val}}\right)}$$

To estimate the parameters of the equation we fitted it against experimental curves from Wessel et al. (2000) (figure 7.20). We fitted all the experimental curves simultaneously and obtained a unique set of kinetic parameters listed in table 7.4. The values of $V_m$ and $K_{eq}$ were estimated as described below.

### Acetolactate Synthase (AHAS)

Acetolactate synthase catalyzes the conversion of ketobutyrate to acetohydroxybutyrate in the pathway of isoleucine biosynthesis and the conversion of pyruvate to acetolactate in the valine biosynthesis pathway.

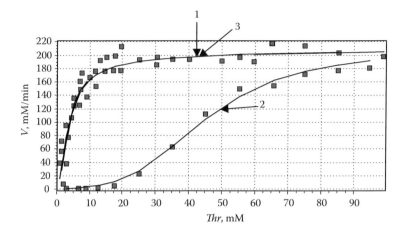

FIGURE 7.20  Steady-state kinetics of TDH. Dependence of the initial rate of product formation on threonine concentration. Symbols correspond to experimental data taken from Wessel et al. (2000), lines designate model curves. 1) Without effectors, $Ile = 0$ mM, $Val = 0$ mM. 2) With isoleucine as inhibitor, $Ile = 0.05$ mM, $Val = 0$ mM. 3) With both effectors, $Ile = 0.05$ mM, $Val = 0.5$ mM.

Three different enzymes catalyzing the formation of acetohydroxy acids have been found (Barak, Chipman, and Gollop 1987; Umbarger 1996). Each isozyme is sensitive to inhibition by isoleucine and valine but at a different level. To simplify the model we have not considered all three isozymes and have taken into account AHAS I and AHAS III only. These isozymes show maximal differences in their sensitivity to the inhibitors and enable us to explore the possible differences in regulation in the framework of the whole model. The rate equations of reactions 3 and 9 (see figure 7.18) catalyzed by AHAS have been taken in the form of (7.15) and (7.16), respectively. The rate equations for AHAS I and AHAS III are the same in form but differ in values of kinetic parameters only:

$$v_{AHAS1} = \frac{\dfrac{V_{m1}^{AHAS}}{F_1 K_{m1}^{Pyr} K_m^{Kb}}\left(Kb \cdot Pyr - \dfrac{Ahb}{K_{eq}^{AHAS}}\right)}{1 + \dfrac{G_1}{F_1} + \dfrac{G_2}{F_2} + \dfrac{Ile}{K_i^{Ile}} + \dfrac{Val}{K_i^{Val}}}, \qquad v_{AHAS2} = \frac{\dfrac{V_{m2}^{AHAS}}{F_2 K_{m2}^{Pyr} K_m^{Al}}\left(Pyr^2 - \dfrac{Al}{K_{eq}^{AHAS}}\right)}{1 + \dfrac{G_1}{F_1} + \dfrac{G_2}{F_2} + \dfrac{Ile}{K_i^{Ile}} + \dfrac{Val}{K_i^{Val}}}$$

where

$$F_1 = 1 + \frac{Pyr}{K_{m1}^{Pyr}}, \quad F_2 = 1 + \frac{Pyr}{K_{m2}^{Pyr}}, \quad G_1 = \frac{Kb}{K_m^{Kb}} + \frac{Ahb}{K_m^{Ahb}}, \quad G_2 = \frac{Pyr}{K_{m2}^{Pyr}} + \frac{Al}{K_m^{Al}}$$

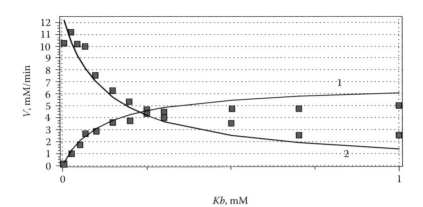

FIGURE 7.21   Dependence of acetohydroxybutyrate (curves 1) and acetolactate (curves 2) formation catalyzed by AHAS I on ketobutyrate concentration with concentration of pyruvate of 10 mM. Symbols correspond to experimental data taken from Barak, Chipman, and Gollop (1987), lines designates model curves.

To estimate values of parameters of these equations we fitted them against experimentally measured curves published in Engel et al. (2000), Vyazmensky et al. (1996) and Bar-Ilan (2001). First, we have evaluated Michaelis constants for AHAS I. We fitted four experimental curves (figure 7.21a and figure 7.21b) simultaneously and obtained a unique set of kinetic parameters listed in table 7.4. Then, we fitted experimental data enabling us to evaluate the inhibition constants of valine (figure 7.22a) and isoleucine (figure 7.22b) listed in table 7.4. By the same procedure we estimated the kinetic parameters of AHAS III (figure 7.23a, figure 7.23b and figure 7.24a, figure 7.24b). The values of the parameters are listed in table 7.4.

*Acetohydroxy Acid Isomeroreductase (IR)*

Acetohydroxy acid isomeroreductase catalyzes the conversion of acetohydroxybutyrate to dihydroxy-methylvalerate in the isoleucine pathway and the conversion of acetolactate (*Al*) to dihydroxy-isovalerate (*Dhi*) in the valine pathway. These reactions are coupled with NADPH reduction. The rate equations of IR-catalyzed reactions have been taken in the form

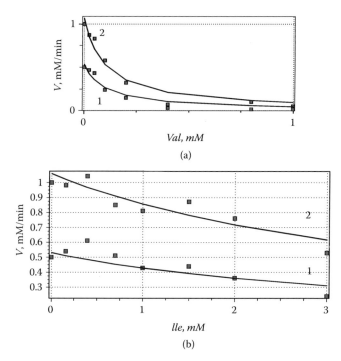

FIGURE 7.22 Dependence of acetohydroxybutyrate (curves 1) and aceto-lactate (curves 2) formation on valine (a) and isoleucine (b) concentration. Experimental data taken from Barak, Chipman, and Gollop (1987).

of (7.13) for the isoleucine pathway and in the form of (7.14) for the valine pathway:

$$v_{IR1} = \frac{\dfrac{V_{m1}^{IR}}{F_1 K_{m1}^{NADPH} K_m^{Ahb}}\left(NADPH \cdot Ahb - \dfrac{NADP \cdot Dhv}{K_{eq}^{IR}}\right)}{1 + \left(\dfrac{NADPH}{K_{m1}^{NADPH}} + \dfrac{Dhv}{K_m^{Ahb}}\right)\left(1 + \dfrac{Ahb}{F_1 K_m^{Ahb}} \dfrac{NADP}{K_{m1}^{NADP}}\right) + \left(\dfrac{NADPH}{K_{m2}^{NADPH}} + \dfrac{Dhi}{K_m^{Al}}\right)\left(1 + \dfrac{Al}{F_2 K_m^{Al}} \dfrac{NADP}{K_{m2}^{NADP}}\right)}$$

$$v_{IR2} = \frac{\dfrac{V_{m2}^{IR}}{F_2 K_{m2}^{NADPH} K_m^{Al}}\left(NADPH \cdot Al - \dfrac{NADP \cdot Dhi}{K_{eq}^{IR}}\right)}{1 + \left(\dfrac{NADPH}{K_{m1}^{NADPH}} + \dfrac{Dhv}{K_m^{Ahb}}\right)\left(1 + \dfrac{Ahb}{F_1 K_m^{Ahb}} \dfrac{NADP}{K_{m1}^{NADP}}\right) + \left(\dfrac{NADPH}{K_{m2}^{NADPH}} + \dfrac{Dhi}{K_m^{Al}}\right)\left(1 + \dfrac{Al}{F_2 K_m^{Al}} \dfrac{NADP}{K_{m2}^{NADP}}\right)}$$

where

$$F_1 = 1 + \frac{Ahb}{K_m^{Ahb}} + \frac{NADP}{K_{m1}^{NADP}}, \quad F_2 = 1 + \frac{Al}{k_m^{Al}} + \frac{NADP}{K_{m2}^{NADP}}$$

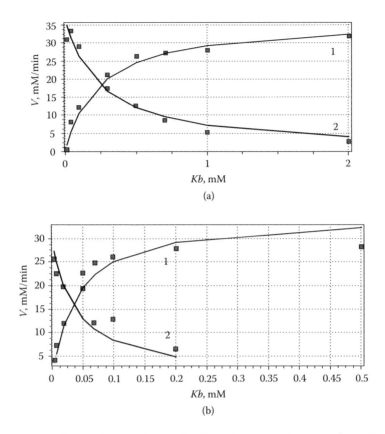

FIGURE 7.23  Dependence of acetohydroxybutyrate (curves 1) and ace-tolactate (curves 2) formation catalyzed by AHAS III on ketobutyrate concentration mM (a) and with concentration of pyruvate 2 mM (b). Experimental data taken from Barak, Chipman, and Gollop (1987).

We have not found experimental curves describing IR kinetics and have used the values of kinetic constants listed in the database BRENDA and in published papers (Aulabaugh and Schloss 1990; Chunduru, Mrachko, and Calvo 1998; Madhavi et al. 1997). The values of the parameters are listed in table 7.4.

*Dihydroxy-Acid Dehydratase (DHAD)*

Dihydroxy-acid dehydratase catalyzes the conversion of dihydroxy-methylvalerate to keto-methylvalerate in the isoleucine pathway and the conversion of dihydroxy-isovalerate to keto-isolvalerate in the valine path-way. The rate equations of DHAD-catalyzed reactions have been taken in

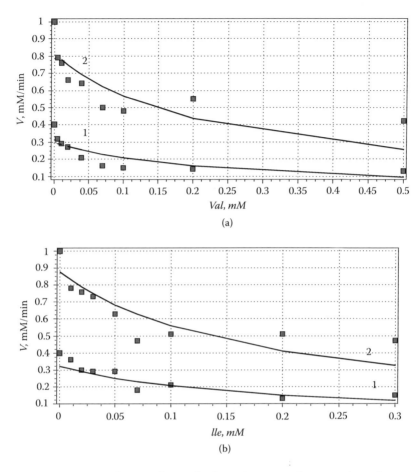

FIGURE 7.24 Dependence of acetohydroxybutyrate (curves 1) and aceto-lactate (curves 2) formation on valine (a) and isoleucine (b) concentration. Experimental data taken from Barak, Chipman, and Gollop (1987).

the form of (7.17) for the isoleucine pathway and in the form of (7.18) for the valine pathway:

$$v_{DHAD1} = \dfrac{\dfrac{V_{m1}^{DHAD}}{K_m^{Dhv}}\left(Dhv - \dfrac{Ktv}{K_{eq}^{DHAD}}\right)}{1 + \dfrac{Dhv}{K_m^{Dhv}} + \dfrac{Ktv}{K_m^{Ktv}} + \dfrac{Dhi}{K_m^{Dhi}} + \dfrac{Kti}{K_m^{Kti}}}, \qquad v_{DHAD2} = \dfrac{\dfrac{V_{m2}^{DHAD}}{K_m^{Dhi}}\left(Dhi - \dfrac{Kti}{K_{eq}^{DHAD}}\right)}{1 + \dfrac{Dhv}{K_m^{Dhv}} + \dfrac{Ktv}{K_m^{Ktv}} + \dfrac{Dhi}{K_m^{Dhi}} + \dfrac{Kti}{K_m^{Kti}}}$$

To estimate the parameters of the equation we fitted (see figure 7.25 for fitting quality) these rate equations against experimental data published

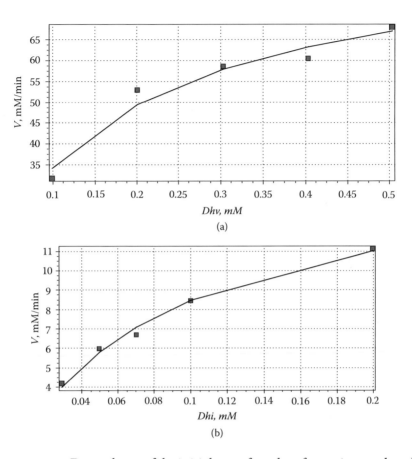

FIGURE 7.25 Dependence of the initial rate of product formation catalyzed by DHAD on dihydroxy-methylvalerate (a) and dihydroxy-isovalerate (b). Experimental data taken from Myers (1961).

in Limberg, Klaffke, and Thiem (1995) and Myers (1961). The values of the parameters are listed in table 7.4.

### Branched-Chain Amino Acid Transaminase (BCAT)

Branched-chain amino acid transaminase catalyzes the conversion of keto-methylvalerate to isoleucine and the conversion of keto-isovalerate to valine. The reaction is coupled with the conversion of glutamate (*Glt*) to ketoglutarate (*Kgl*). The rate equations of BCAT-catalyzed reactions have been taken in the form of (7.13) for the isoleucine pathway and in the form of (7.14) for the valine pathway:

$$v_{BCAT1} = \frac{\frac{V_{m1}^{BCAT}}{F_1 K_m^{Ktv} K_m^{Glt}} \left( Ktv \cdot Glt - \frac{Ile \cdot Kgl}{K_{eq}^{BCAT}} \right)}{1 + \left( \frac{Kgl}{K_{ml}^{Kgl}} + \frac{Ktv}{K_m^{Ktv}} \right) \left( 1 + \frac{Ile}{F_1 K_m^{II}} \frac{Glt}{K_{m1}^{Glt}} \right) + \left( \frac{Kgl}{K_{m2}^{Kgl}} + \frac{Kti}{K_m^{Kti}} \right) \left( 1 + \frac{Val}{F_2 K_m^{II}} \frac{Glt}{K_{m2}^{Glt}} \right)}$$

$$v_{BCAT2} = \frac{\frac{V_{m2}^{BCAT}}{F_2 K_m^{Ktv} K_m^{Glt}} \left( Kti \cdot Glt - \frac{Val \cdot Kgl}{K_{eq}^{BCAT}} \right)}{1 + \left( \frac{Kgl}{K_{ml}^{Kgl}} + \frac{Ktv}{K_m^{Ktv}} \right) \left( 1 + \frac{Ile}{F_1 K_m^{II}} \frac{Glt}{K_{m1}^{Glt}} \right) + \left( \frac{Kgl}{K_{m2}^{Kgl}} + \frac{Kti}{K_m^{Kti}} \right) \left( 1 + \frac{Val}{F_2 K_m^{II}} \frac{Glt}{K_{m2}^{Glt}} \right)}$$

where

$$F_1 = 1 + \frac{Ktv}{K_m^{Ktv}} + \frac{Kgl}{K_{ml}^{Kgl}}, \quad F_2 = 1 + \frac{Kti}{K_m^{Kti}} + \frac{Kgl}{K_{m2}^{Kgl}}$$

We have not found kinetic experimental data for BCAT and have used the values of kinetic constants listed in data base BRENDA and in published papers (Hall et al. 1993; Inoue et al. 1988; Lee-Peng, Hermodson, and Kohlhaw 1979). The values of the parameters are listed in table 7.4.

### NADP Recycling and Effluxes

The rate equations describing kinetics of effluxes and NADP recycling have been written in the simplest form as the product of appropriate constants and concentrations.

$$v_{nadp} = V_m^{NADP} NADP$$

$$v_{efflux1} = V_{efflux1} Ile$$

$$v_{efflu2} = V_{efflux2} Val$$

Parameters of $V_m$ were evaluated by the procedure described below.

### Evaluation of Maximal Reaction Rates

To estimate maximal reaction rates we used an approach described in Chassagnole (2002). According to this approach, at steady state each rate equation corresponding to a reaction catalyzed by enzyme $i$ is given by:

$$\tilde{v}_i = V_i^{max} f_i(\tilde{C}_i, \tilde{P}_i), \tag{7.19}$$

where $\tilde{P}_i$ is the parameter vector and $\tilde{C}_i$ is the steady-state concentration vector of the metabolites involved in the reaction. From equation (7.19) it follows that the maximal rate can be calculated in the following manner:

$$V_i^{max} = \frac{\tilde{v}_i}{f_i(\tilde{C}_i, \tilde{P}_i)}.$$

The stationary rates are estimated from the values of stationary fluxes taken from Holms (1986). The value of the stationary rate was calculated by the following expression:

$$\tilde{v}_i = 5.56 f_i$$

where $f_i$ is the value of stationary flux.

TABLE 7.5 Values of Steady-state Concentrations, Maximal Rates and Equilibrium Constants of Isoleucine and Valine Biosynthesis Pathway in *Escherichia coli*

| Steady-State Concentrations (mM) | Maximal Rates (mM/min) | Equilibrium Constants |
|---|---|---|
| OA = 1 | $V_{influx1} = 1$ | $K_{eq}^{in1} = 1.1$ |
| Pep = 2.8 (Chassagnole 2002) | $V_{influx2} = 1.2$ | $K_{eq}^{in2} = 0.9$ |
| Thr = 0.1Kb = 0.1 | $V_m^{TDH} = 10$ | $K_{eq}^{TDH} = 10$ |
| Pyr = 2.8 (Chassagnole 2002) | $V_{m1}^{AHAS} = 0.11$ | $K_{eq1}^{AHASI} = 10$ |
| Ahb = 0.05 | $V_{m2}^{AHAS} = 0.11$ | $K_{eq1}^{AHASIII} = 10$ |
| Al = 0.05 | $V_{m1}^{IR} = 0.34$ | $K_{eq2}^{AHASI} = 0.1$ |
| Dhb = 0.03 | $V_{m2}^{IR} = 0.34$ | $K_{eq2}^{AHASIII} = 0.1$ |
| Dhi = 0.03 | $V_{m1}^{DHAD} = 0.03$ | $K_{eq1}^{IR} = 100$ |
| Ktv = 0.02 | $V_{m2}^{DHAD} = 0.03$ | $K_{eq2}^{IR} = 3$ |
| Kti = 0.02 | $V_{m1}^{BCAT} = 0.55$ | $K_{eq1}^{IR} = 3$ |
| Ile = 0.001 | $V_{m2}^{BCAT} = 0.55$ | $K_{eq2}^{DHAD} = 3$ |
| Val = 0.001 | $V_{efflux2} = 1$ | $K_{eq1}^{BCAT} = 10$ |
| NADP = 0.19 (Chassagnole 2002) | $V_{efflux2} = 1.2$ | $K_{eq2}^{BCAT} = 10$ |
| NADPH = 0.06 (Chassagnole 2002) | $V_m^{NADP} = 0.1$ | |
| Glu = 0.1 | | |
| α Kg = 0.1 | | |

The steady-state concentrations were chosen according to the *in vivo* concentration of metabolites in the *E. coli* cell. The equilibrium constants were evaluated to maintain the real steady-state concentrations of metabolites. Steady-state concentrations, maximal rates and equilibrium constants are collected in table 7.5.

## CHAPTER 8

# Modelling of Mitochondrial Energy Metabolism

This chapter illustrates how the principles of kinetic modelling described in detail in chapters 4 and 6 can be applied to collect all the available information on energy metabolism of whole organelles, such as mitochondria, to construct a kinetic model and predict the response of the organelle to changes in external conditions. We present here a kinetic model describing the operation of oxidative phosphorylation in mitochondria respiring on succinate. In the framework of the model we have simulated not only biochemical reactions but reactions coupled to production and consumption of electric potential difference as well as transport processes.

## OXIDATIVE PHOSPHORYLATION AND SUPEROXIDE PRODUCTION IN MITOCHONDRIA

It is well known that mitochondria convert organic materials into cellular energy in the form of ATP. This is done by oxidising different substrates such as pyruvate, succinate and glutamate that are produced in the cytosol. This process, known as aerobic respiration, is dependent on the presence of oxygen. Respiration, i.e., transfer of electrons from the oxidized substrates to a terminal electron acceptor—oxygen—is catalyzed by enzymes of the mitochondrial Krebs cycle and the respiratory chain. The mitochondrial Krebs cycle, also known as the tricarboxylic acid cycle (TCA), converts different mitochondrial substrates to $CO_2$ and reduced equivalents in the form of NADH. Then, NADH donates its two electrons to the respiratory chain which transfers them to oxygen to form water. Operation of the respiratory chain results in a proton motive force which consists of two parts: an electric potential difference across inner mitochondrial membrane, $\Delta\Psi$, and a transmembrane proton gradient, $\Delta pH$. Both $\Delta\Psi$ and $\Delta pH$ are consumed by reactions involved in ATP synthesis in mitochondria. Thus, the proton motive force is generated by the respiratory chain and is consumed by ATP production processes.

Employing the respiratory chain with $O_2$ as the electron acceptor allows the cell to increase the number of ATP molecules synthesized for a given number of substrate molecules consumed (Skulachev 1988). However, this mechanism for elevating the yield of ATP per redox equivalent is a potential source of danger. Indeed, the set of oxidative-reducing reactions that is usually called 'the respiratory chain' includes redox intermediates that can donate a single electron to molecular oxygen to yield superoxide anion, $O_2^{\bullet-}$ (Skulachev 1997). It is commonly accepted that at least two respiratory chain enzyme complexes are able to form these dangerous intermediates. These are NADH ubiquinone oxido reductase (complex I) and ubiquinol cytochrome $c$ oxido reductase (bc$_1$ complex; Boveris and Chance 1973; Hansford, Hogue, and Mildaziene 1997; Korshunov et al. 1998; Korshunov, Skulachev, and Starkov 1997). Among the intermediates of the reactions catalyzed by the bc$_1$ complex, ubisemiquinone has this property. Adventitious formation of the same species, i.e., ubisemiquinone, has been proposed as being responsible for $O_2^{\bullet-}$ production with complex I (Hansford, Hogue, and Mildaziene 1997; Korshunov, Skulachev, and Starkov 1997). Because of the electrogenic nature of the reactions surrounding semiquinone, its concentration and, consequently, the rate of superoxide production may depend on the membrane potential ($\Delta\Psi$); i.e., on the energy state of the mitochondria.

Several models of energy conversion in mitochondria have been proposed. Bohnensack (1982), Korzeniewski and Froncisz (Korzeniewski 1996, 1998; Korzeniewski and Froncisz 1991) and Westerhoff and Van Dam (1987; Van Dam et al. 1980) developed quantitative models of oxidative phosphorylation in mitochondria describing the State 3–State 4 transition. Earlier, Kholodenko (1988) developed a kinetic model of mitochondria respiring on succinate. In this model, the electric potential across the membrane was determined by $\Delta\bar{\mu}_H$ generators and $\Delta\bar{\mu}_H$ consumers but the kinetic description of enzymes of the respiratory chain was semiempirical. To understand how generation of superoxide with bc$_1$ complex correlates to the transmembrane electric potential, a kinetic model of the $bc_1$ complex, considered as a single enzyme (in isolation from the rest of the oxidative phosphorylation machinery and, consequently, assuming $\Delta\Psi$ as a parameter), has been developed (Demin, Kholodenko, Skulachev 1998). In this chapter, we develop a kinetic model of mitochondria respiring on succinate that enables us to examine how the $bc_1$-linked superoxide generation may vary with $\Delta\Psi$ when both the processes of $\Delta\bar{\mu}_H$ generation and $\Delta\bar{\mu}_H$ consumption are taken into account. This model is based on existing experimental data. It should allow us to predict possible relationships between the energy state of mitochondria and the ability of the $bc_1$ complex to generate superoxide on the basis of what is known about mitochondria. We shall show that the rate of superoxide generation is determined by many interacting molecular processes and may only be understood by employing a kinetic model.

## DEVELOPMENT OF KINETIC MODELS

Our model has two compartments; i.e., it describes the case in which all intermediates can be divided into two groups depending on their location. The first group consists of compounds located in the inner mitochondrial membrane; these are all intermediates of the respiratory chain. The second group comprises the metabolites localized in the mitochondrial matrix; i.e., $ATP_n^{4-}, ADP_n^{3-}, H_n^+, K_n^+, H_2PO_4^-, HPO_4^{2-}, MgHPO_4, MgADP_n^-, MgATP_n^{2-}$.

Let us first consider the set of reduction-oxidation reactions located inside the inner mitochondrial membrane which are usually referred to as the respiratory chain. These reactions are catalyzed by three membrane protein complexes: NADH:ubiquinone oxidoreductase (complex I), ubiquinol: cytochrome $c$ oxidoreductase (complex III or $bc_1$ complex) and cytochrome $c$ oxidase (complex IV). Here, we will focus on the functioning of the $bc_1$ complex and on its possible role in superoxide generation. Reactions catalyzed by other enzymes of the respiratory chain will not be considered in any detail.

The mitochondrial $bc_1$ complex is a $\Delta\bar{\mu}_H$ generator which couples two processes: electron transport from ubiquinol to cytochrome $c$ and proton transfer out of the mitochondrial matrix to outside the mitochondrion. The most widely accepted model for the reduction-oxidation (redox) reactions catalyzed by the $bc_1$ complex is the Q-cycle (figure 8.1). The $bc_1$ complex has two catalytic sites with respect to ubiquinone usually called the $i$ (inside) and $o$ (outside) sites, respectively (Skulachev 1988).

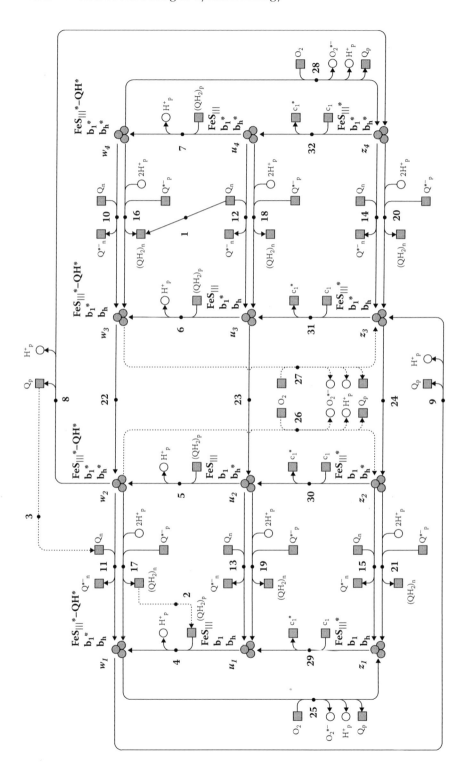

FIGURE 8.1 Scheme of reactions of the Q-cycle. The kinetic model implemented for $bc_1$ complex. The $bc_1$ complex is considered to be in any one of twelve states depending on the redox states of the $b$ cytochromes (subscripts 1–4), the redox state of the $FeS_{III}$ centre ($u$ versus $z$ or $w$) and the consideration that semiquinone can be bound to $FeS_{III}$ centre ($w$ versus $u$). Transitions between these states that are the reduced part of the main electron transfer pathway are indicated by full arrows, proton release or uptake and reduction or oxidation of ubiquinone or cytochrome $c_1$ being indicated alongside. Dotted arrows refer to reactions engaged in oxygen free radical production. Subscripts $n$ and $p$ refer to matrix and intermembrane sides of the membrane, respectively.

In our model, electron supply to the Q-cycle results in the reduction of ubiquinone, $Q$, to ubiquinol, $QH_2$, as catalyzed by succinate dehydrogenase (reaction 1). The reaction occurs at the matrix $n$ side of the inner mitochondrial membrane. Ubiquinol diffuses to the other $p$ side of the membrane (reaction 2). After binding to the site $o$ of the bc$_1$ complex, the ubiquinol donates one electron to the non-heme iron-sulfur cluster (ISC), $FeS_{III}$, which then releases one proton into the intermembrane space and forms the complex $FeS_{III}^r - QH^\bullet$ (reactions 4, 5, 6 or 7 depending on the preexisting redox states of the FeS centre and the hemes of cytochrome $b$). If the low potential heme of cytochrome $b$ is reduced, semiquinone remains bound to the iron sulfur centre until heme b$_1$ becomes oxidized (due to reactions 10, 16 and 22) and, consequently, able to accept a second electron from semiquinone. Subsequently, a second proton is released and the bound semiquinone anion donates an electron to heme $b_1$ and is converted to free ubiquinone (reactions 8 or 9). Then, ubiquinone diffuses or flips to the site $n$ of the bc$_1$ complex (reaction 3). The next step of the Q-cycle is a transmembrane electron transfer between the two hemes of cytochrome $b$: from the low midpoint-potential heme $b_1$ to the high potential heme $b_h$ (reactions 22–24). Subsequently, reduced heme $b_h$ reduces ubiquinone at site $n$ with concomitant formation of the semiquinone $Q_n^{\bullet-}$ (reactions 10–15). This semiquinone takes a second electron from heme $b_h$ to become ubiquinol, thereby absorbing two protons out of the mitochondrial matrix (reactions 16–21). In parallel to these events, the electron captured by the iron-sulfur cluster is transferred to cytochrome c$_1$ (reaction 29–32) and further to cytochrome $c$ (reaction 33 in table 8.1). Superoxide anion formation through a one-electron reduction of $O_2$ by the complex $FeS_{III}^r - QH^\bullet$ is a side reaction that is of particular importance for the issue we address here (reactions 25–28). This reaction also results in ubiquinone and one proton is released into the intermembrane space.

The oxidation-reduction reactions catalyzed by cytochrome $c$ oxidase are approximated by four (overall) processes. The first involves release of two $H_2O$ molecules formed in a previous catalytic cycle, donation of three cytochrome $c$ electrons to the cytochrome $c$ oxidase intermediate designated by $Y$, and adsorption of three Mitchell protons out of the mitochondrial matrix (reaction 34 in table 8.1). This reaction also results in transmembrane pumping of one Wikstrom proton. The second reaction is di-oxygen binding (reaction 35). The third process is oxidation of one cytochrome $c$ molecule accompanied by adsorption of one proton and transmembrane translocation of two protons (reaction 36).

TABLE 8.1    Chemical Reactions and Corresponding Rate Equations

| N | Reaction | Rate Equation |
|---|----------|---------------|
| 1 | $Q_n = (QH_2)_n$ | Equation (8.10) |
| 2 | $(QH_2)_n = (QH_2)_p$ | Equation (8.17) |
| 3 | $Q_p = Q_n$ | Equation (8.17) |
| 4–7 | $(QH_2)_p + u_i = w_i + H_p^+$, $i = 1,\ldots,4$ | $k_{(i+3)}.\exp(-d_1.\alpha.F.\Delta\Psi/RT).(QH_2)_p.u_i - k_{-(i+3)}.\exp((1-d_1).\alpha.F.\Delta\Psi/RT).w_i.\,H_p^+$, $i=1,\ldots,4$ |
| 8 | $w_2 = z_4 + H_p^+ + Q_p$ | $k_8.\exp(-d_2.\alpha.F.\Delta\Psi/RT).w_2 - k_{-8}.\exp((1-d_2).\alpha.F.\Delta\Psi/RT).\,H_p^+.z_4.Q_p$ |
| 9 | $w_1 = z_3 + H_p^+ + Q_p$ | $k_9.\exp(-d_2.\alpha.F.\Delta\Psi/RT).w_1 - k_9.\exp((1-d_2).\alpha.F.\Delta\Psi/RT).\,H_p^+.z_3.Q_p$ |
| 10–11 | $w_{2i} + Q_n = Q_n^{\bullet-} + w_{2i-1}$, $i = 2,1$ | $k_{(12-i)}.w_{2i}.Q_n - k_{-(12-i)}.\,Q_n^{\bullet-}.w_{2i-1}$, $i = 2,1$ |
| 12–13 | $u_{2i} + Q_n = Q_n^{\bullet-} + u_{2i-1}$, $i = 2,1$ | $k_{(14-i)}.u_{2i}.Q_n - k_{-(14-i)}.\,Q_n^{\bullet-}.u_{2i-1}$, $i = 2,1$ |
| 14–15 | $z_{2i} + Q_n = Q_n^{\bullet-} + z_{2i-1}$, $i=2,1$ | $k_{(16-i)}.z_{2i}.Q_n - k_{-(16-i)}.\,Q_n^{\bullet-}.z_{2i-1}$, $i = 2,1$ |
| 16–17 | $w_{2i} + Q_n^{\bullet-} + 2\,H_n^+ = w_{2i-1} + (QH_2)_n$, $i = 2,1$ | $k_{(18-i)}.\exp(-d_4.\gamma.F.\Delta\Psi/RT).w_{2i}.\,Q_n^{\bullet-}.(\,H_n^+)^2 - k_{-(18-i)}.\,\exp((1-d_4).\gamma.F.\Delta\Psi/RT).w_{2i-1}.\,(QH_2)_n$, $i =2,1$ |
| 18–19 | $u_{2i} + Q_n^{\bullet-} + 2\,H_n^+ = u_{2i-1} + (QH_2)_n$, $i = 2,1$ | $k_{(20-i)}.\exp(-d_4.2.\gamma.F.\Delta\Psi/RT).u_{2i}.\,Q_n^{\bullet-}.(\,H_n^+)^2 - k_{-(20-i)}.\,\exp((1-d_4).2.\gamma.F.\Delta\Psi/RT).u_{2i-1}.\,(QH_2)_n$, $i = 2,1$ |
| 20–21 | $z_{2i} + Q_n^{\bullet-} + 2\,H_n^+ = z_{2i-1} + (QH_2)_n$, $i = 2,1$ | $k_{(22-i)}.\exp(-d_4.2.\gamma.F.\Delta\Psi/RT).z_{2i}.\,Q_n^{\bullet-}.(\,H_n^+)^2 - k_{-(22-i)}.\,\exp((1-d_4).2.\gamma.F.\Delta\Psi/RT).z_{2i-1}.\,(QH_2)_n$, $i = 2,1$ |
| 22 | $w_3 = w_2$ | $k_{22}.\exp(-d_3.\beta.F.\Delta\Psi/RT).w_3 - k_{-22}.\exp((1-d_3).\beta.F.\Delta\Psi/RT).w_2$ |
| 23 | $u_3 = u_2$ | $k_{23}.\exp(-d_3.\beta.F.\Delta\Psi/RT).u_3 - k_{-23}.\exp((1-d_3).\beta.F.\Delta\Psi/RT).u_2$ |
| 24 | $z_3 = z_2$ | $k_{24}.\exp(-d_3.\beta.F.\Delta\Psi/RT).z_3 - k_{-24}.\exp((1-d_3).\beta.F.\Delta\Psi/RT).z_2$ |
| 25–28 | $w_i + O_2 = z_i + O_2^{\bullet-} + Q_p + H_p^+$, $i = 1,\ldots,4$ | $k_{(24+i)}.w_i.O_2 - k_{-(24+i)}.z_i.\,O_2^{\bullet-}.Q_p.\,H_p^+$, $i = 1,\ldots,4$ |
| 29–32 | $z_i + c_1^{ox} = u_i + c_1^r$, $i = 1,\ldots,4$ | $k_{(28+i)}.z_i.\,c_1^{ox} - k_{-(28+i)}.u_i.\,c_1^r$, $i = 1,\ldots,4$ |
| 33 | $c_1^r + c^{ox} = c_1^{ox} + c^r$ | $k_{33}.\,c_1^r.\,c^{ox} - k_{-33}.\,c_1^{ox}.\,c^r$ |
| 34 | $3\,c^r + Y + 4\,H_n^+ = 3\,c^{ox} + Yr + H_p^+ + 2H_2O$ | $k_{34}.\exp(-d_5.4.F.\Delta\Psi/RT).(\,c^r\,)^3.Y.(\,H_n^+\,)^4 - k_{-34}.\exp((1-d_5).4.F.\Delta\Psi/RT).(\,c^{ox}\,)^3.Yr.\,H_p^+$ |
| 35 | $Yr + O_2 = YO$ | $k_{35}*Yr*O_2$ |

(*Continued*)

TABLE 8.1   Chemical Reactions and Corresponding Rate Equations (Continued)

| N | Reaction | Rate Equation |
|---|----------|---------------|
| 36 | $c^r + YO + 3H_n^+ = c^{ox} + YOH$ $+2H_p^+$ | $k_{36}.\exp(-d_5.3.F.\Delta\Psi/RT). c^r .YO.(H_n^+)^3 - k_{-36}.$ $\exp((1-d_5).3.F.\Delta\Psi/RT). c^{ox} .YOH.( H_p^+ )^2$ |
| 37 | $YOH + H_n^+ = Y + H_p^+$ | $k_{37}.\exp(-d_5.F.\Delta\Psi/RT).YOH. H_n^+ - k_{-37}.$ $\exp((1-d_5).F.\Delta\Psi/RT).Y. H_p^+$ |
| 38 | $H_p^+ = H_n^+$ | Equation (8.13) |
| 39 | $ADP_n + 3 H_p^+ + P_n^- =$ $ATP_n + 3 H_n^+$ | Equation (8.11) |
| 40 | $ADP_p + ATP_n = ATP_p + ADP_n$ | Equation (8.12) |
| 41 | $P_p^- + H_p^+ = P_n^- + H_n^+$ | Equation (8.15) |
| 42 | $K_p^+ = K_n^+$ | Equation (8.14) |
| 43 | $K_n^+ + H_p^+ = K_p^+ + H_n^+$ | Equation (8.16) |

The fourth reaction completes the catalytic cycle and produces translocation of one proton outside the mitochondrial matrix (reaction 37). This
simple description based on recent structural data (Tsukihara et al. 1995,
1996; Yoshikawa et al. 1998) was developed in accordance with the mechanistic model of cytochrome $c$ oxidase operation suggested by Michel
(1998). This simple mechanism involves three electrogenic steps. Reaction
34 results in the transmembrane transfer of four charges: translocation
of one Wikstrom proton from the mitochondrial matrix and transfer of
three electrons accompanied by adsorption of three protons. As a result
of reaction 36, three charges are transferred across the membrane: two
Wikstrom protons and one electron accompanied by one Mitchell proton. Reaction 37 is coupled to the translocation of one proton from the
mitochondrial matrix to the outside. We assume that electrons donated
by cytochrome $c$ are accepted by cytochrome oxidase residues which are
located halfway through the membrane. To balance these electrons, protons adsorbed out of the matrix should be transferred through a hydro-
phobic domain of the enzyme as well. Therefore, both electron transfer
and proton translocation are assumed to contribute equally to the total
electrogenesis of cytochrome $c$ oxidase.

As in Demin, Westerhoff, Kholodenko (1998), we assume that there are
four electrogenic steps in the Q-cycle: the first corresponds to intramembrane electron transfer from heme $b_l$ to heme $b_h$ (reactions 22–24) and the

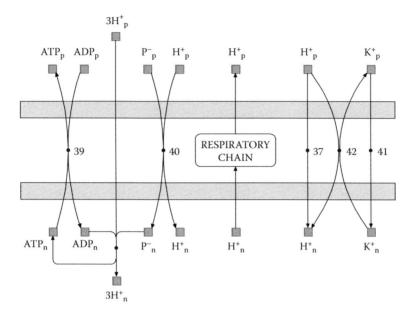

FIGURE 8.2   Scheme of ATP synthesis processes.

others correspond to proton translocations in the reactions of ubiquinone reduction (reactions 16–21) and ubiquinol oxidation (reactions 4–9). The process of intramembrane electron transfer represents 80 percent of total electrogenesis and the other 20 percent is represented by proton translocation steps (Drachev et al. 1989; Semenov 1993). In addition, we take account of the fact that proton translocation in the reactions catalyzed by cytochrome $c$ oxidase is also electrogenic.

In addition to these $\Delta\bar{\mu}_H$ generators (bc$_1$ complex and cytochrome $c$ oxidase), the model includes the following $\Delta\bar{\mu}_H$ consumers (see figure 8.2):

1. ATP synthesis, catalyzed by ATP-ase/ATP-synthase (reaction 38);

2. Electrogenic translocation of adenine nucleotides, catalyzed by the adenine nucleotide translocator (reaction 39);

3. Electroneutral symport of inorganic phosphate and a proton, as catalyzed by the phosphate carrier (reaction 40);

4. H$^+$ leakage (reaction 37);

5. K$^+$ leakage (reaction 41);

6. Electroneutral exchange of K$^+$ for H$^+$ (reaction 42).

The mitochondrial matrix buffers changes in proton concentration. We approximate the buffering capacity of the mitochondrial matrix by three substances $B_1$, $B_2$ and $B_3$ with pK for protons ranging from 6.7 to 8.7:

$$B_j + H_n^+ = B_j H, \quad K_B^j = \frac{B_j \cdot H_n^+}{B_j H}, \quad K_B^1 = 10^{-6.7} M, \quad K_B^2 = 10^{-7.7} M,$$

(8.1)

$$K_B^3 = 10^{-8.7} M, \quad B_j + B_j H = B_j^{tot} = 80 mM, \quad j = 1,2,3$$

These values of $K_B^j$ and the total concentration of the buffer $B_j$, $B_j^{tot}$, are chosen to meet experimental data on the matrix buffer capacity (Harris and Bangham 1972):

$$\sum_{j=1}^{3} \frac{dB_i^-}{dpH_n^+} = \sum_{j=1}^{3} \frac{B_j^{tot} \cdot K_B^j \cdot H_n^+}{\left(K_B^j + H_n^+\right)^2} \approx \frac{60}{\ln 10}$$

The model considers the following ion balances:

$$P_n^{2-} + H_n^+ = P_n^-, \quad K_{eq} = \frac{P_n^{2-} \cdot H_n^+}{P_n^-} = 1.76 \cdot 10^{-4} mM$$

(8.2)

$$P_n^{2-} + Mg_n^{2+} = MP, \quad K_{eq} = \frac{P_n^{2-} \cdot Mg_n^{2+}}{MP} = 12 mM$$

(8.3)

$$ATP_n^{4-} + Mg_n^{2+} = MATP_n, \quad K_{eq} = \frac{ATP_n^{4-} \cdot Mg_n^{2+}}{MATP_n} = 0.08 mM$$

(8.4)

$$ADP_n^{3-} + Mg_n^{2+} = MADP_n, \quad K_{eq} = \frac{ADP_n^{3-} \cdot Mg_n^{2+}}{MADP_n} = 0.8 mM$$

(8.5)

where $P_n^{2-}, P_n^-, MP, MATP_n$ and $MATP_n$ are concentrations of matrix $HPO_4^{2-}$, $H_2PO_4^-$, $MGHPO_4$, $MgATP_n^{2-}$ and $MgADP_n^-$, respectively. Values of equilibrium constants are taken in accordance with Alberty (1969) and Lawson and Veech (1979). The reactions of ion binding and dissociation are taken to be at equilibrium. The system of these rapid reactions (equations (8.1)–(8.5)) has five moiety conservations:

$$P_n^{2-} + P_n^- + MP = P_n,$$

$$H_n^+ + \sum_{j=1}^{3} B_j H + P_n^- = H_n^*$$

$$ATP_n^{4-} + MATP_n = ATP_n,$$

$$ADP_n^{3-} + MADP_n = ADP_n, \tag{8.6}$$

where $P_n, H_n^*, ATP_n, ADP_n$ stand for the total concentrations of matrix phosphate, proton, adenosine triphosphate, adenosine diphosphate and magnesium. The values $P_n, H_n^*, ATP_n, ADP_n$ are the variables resulting from the processes of ATP synthesis and $\Delta\bar{\mu}_H$ production (consumption).

The processes considered in the model are listed in table 8.1. They can be described by the following system of differential equations:

$$dQ_p/d_t = v_8 + v_9 + v_{25} + v_{26} + v_{27} + v_{28} - v_3,$$
$$dQ_n/d_t = v_3 - v_1 - v_{10} - v_{11} - v_{12} - v_{13} - v_{14} - v_{15},$$
$$dQ_n^{-\bullet}/d_t = v_{10} + v_{11} + v_{12} + v_{13} + v_{14} + v_{15} - v_{16} - v_{17} - v_{18} - v_{19} - v_{20} - v_{21},$$
$$d(QH_2)_n/d_t = v_1 - v_2 + v_{16} + v_{17} + v_{18} + v_{19} + v_{20} + v_{21},$$
$$d(QH_2)_p/d_t = v_2 - v_4 - v_5 - v_6 - v_7,$$
$$dw_1/d_t = v_4 + v_{11} + v_{17} - v_9 - v_{25},$$
$$dw_2/d_t = v_5 + v_{22} - v_8 - v_{11} - v_{17} - v_{26},$$
$$dw_3/d_t = v_6 + v_{10} + v_{16} - v_{22} - v_{27},$$
$$dw_4/d_t = v_7 - v_{10} - v_{16} - v_{28},$$
$$du_1/d_t = v_{13} + v_{19} + v_{29} - v_4,$$
$$du_2/d_t = v_{23} + v_{30} - v_5 - v_{13} - v_{19},$$
$$du_3/d_t = v_{12} + v_{18} + v_{31} - v_6 - v_{23},$$
$$du_4/d_t = v_{32} - v_7 - v_{12} - v_{18},$$
$$dz_1/d_t = v_{15} + v_{21} + v_{25} - v_{29},$$
$$dz_2/d_t = v_{24} + v_{26} + v_{30} - v_{15} - v_{21},$$
$$dz_3/d_t = v_9 + v_{14} + v_{20} + v_{27} - v_{24} - v_{31},$$
$$dz_4/d_t = v_8 + v_{28} - v_{14} - v_{20} - v_{32},$$
$$dc_1^r/d_t = v_{29} + v_{30} + v_{31} + v_{32} - v_{33},$$
$$dc_1^{ox}/d_t = v_{33} - v_{29} - v_{30} - v_{31} - v_{32},$$
$$dc^r/d_t = v_{33} - 3v_{34} - v_{35},$$
$$dc^{ox}/d_t = 3v_{34} + v_{35} - v_{33},$$
$$dY/d_t = v_{34} - v_{35},$$
$$dYr/d_t = v_{35} - v_{36},$$
$$dYO/d_t = v_{36} - v_{37},$$

$dYOH/d_t = v_{37} - v_{34},$

$dH_n^*/d_t = v_{38} + 3v_{39} + v_{41} + v_{43} - 2(v_{16} + v_{17} + v_{18} + v_{19} + v_{20} + v_{21}) - 4v_{34}$
$\qquad - 3v_{36} - v_{37},$

$dP_n/d_t = v_{41} - v_{39},$

$dK_n^+/d_t = v_{42} - v_{43},$

$dADP_n/d_t = v_{40} - v_{39},$

$dATP_n/d_t = v_{39} - v_{40},$

$(c_M/F) \times d(\Delta\Psi)/d_t = 2(v_{16} + v_{17} + v_{18} + v_{19} + v_{20} + v_{21}) + 4v_{34} + 3v_{36} + v_{37}$
$$\qquad\qquad\qquad\qquad - v_{38} - 3v_{39} - v_{40} - v_{42}, \tag{8.7}$$

Here, $v_i$ is the rate of $i$th reaction, $\Delta\Psi > 0$ is the electric potential difference, $c_M$ is the apportioned electric capacity of mitochondrial membrane, $c_M = 7.8 \cdot 10^{-3}$ Farad/(g mit. prot.; Reich and Rohde 1983) or, using the data on an area of the membrane corresponding to 1 g of mitochondrial protein (Schwerzman et al. (1986)), one obtains $c_M = 5 \cdot 10^{-4}$ Farad/m$^2$), $F$ is the Faraday constant and superscripts $n$ and $p$ refer to the mitochondrial matrix (matrix side of inner mitochondrial membrane) and the extra-mitochondrial space, respectively. $Y$, $Yr$, $YO$ and $YOH$ are intermediate forms of the cytochrome $c$ oxidase. $ATP_n$, $ADP_n$, $P_n$ are concentrations of adenine nucleotides and inorganic phosphate in the mitochondrial matrix (see equation (8.6) for details). $K_n^+$ is the intra-mitochondrial potassium concentration. The different states of the cytochrome bc$_1$ complex ($u_i$, $z_i$, $w_i$, $i$ = 1, 2, 3, 4) differ in terms of the redox states of the iron sulfur cluster and hemes $b_1$ and $b_h$: $u_i$, $i$ = 1, 2, 3, 4 correspond to states where the iron sulfur cluster is oxidized; $z_i$, $i$ = 1, 2, 3, 4 correspond to states where the iron sulfur cluster is reduced; $w_i$, $i$ = 1, 2, 3, 4 correspond to states where the reduced iron sulfur cluster forms a complex with semiquinone at site $o$; $u_1$, $z_1$, $w_1$ and $u_4$, $z_4$, $w_4$ correspond to states where both hemes are either oxidized or reduced (the dot or asterisk refers to an electron); $u_3$, $z_3$, $w_3$ correspond to states with heme $b_1$ reduced and heme $b_h$ oxidized; and $u_2$, $z_2$, $w_2$ correspond to the state with heme $b_h$ reduced and heme $b_1$ oxidized. Concentrations of succinate, fumarate, oxygen, magnesium, extra-mitochondrial adenine nucleotides ($ATP_p$ and $ADP_p$), external inorganic phosphate ($P_p$) and potassium ($K_p^+$) and external pH ($pH_p$) are assumed constant (see table 8.2). Stoichiometric coefficients placed before terms $v_{34}$, $v_{35}$, $v_{36}$ and $v_{38}$ in the last equation of system (8.7) result from the fact that cytochrome $c$ oxidase extracts eight protons (and, consequently, eight positive charges) out of the mitochondrial matrix per one molecule of oxygen oxidised and from the current point of view that synthesis of one molecule of ATP is

TABLE 8.2    Boundary Conditions of the Model

| Compartment | Boundary Conditions |
| --- | --- |
| Extra-mitochondrial space | $O_2 = 6\,\mu M$ (Jones 1986), $O_2^\infty = 0.001\mu M$; $H_p^+ = 10^{-4.2}\,mM$, $P_p^- = 3mM$, $K_p^+ = 140mM$, $ATP_p = 3\,mM$ |
| Inner mitochondrial membrane | $Q_{tot} = 2\,nmol/(mg\ prot.)$, $b_{tot} = 0.325\,nmol/(mg\ prot.$; Green and Wharton 1963; Srere 1981) $Y_{tot} = c_{tot} = 0.325\,nmol/(mg\ prot.$; Green and Wharton 1963) |
| Mitochondrial matrix | $A_{tot} = 10\,mM$, $q_{tot} = 80\,mM$ |

coupled to translocation of three protons (and, consequently, three positive charges) from the intermembrane space to the mitochondrial matrix, respectively.

The system of differential equations (8.7) reflects the stoichiometry of the set of reactions depicted in figure 8.1 and figure 8.2. $Q_{tot}$, $b_{tot}$, $c_{tot}^1$, $Y_{tot}$, $c_{tot}$, $A_{tot}$, $q_{tot}$ stand for the total concentrations of quinone, $bc_1$ complex, cytochrome $c_1$, cytochrome $c$ oxidase (all the species are in the membrane phase), cytochrome $c$, adenine nucleotides and membrane impermeant charges (in the mitochondrial matrix). Summing up the first nine equations of the system (8.7), it can be shown that $dQ_{tot}/d_t = 0$. This reflects the fact that $Q_{tot}$ does not change with time:

$$Q_n + (QH_2)_n + Q_p + (QH_2)_p + Q_n^{-\bullet} + w_1 + w_2 + w_3 + w_4 = Q_{tot} = \text{const}$$

Similarly, the other equations of the system (8.7) imply the following moiety conservations: $w_1 + w_2 + w_3 + w_4 + u_1 + u_2 + u_3 + u_4 + z_1 + z_2 + z_3 + z_4 = b_{tot}$,

$$c_1^r + c_1^{ox} = c_{tot}^1$$

$$c^r + c^{ox} = c_{tot}$$
$$ADP_n + ATP_n = A_{tot},$$

$$H_n^* + (c_M/F) \times (\Delta\Psi) + K_n^+ - P_n - ATP_n = q_{tot}$$

(or using equation (8.6)):

$$H_n^+ + \sum_{j=1}^{3} B_j H - P_n^{2-} + (c_M/F) \times (\Delta\Psi) + K_n^+ - MP - ATP_n^{4-} - MATP_n = q_{tot})$$

$$Y + Yr + YO + YOH = Y_{tot} \qquad (8.8)$$

The seven conservation laws (8.8) reduce the number of independent differential equations of system (8.7) from thirty to twenty-three.

At steady state, fluxes and metabolite concentrations do not change with time; i.e., all derivatives with respect to time equal zero. Consequently, steady-state fluxes have to satisfy the following relationships (see the kinetic scheme, figure 8.1 and figure 8.2):

$$v_2 = v_3 = 2v_1 - v_{25} - v_{26} - v_{27} - v_{28},$$

$$v_4 = v_{13} + v_{15} + v_{19} + v_{21} + v_{25},$$

$$v_5 = v_{23} + v_{24} + v_{26} - v_{13} - v_{15} - v_{19} - v_{21},$$

$$v_6 = v_1 + v_{11} + v_{12} + v_{13} + v_{14} + v_{15} - v_{16} - v_{23} - v_{24} - v_{25} - v_{26} - v_{28},$$

$$v_7 = v_1 + v_{16} - v_{11} - v_{12} - v_{13} - v_{14} - v_{15} - v_{25} - v_{26} - v_{27},$$

$$v_8 = 2(v_1 - v_{25} - v_{26} - v_{27} - v_{28}) - v_{11} - v_{13} - v_{15} - v_{17} - v_{19} - v_{21}$$

$$v_9 = v_{11} + v_{13} + v_{15} + v_{17} + v_{19} + v_{21},$$

$$v_{10} = v_1 - v_{11} - v_{12} - v_{13} - v_{14} - v_{15} - v_{25} - v_{26} - v_{27} - v_{28},$$

$$v_{20} = v_1 - v_{16} - v_{17} - v_{18} - v_{19} - v_{21} - v_{25} - v_{26} - v_{27} - v_{28},$$

$$v_{22} = 2(v_1 - v_{25} - v_{26} - v_{27} - v_{28}) - v_{23} - v_{24},$$

$$v_{29} = v_{15} + v_{21} + v_{25},$$

$$v_{30} = v_{24} + v_{26} - v_{15} - v_{21};$$

$$v_{31} = v_1 + v_{11} + v_{13} + v_{14} + v_{15} - v_{16} - v_{18} - v_{24} - v_{25} - v_{26} - v_{28},$$

$$v_{32} = v_1 + v_{16} + v_{18} - v_{11} - v_{13} - v_{14} - v_{15} - v_{25} - v_{26} - v_{27},$$

$$v_{33} = 4v_{34} = 4v_{35} = 4v_{36} = 4v_{37} = 2v_1 - v_{25} - v_{26} - v_{27} - v_{28},$$

$$v_{39} = v_{41} = v_{40},$$

$$v_{43} = v_{42} = 6v_1 - v_{38} - 4v_{40} - 4(v_{25} - v_{26} - v_{27} - v_{28}) \qquad (8.9)$$

Relationships (8.9) demonstrate that all steady-state fluxes can be expressed in terms of the independent rates $v_1$, $v_{11}$, $v_{12}$, $v_{13}$, $v_{14}$, $v_{15}$, $v_{16}$, $v_{17}$, $v_{18}$, $v_{19}$, $v_{21}$, $v_{23}$, $v_{24}$, $v_{25}$, $v_{26}$, $v_{27}$, $v_{28}$, $v_{38}$, $v_{40}$. In particular, the rate of oxygen consumption ($v_{35}$) equals $(2v_1 - v_{25} - v_{26} - v_{27} - v_{28})/4$, where $v_1$ is the rate of succinate consumption and the sum $v_{25} + v_{26} + v_{27} + v_{28}$ reflects bypass of interheme electron transfer by direct one-electron oxygen reduction.

## DESCRIPTION OF INDIVIDUAL PROCESSES OF THE MODEL

Now, we consider how the rates of the reactions depend on the concentrations of their substrates and products. Our model assumes that succinate is in excess. In accordance with data in Grivennikova and Vinogradov (1982) and Kotlyar and Vinogradov (1984), the rate of the succinate dehydrogenase reaction depends only on the ratio of ubiquinol concentration to ubiquinone,

$$v_1 = \frac{V \cdot Q/(Q + QH_2)}{K + Q/(Q + QH_2)}, \qquad (8.10)$$

where $V = 256$ mM/min and $K = 0.6$. Most notably, we ignore any direct sensitivity of succinate dehydrogenase to membrane potential other than through the $QH_2/Q$ ratio.

The rate equation of the ATP-synthase reaction is based on a minimal kinetic scheme for ATP synthesis-hydrolysis (Boork and Wennestrom 1984; Kholodenko 1988; Skulachev and Kozlov 1997):

$$v_{39} = \frac{V_A}{K_D \cdot K_P} \cdot e^{n_A \cdot \delta \cdot \varphi} \cdot \frac{(H_p/K_H^p)^{n_A} \cdot [ADP_n \cdot P_n - \tilde{K}_{eq} \cdot e^{-n_A \cdot \varphi} \cdot (H_n/H_p)^{n_A} \cdot ATP_n]}{1 + \frac{ADP_n \cdot P_n}{K_D \cdot K_P} \cdot \left(\frac{H_p}{K_H^p}\right)^{n_A} + \frac{ATP_n}{K_T} \cdot \left(\frac{H_n}{K_H^n \cdot e^{\delta_n \varphi}}\right)^{n_A}}$$

(8.11)

Here, $\phi = F \cdot \Delta\Psi/RT$, $\tilde{K}_{eq} = \tilde{K} \cdot e^{-\Delta G_0'/RT}$ $\Delta G \cdot 0 = -33.5$ kJ/mol (see Rosing and Slater 1972) and concentrations are in mM. Kinetic constants $K_D$, $K_T$, $K_P$ and apparent equilibrium constant $\tilde{K}_{eq}$ depend on the concentration of free magnesium, $Mg_n$ ($K_P$ and $\tilde{K}_{eq}$ depend also on $pH_n$; Kholodenko 1988):

$$K_D = \frac{0.8 + Mg_n}{180 \cdot Mg_n}, \quad K_T = \frac{0.08 + Mg_n}{1.08 \cdot Mg_n},$$

$$\tilde{K}_{eq} = K_P = \frac{1 + (1.76 \cdot 10^{-4} / H_n) \cdot (1 + Mg_n/12)}{2.82}$$

$V_A = 0.12$ μM/min, $n_A = 3$, $K_H^p = 3 \cdot 10^{-5} mM$, $K_H^n = 10^{-6} mM$, $\delta = 0.9$, $\delta_n = 0.1$.

The rate equation for adenine nucleotide translocation has been derived in Kholodenko (1988) based on the kinetic data reported in Barbour and Chan (1981), Dupont, Brandolin, and Vignais (1982) and Kramer and Klingenberg (1982):

$$v_{40} = \frac{V_{T/D} \cdot e^{(4\delta_m^T - 3\delta_m^D)\varphi} \cdot (ADP_p \cdot ATP_n - K_{T/D} \cdot e^{-\varphi} \cdot ADP_n \cdot ATP_p)}{\left[\left(K_D^p + ADP_p + \frac{K_D^p \cdot e^{-4\delta_p^D \varphi}}{K_T^p} \cdot ATP_p\right) \cdot (K_T^n + ATP_n + K_T^n \cdot (D_n/K_D^n + I/K_I))\right]}$$

(8.12)

Although all experiments (see Barbour and Chan 1981; Dupont, Brandolin, and Vignais 1982; Kholodenko 1988; Kramer and Klingenberg 1982) to determine Michaelis constants of adenine nucleotide translocator were carried out in a magnesium-free medium, the kinetic constants $K_{T/D}$, $K_D^p, K_T^p, K_D^n, K_T^n$ depend on the concentration of free magnesium both outside ($Mg_p$) and inside ($Mg_n$) the mitochondrion (Kholodenko 1988). This dependence is due to the ability of nucleotides to bind free magnesium and thereby to avoid translocation. The adenine nucleotide translocator operates with free nucleotides. Accordingly:

$$K_{T/D} = \frac{(0.8 + Mg_p) \cdot (0.08 + Mg_n)}{(0.8 + Mg_n) \cdot (0.08 + Mg_p)}, \quad K_D^p = 0.007 \cdot (1 + Mg_p/0.8),$$

$$K_T^p = 0.007 \cdot (1 + Mg_p/0.08),$$

$$K_D^n = 0.06 \cdot (1 + Mg_n/0.8), \quad K_T^n = 0.0015 \cdot (1 + Mg_n/0.8),$$

$$V_T/D = 90 \text{ mM/min}, \quad I = 0.15 \text{ mM}, \quad K_I = 0.001 \text{ mM},$$
$$\delta T_p = 0.15, \quad -4\delta T_m + 3\delta D_m = 0.13.$$

The dependence of the proton leak rate on membrane potential (Brown and Brand 1986; Kholodenko 1988; Nicholls 1974) has been obtained in the framework of Eyring's approach to describe ion transport through a channel with a three-peaks potential energy profile (Markin and Chizmadgev 1974):

$$v_{38} = \frac{V_H \cdot e^{\delta_H \varphi} \cdot (H_p - e^{-\varphi} \cdot H_n)}{\left(1 + \frac{H_p \cdot e^{\delta_H^p \varphi}}{K_H^p}\right) \cdot \left(1 + \frac{H_n \cdot e^{-\delta_H^n \varphi}}{K_H^n}\right)}, \tag{8.13}$$

The parameters $\delta_H, \delta_H^p, \delta_H^n, K_H^p, K_H^n$ were chosen to provide the best quantitative coincidence of proton leak dependence on both $\Delta\Psi$ and $\Delta$pH with experimental data (Brown and Brand 1986):

$$\delta_H = 0.87, \quad \delta_H^p = 0.1, \quad \delta_H^n = 0.1, \quad K_H^p = 10^{-5} mM, \quad K_H^n = 0.4 \cdot 10^{-5} mM,$$

$$V_H = 1500 \text{ min}^{-1}.$$

To describe the rate of electrogenic potassium transfer through the mitochondrial membrane we used an expression similar to that of the rate of proton leak:

$$v_{42} = \frac{V_K \cdot e^{\delta_k \varphi} \cdot (K_p - e^{-\varphi} \cdot K_n)}{\left(1 + \frac{K_p \cdot e^{\delta_k^p \varphi}}{K_K^p}\right) \cdot \left(1 + \frac{K_n \cdot e^{-\delta_k^n \varphi}}{K_K^n}\right)}, \tag{8.14}$$

where $\delta_K = 0.8$, $\delta_K^p = 0.1$, $\delta_K^n = 0.1$, $K_K^p = 254 mM$, $K_K^n = 66 mM$, $V_K = 0.005$ $min^{-1}$.

Symport of inorganic phosphate and a proton (reaction 21), as catalyzed by the phosphate carrier, is a rapid, reversible process that is near thermodynamic equilibrium. The rate of this process is approximated by:

$$v_{41} = V_P \cdot \left(1 - \frac{P_p^- \cdot H_p}{P_n^- \cdot H_n}\right),$$ (8.15)

In accordance with previous studies (La Noue and Schoolwerth 1984; Ligeti et al. 1985), the maximal rate of the phosphate carrier exceeds the rate of phosphate consumption by ATP synthase even at the maximal rate of oxidative phosphorylation, $V_p = 20$ μmol/[min*(mg prot.)].

The rate of electroneutral exchange of potassium ions for protons, catalyzed by the K$^+$/H$^+$-exchanger (Brandolin, Dupont, and Vignais 1982; Garlid 1980; Martin, Beavis, and Garlid 1984; Nakashima, Dordick, and Garlid 1982), is described by the following Michaelis-Menten expression (Kholodenko 1988):

$$v_{43} = \frac{V_{K/H} \cdot \left(K_n^+ \cdot H_p - K_p^+ \cdot H_n\right)}{K_M + \left(K_n^+ + c_1 \cdot K_p^+\right) \cdot \left(H_p + c_2 \cdot H_n\right)}$$ (8.16)

where $V_{K/H} = 3500$ min$^{-1}$, $K_M = 0.9$ (mM)$^2$, $c_1 = 100$, $c_2 = 0.01$.

The scheme of Q-cycle reactions depicted in figure 8.1 includes two diffusion processes: ubiquinol transport from site $i$ to site $o$ and ubiquinone transport from site $o$ to site $i$. The rates of these processes were approximated as follows:

$$\begin{aligned} v_2 &= k_d[(QH_2)_n - (QH_2)_p], \\ v_3 &= k_d(Q_p - Q_n), \end{aligned}$$ (8.17)

where the value of the rate constant transmembrane diffusion of quinones was chosen in accordance with Kingsley and Feigenson (1981): $k_d = 1380$ min$^{-1}$.

The Q-cycle reactions (see table 8.1 and figure 8.1) are assumed to follow mass action kinetics. In order to give a complete description of each reaction, it suffices to know its equilibrium constant and the kinetic constant of the forward reaction. Given the value of the equilibrium constant, the kinetic constant of the direct reaction ($k_i$) was then expressed in terms of the kinetic constant of the reverse reaction ($k_{-i}$) as follows:

$$k_i = k_{-i} \cdot K_{eq}^i \qquad (8.18)$$

Equilibrium constants of redox reactions were determined in accordance with experimental data on midpoint redox potentials:

$$K_{eq} = \exp\left( \frac{\Delta E_m'}{RT/nF} \right) \qquad (8.19)$$

Here, $T$ is the absolute temperature, $F$ is the Faraday constant, $R$ is the gas constant, $\Delta E_m'$ is the difference of the midpoint potentials with respect to the standard hydrogen electrode at pH $= 7$ and $n$ is the number of electrons transferred.

In the present model, the $\Delta E_m'$ values for the steps of the Q-cycle (see figure 8.1) in which states with the complex $FeS_{III}^r - QH^\bullet$ are the substrate or the product cannot be obtained from experimental data. As data are only available for the free quinone forms, only the equilibrium constant for the overall reaction, which includes both the oxidation-reduction step and the dissociation/association steps, is known. Therefore, the choice of either equilibrium constant was arbitrary, provided that the product of the two constants was determined by $\Delta E_m$ of the overall reaction. Using equations (8.18)–(8.19) we can determine either the forward or the reverse rate constants. The other one still then remains to be determined. However, as for reactions 4–7, 16–24 and 29–33 there are additional kinetic data on rise times, $t_{1/2}$, of the time course of transmembrane electric potential difference measured in Drachev et al. (1989), Semenov (1993) and Rich (1984). The rise time, $t_{1/2}$, of each reaction can be approximately expressed in terms of the rate constant of the reaction:

$$\frac{1}{(k_i + k_{-i}) \cdot c_{tot}} \approx t_{1/2}^i = 0.2ms, \quad i = 29-32,$$

$$\frac{1}{(k_{33} + k_{-33}) \cdot c_{tot}} \approx t_{1/2}^{33} = 0.15ms,$$

$$\frac{1}{k_i \cdot \bar{H}_n + k_{-i}} \approx t_{1/2}^i = 40ms, \quad i = 16-21,$$

$$\frac{1}{k_i + k_{-i}} \approx t^i_{1/2} = 7ms, \quad i = 22-24,$$

$$\frac{1}{k_i \cdot b_{tot} + k_{-i} \cdot H^+_p} \approx t^i_{1/2} = 7ms, \quad i = 4-7, \qquad (8.20)$$

Here, $H^+_p = 10-4.2$ mM is the intermembrane proton concentration; $b_{tot}$, $c^1_{tot}$ $c_{tot}$ are total concentrations of the IRC-cytochrome $b$ complex, cytochrome $c_1$ and cytochrome $c$; $\bar{H}_n = 10-4.6$ mM is the averaged concentration of matrix protons.

Equations (8.18)–(8.20) enable us to determine both direct and reverse rate constants of the reactions 2, 4–7, 16–24 and 29–33 (the numerical values are given in table 8.3). As for the rest of the Q-cycle reactions (8–15 and 25–28) their direct rate constants are expressed in terms of the reverse rate constants (see equation (8.18)). The latter are free parameters of our model.

The reactions 34–37 that describe the catalytic cycle of cytochrome $c$ oxidase are assumed to follow mass action kinetics (see table 8.1). From recent kinetic data on the rates of transitions between experimentally distinguishable intermediates of cytochrome $c$ oxidase (Brzezinski and Adelroth 1998; Paula et al. 1999; Wrigglesworth et al. 1988), we can estimate rise times, $t_{1/2}$, for the reactions 34–37. Employing the logic applied above to estimate kinetic constants of the Q-cycle reactions, we can calculate kinetic constants of reactions 34, 36 and 37 according to the following equations:

$$\frac{1}{\left[ k_{34} \cdot \left( \bar{H}^+_n \right)^4 \cdot O_2 + k_{-34} \cdot H^+_p \right] \cdot (c_{tot})^3} \approx t^{34}_{1/2} = 50.9 \mu s$$

$$\frac{1}{\left[ k_{36} \cdot \left( \bar{H}^+_n \right)^3 + k_{-36} \cdot \left( H^+_p \right)^2 \right] \cdot c_{tot}} \approx t^{35}_{1/2} = 1.49ms$$

$$\frac{1}{k_{37} \cdot \bar{H}^+_n + k_{-37} \cdot H^+_p} \approx t^{36}_{1/2} = 1.2ms \qquad (8.21)$$

Any step including electric charge transfer across the membrane may produce membrane potential ($\Delta\Psi$). This ($\Delta\Psi$) in turn affects the rate and equilibrium constants of all electrogenic steps (Boork and Wennestrom 1984; Reynolds, Johnson, and Tanford 1985):

$$K_{eq}(\Delta\Psi) = \exp[-\alpha \cdot \Delta\Psi/(RT/F)] \cdot K_{eq},$$
$$k_+(\Delta\Psi) = \exp[-\delta \cdot \alpha \cdot \Delta\Psi/(RT/F)] \cdot k_+,$$
$$k_-(\Delta\Psi) = \exp[(1-\delta) \cdot \alpha \cdot \Delta\Psi/(RT/F)] \cdot k_-, \qquad (8.22)$$

TABLE 8.3 Parameter Values of Reaction Rates

| N | Midpoint Redox Potential, Em (mV), (ref) | Equilibrium Constant at $\Delta\Psi = 0$ | Rate Constant mM/min or min$^{-1}$ | Free Parameters | Other Parameters |
|---|---|---|---|---|---|
| 4–7 | $E(QH|QH_2^\cdot) = 280$ (Trumpower 1981), $E(Fes_{III}^{ox}|Fes_{III}^r) = 280$ (Skulachev 1988) | $K_i = 10^{-4}, i = 4,...,7$ | $k_i = 476, k_{-i} = 476 \cdot 10^4, i = 4,...,7$ | | $\alpha = 0.1$, $d_1 = 0.5$ |
| 8–9 | $E(Q|Q_p^\cdot) = -160$ (Bowyer and Trumpower 1981), $E(b_l^{ox}|b_l^r) = -40$ (Skulachev 1988) | $K_i = 108 \cdot 10^{-4}, i = 8,9$ | $k_i = 7668, k_{-i} = 71 \cdot 10^4, i = 8,9$ | | $d_2 = 0.5$ |
| 10–15 | $E(Q|Q_n^\cdot) = 70$ (Markin and Chizmadgev 1974), $E(b_h^{ox}|b_h^r) = 40$ (Skulachev 1988) | $K_i = 3.2, i = 10,...,15$ | $k_i = s_1 \cdot 3.2, k_{-i} = s_1, i = 10,...,15$ | $s_1 = 5000$ | |
| 16–21 | $E(Q_n^\cdot|QH_2) = 170$ (Ligeti et al. 1985), $E(b_h^{ox}|b_h^r) = 40$ (Skulachev 1988) | $K_i = 5.07 \cdot 10^8, i = 16,...,21$ | $k_i = 7605 \cdot 10^8, k_{-i} = 1500, i = 16,...,21$ | | $\gamma = 0.1$, $d_4 = 0.5$ |
| 22–24 | $E(b_h^{ox}|b_h^r) = 40$ (Skulachev 1988), $E(b_l^{ox}|b_l^r) = -40$ (Skulachev 1988) | $K_i = 22.6, i = 22,...,24$ | $k_i = 8224, k_{-i} = 364, i = 22,...,24$ | | $\beta = 0.8$, $d_3 = 0.5$ |
| 25–28 | $E(Q|Q_p^\cdot) = -160$ (Wrigglesworth et al. 1988), $E(O_2|O_2^\cdot) = -150$ (Skulachev 1988) | $K_i = 1.5 \cdot 10^{-4}, i = 25,...,28$ | $k_i = 1.5 \cdot s_2, k_{-i} = s_2 \cdot 10^4, i = 25,...,28$ | $s_2 = 0.1$ | |
| 29–32 | $E(FeS_{III}^{ox}|FeS_{III}^r) = 280$ (Skulachev 1988), $E(c_1^{ox}|c_1^r) = 245$ (Rich 1984) | $K_i = 0.8, i = 29,...,32$ | $k_i = 82,305, k_{-i} = 102,881, i = 29,...,32$ | | |
| 33 | $E(c_1^{ox}|c_1^r) = 245$(Rich 1984), $E(c^{ox}|c^r) = 255$ (Skulachev 1988) | $K_{33} = 1.5$ | $k_{33} = 148,148, k_{-33} = 98,765$ | | |
| 34 | | | $k_{34} = 2.8 \cdot 10^{30}, k_{-34} = 275.8$ | | $d_5 = 0.5$ |
| 35 | | | $k_{35} = 1.5 \cdot 10^8 \text{ M}^{-1} \cdot \text{s}^{-1}$ | | |
| 36 | | | $k_{35} = 4.6 \cdot 10^{22}, k_{-35} = 4.6 \cdot 10^7$ | | $d_6 = 0.5$ |
| 37 | | | $k_{36} = 3.5 \cdot 10^{14}, k_{-36} = 3.5 \cdot 10^6$ | | |

Here $\alpha^*\Delta\Psi$ is the membrane potential difference that drops along the path of an electron in a particular step and $\delta$ is the relative fraction of the membrane potential difference ($\alpha^*\Delta\Psi$), which affects the rate of the forward reaction.

Table 8.2 and table 8.3 summarize the rate equations and the parameter values for the model considered. The parameters include all the rate and equilibrium constants of steps and conserved moieties. Experimental data on midpoint redox potentials, required for estimation of the equilibrium constants, are given as well. All results presented in the next section were obtained from numeric solutions of the system (8.7) of differential equations. Since we are only interested in studying steady-state characteristics of the model, the system of algebraic equations obtained by setting the right-hand side of equation (8.7) to zero was solved. This led to steady-state values of the model variables.

## MODEL PREDICTIONS

In this section we address the question of whether one should expect, on the basis of present kinetic knowledge, that $bc_1$-linked superoxide production is determined by any single property of mitochondrial energy metabolism. We consider respiratory rate, the ubisemiquinone concentration, membrane potential and oxygen tension as potentially important factors. We calculate the variation of superoxide production with the variation in each of these properties; i.e., along the different transitions of mitochondrial state resulting from workload variation, uncoupler and inhibitor titration. The simulations presented below can be divided in two sets. The first one consists of model-derived dependencies that reproduce experimental data obtained earlier. These results serve to validate our model and justify predictions that we refer to as the second set of simulations.

Firstly, we reproduce numerically the experimental data on titration of mitochondria respiring on succinate with uncoupler at State 4 (Boveris and Chance 1973; Korshunov, Skulachev, and Starkov 1997). Figure 8.3 shows that partial uncoupling resulting in a less than 20 percent decrease in $\Delta\Psi$ level (solid line) was accompanied by approximately fourfold decrease in $O_2^{\cdot-}$ production (dotted line) and by the same increase in oxygen consumption (dashed line). These calculated dependencies correspond closely to those obtained experimentally through titration with uncoupler SF6847 (see figure 8.3 in Korshunov, Skulachev, and Starkov 1997).

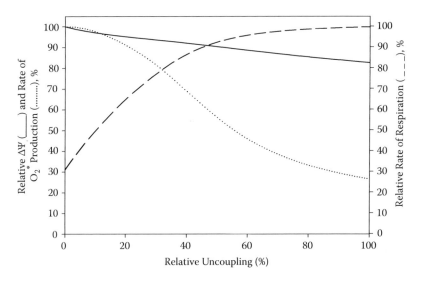

FIGURE 8.3  Effect of uncoupling on respiration (dashed line), $\Delta\Psi$ (solid line) and rate of superoxide production (dotted line). Increase in H$^+$ permeability of the inner mitochondrial membrane was due to increase in the maximal rate of proton leak, $V_H$.

The simplest hypothesis that could be posed on the basis of figure 8.3 is that $O_2^{\bullet-}$ production is governed by respiratory rate in such a way that, the more respiration, the less superoxide is produced by the bc$_1$ complex. However, experimental data (Korshunov, Skulachev, and Starkov 1997; Liu 1997; Liu and Huang 1996) and numerical simulations derived from our model reject this hypothesis. Indeed, paralleling the experimental work of Korshunov, Skulachev, and Starkov (1997), we simulate titration of the succinate oxidising mitochondria with malonate at State 4. In the framework of our model, this can be mimicked by decreasing the succinate dehydrogenase activity. Figure 8.4 shows that both the respiratory rate (solid line) and $O_2^{\bullet-}$ production (dotted line) diminish with the decrease in $\Delta\Psi$ brought about by malonate titration.

Figure 8.3 and figure 8.4 taken together might suggest that what determines intensity of bc$_1$-linked superoxide generation is the magnitude of $\Delta\Psi$. Indeed, in the simulated malonate titrations (figure 8.4), a decrease in $\Delta\Psi$ from 185 mV to 170 mV results in about a thirty-fold decrease in $O_2^{\bullet-}$ production. The same effect was found experimentally (see figure 8.4 in Korshunov, Skulachev, and Starkov 1997).

Figure 8.3 and figure 8.4 exhaust the set of experimentally tested simulations. All other results of kinetic modelling presented in this section are

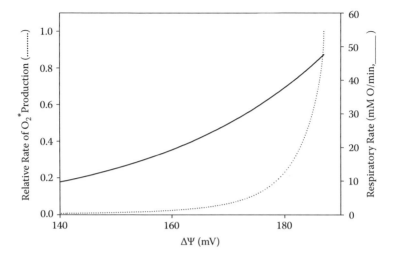

FIGURE 8.4   Variation of relative rate of superoxide production (solid line) and the respiration rate (dashed line) with membrane potential. $\Delta\Psi$ was varied by decreasing activity of succinate dehydrogenase.

predictions that will have to be tested experimentally. The first of these predictions is the dependence of the $bc_1$-linked superoxide production on $\Delta\Psi$ level when varying the workload.

We varied the external ADP concentration from 20 to 0 μM; i.e., decreased the external workload. This variation gave rise to an increase in the electric potential from 145 to 185 mV (figure 8.5a). Figure 8.5b (solid line) shows how respiration rate increased with decreasing electric potential. These results are in line with experimental data (Kunz et al. 1981; Letko et al. 1980; Van Dam et al. 1980). Figure 8.5b (dashed line) shows the predicted dependence of superoxide production rate on $\Delta\Psi$. Superoxide generation increased dramatically with increasing $\Delta\Psi$ in the range from 155 to 180 mV. A similar behaviour was reported in Demin et al. (1998, 1998, 2001) where the ability of $bc_1$ complex itself to generate superoxide radicals was considered. Here we show that the effect is retained in the intact system making ATP. The variation of superoxide production rate with $\Delta\Psi$ was not the same as in the case of inhibition of succinate dehydrogenase.

The other possibility to decrease superoxide generation is to regulate the ion content of the extra-mitochondrial medium. Indeed, a decrease in the extra-mitochondrial phosphate concentration (figure 8.6a) or an increase in the extra-mitochondrial potassium concentration (not shown) resulted

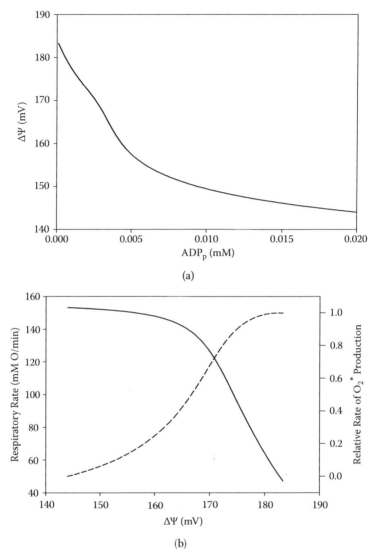

FIGURE 8.5 Results of modelling of coupled mitochondria: a: dependence of electric potential difference on external ADP concentration; b: variation of respiratory rate (solid line) and relative rate of superoxide production (dashed lane) with $\Delta\psi$.

in a simulated decrease of the superoxide production rate at the same electric potential values. We also calculated the rate of superoxide formation as a function of the phosphate concentration (figure 8.6c) and found that the dependence slightly increased with external phosphate. Yet, addition of

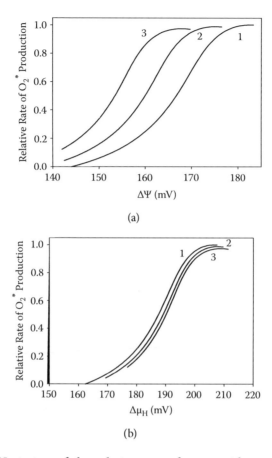

(a)

(b)

FIGURE 8.6   Variation of the relative rate of superoxide production with $\Delta\Psi$ (a) and $\Delta\bar{\mu}_H$ (b): curves 1, 2 and 3 correspond to 3 mM, 1 mM and 0.3 mM of phosphate, respectively; $\Delta\Psi$ was varied by modulating $ADP_p$ concentration. Dependencies of relative rate of superoxide production (c), $\Delta\Psi$ (d, solid lane), $\Delta\bar{\mu}_H$ (d, dashed lane) and $\Delta pH$ (d, dotted lane) on external phosphate concentration at $ADP_p = 0$ μM.

phosphate increased the membrane potential (figure 8.6d), which, according to figure 8.6a, should lead to an increase in $O_2^*$ production rate.

However, the rate of superoxide production need not solely depend on the electric component of $\Delta\bar{\mu}_{H^+}$. It might also depend on $\Delta pH$. Figure 8.6d shows that the increase in $\Delta\Psi$ caused by adding phosphate was accompanied by a decrease in $\Delta pH$, even such that $\Delta\bar{\mu}_{H^+}$ only slightly decreased

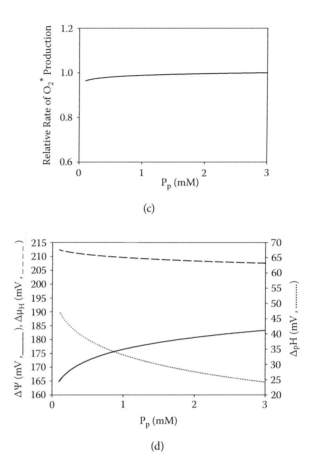

FIGURE 8.6 (Continued).

with external phosphate. If the dependence of superoxide anion production on $\Delta\Psi$ and $\Delta pH$ had been equal, and its dependence on $\Delta\bar{\mu}_{H^+}$ unique and if $\Delta\bar{\mu}_{H^+}$ had kept the same magnitude for the various phosphate concentrations, we would have found independence of superoxide anion production on phosphate. Then plotting $J_{o_2^-}$ versus $\Delta\bar{\mu}_{H^+}$ should have produced the same line for all selected concentrations of phosphate. What actually happened was that $\Delta\bar{\mu}_{H^+}$ decreased with $P_p$ and, consequently, dependencies of $J_{o_2^-}$ on $\Delta\bar{\mu}_{H^+}$ at three selected $P_p$ values did not coincide with each other (figure 8.6b). This was due to the fact that transfer of $\Delta pH$ into $\Delta\Psi$ with increase of extra-mitochondrial phosphate concentration was accompanied by unbalanced changes of $\Delta pH$ and $\Delta\Psi$ (figure 8.6d), so that total $\Delta\bar{\mu}_{H^+}$ did decrease. The transfer $\Delta pH$ into $\Delta\Psi$ moved the dependence of superoxide production rate on potential as a whole to higher $\Delta\Psi$ values (figure 8.6a) but the phosphate elevation decreased $\Delta pH$ more than it increased $\Delta\Psi$. This leads to a reverse order of curves when we consider the rate of superoxide production as a function of $\Delta\bar{\mu}_{H^+}$ (figure 8.6b) as compared with the rate as a function of $\Delta\Psi$ (figure 8.6a). The same phenomenon was observed when the potassium concentration was varied.

Since, in the framework of the Q-cycle scheme depicted in figure 8.1, only ubisemiquinone bound to the iron sulfur center is able to give one electron to oxygen to form superoxide, the rate of superoxide production is driven by the concentration of the corresponding states of the $bc_1$ complex.

Assuming that they are all equally reactive, the relevant property is the total concentration of semiquinone associated with the Reiske iron-sulfur protein, which we denote as $fQr$:

$$fQr = w_1 + w_2 + w_3 + w_4;$$

In the case of variation of oxygen tension one might expect the dependence of the total concentration of ubisemiquinone bound in site $o$, i.e., $fQr$, to be bell-shaped. Indeed, $fQr$ depends on oxygen concentration in two ways. The first one consists of supplying cytochrome $c$ oxidase with $O_2$ that provides electron flow along the respiratory chain and, consequently, replenishes $fQr$ due to its formation from states of the $bc_1$ complex with an oxidized iron-sulfur cluster (reactions 4–6 in figure 8.1). The second way depletes $fQr$ to form superoxide (reactions 25–28). While the depletion rate is proportional to oxygen concentration due to the nonenzymatic nature of the reaction, the rate of replenishment, i.e., the rate of cytochrome $c$ oxidase, is saturable with oxygen concentration. This means that an increase in oxygen concentration might result in elevation of $fQr$ when the oxygen concentration is below the $K_m$ of cytochrome $c$ oxidase and then a decrease in $fQr$ due to the unsaturable rate of superoxide formation. However, figure 8.7 (solid lane) shows that in the simulation $fQr$ increased monotonically with $O_2$ concentration. This can be explained if we take into account that there is another process of $fQr$ consumption different from superoxide formation. This process, which is presented by reactions 8 and 9 (figure 8.1), really determines the steady-state $fQr$, since

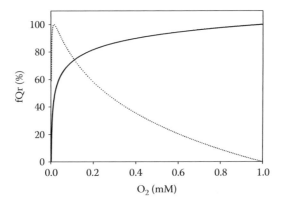

FIGURE 8.7 Dependencies of total ubisemiquinone concentration on $O_2$ concentration at State 4 ($ADP_p = 0$ μM): solid line corresponds to normal rate of Q-cycle operation (parameter values are listed in table 8.3) and dotted line corresponds to greatly reduced Q-cycle operation (the rate constants of reactions 8 and 9 were decreased tenfold and the maximal rate of succinate dehydrogenase, $V_1$, was decreased tenfold). The concentrations of ubisemiquinone at site $o$ of $bc_1$ complex; i.e., $fQr$, were calculated per unit of volume of inner mitochondrial membrane.

steady-state fluxes going via reactions 8 and 9 are many times higher than the steady-state rate of superoxide production. Only when the rate of Q-cycle operation is slowed down by inhibiting reactions 8, 9 and 1 should the expected bell-shaped dependence of $fQr$ on $O_2$ concentration should be observed (figure 8.7, dotted line).

# Application of the Kinetic Modelling Approach to Problems in Biotechnology and Biomedicine

In this chapter we illustrate how kinetic models can be applied to resolve different problems arising in the areas of biotechnology and biomedicine. We present several examples of these types of applications aimed at improving drug safety (Mogilevskaya, Demin, and Goryanin 2006), drug discovery (Goryanin, Demin, and Tobin 2003) and strain improvement (Demin et al. 2005).

## STUDY OF THE MECHANISMS OF SALICYLATE HEPATOTOXIC EFFECT

Understanding of drug side effects is one of the most challenging problems of modern pharmacology (Bjorkman 1998). The problem has two aspects. The first is how to reduce adverse effects of already existing drugs, and the second is how to predict possible side effects of new compounds that are still under development. The pathways involved in toxic drug effects can be examined using knowledge about the regulatory mechanisms of the intracellular

biochemistry. In order to enable this, the strategy of kinetic pathways reconstruction and modelling has been developed (Demin et al. 2004).

Usually, drugs have multiple effects on the intracellular metabolism
(nonspecificity); i.e., more than one enzyme is affected (inhibited or activated) or more than one transporter is involved in utilization and excretion
of the drug. Kinetic modelling enables us to study each effect individually.
At the same time, we can estimate the synergy of individual impacts on
the total adverse effect of the drug. Using this approach, we have studied the mechanisms of the hepatotoxic side effects of acetylsalicylic acid
by assessing individual contributions to the inhibition of mitochondrial
energy metabolism as a whole.

It is known that, during utilization in the liver, salicylates can inhibit β-
oxidation of fatty acids (Fromenty and Pessayre 1995), decrease the pool of
coenzyme A (CoA; Vessey, Hu, and Kelly 1996), inhibit succinate dehydrogenase and $\alpha$-ketoglutarate dehydrogenase (Kaplan, Kennedy, and Davis
1954), and increase the permeability of the inner mitochondrial membrane,
thereby decreasing the proton motive force (Haas et al. 1985; Miyahara and
Karler 1965). One of the reasons for liver injury during aspirin therapy is
the disturbance of hepatocyte energy metabolism, particularly in the Krebs
cycle. In chapter 7, we developed a kinetic model of the mitochondrial Krebs
cycle. In this section, we apply this model to understand which of the modes
of action of aspirin are most critical for the hepatocytes.

We have estimated the contributions for each mode to changes in the
global regulatory properties of mitochondrial energy metabolism. Using
our model, we have studied the influences of salicylates on the quasi-steady-
state flux in the Krebs cycle and examined possible ways to prevent such
changes.

Kinetic Description of the Influence of Salicylates on the Krebs Cycle

Salicylates have multiple influences on the Krebs cycle (see figure 9.1).
These are:

(A) Sequestration of coenzyme A,

(B) Uncoupling of oxidative phosphorylation,

(C) Inhibition of $\alpha$-ketoglutarate and succinate dehydrogenases.

To account for all these impacts in the kinetic model, we have assumed
the intra-mitochondrial salicylate concentration to be constant; i.e., the

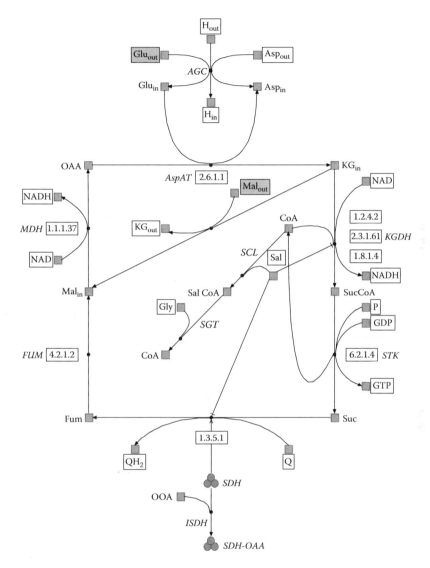

FIGURE 9.1 The scheme of the Krebs cycle oxidising glutamate and malate as substrates with the influence of salicylates.

concentration of salicylates does not change with time. Below we describe each individual salicylate impact on the Krebs cycle in more detail:

Salicylate-induced CoA sequestration results from salicylate activation consisting of two processes (Forman, Davidson, and Webster 1971): reaction of salicyl-CoA formation accompanied by CoA consumption and catalyzed by salicyl-CoA ligase (SCL in figure 9.1) and acylation of glycine accompanied by CoA release and catalyzed by salicyl-CoA-glycine acyltransferase (SGT in figure 9.1). These enzymes catalyzing salicylate activation have been described in the following manner.

It is known (Forman, Davidson, and Webster 1971) that the mechanism of salicylate (Sal) activation in mitochondria coincides with that of fatty acid activation, involving ATP consumption and pyrophosphate release, and is catalyzed by salicyl-CoA ligase (SCL). To describe the functioning of this enzyme we derive the following rate equation:

$$V_{SCL} = \frac{V_f \times Sal \times CoA}{K_m^{Sal} K_i^{CoA} + K_m^{Sal} CoA + K_m^{CoA} Sal + CoA \times Sal}$$

Here, $V_f$ is the maximal rate of salicyl-CoA ligase; $K_m^{Sal}$, $K_m^{CoA}$ are the Michaelis constants for salicylate and CoA; and $K_i^{CoA}$ is the inhibition constant for CoA. Since the concentration of ATP in the mitochondrial matrix is much higher than the corresponding Michaelis constant of salicyl-CoA ligase (Ricks and Cook 1981), we assume that the enzyme is in saturation with respect to ATP. This assumption allows us to ignore dependence of the enzyme on ATP concentration. Parameters whose values are known from the literature are $V_f$ (Forman, Davidson, and Webster 1971) and $K_m^{Sal}$ (Vessey, Hu, and Kelly 1996). Because of the lack of experimental data on dependence of salicyl-CoA ligase on CoA, we have assumed that both the value of the Michaelis constant and the value of the inhibition constant with respect to CoA for salicyl-CoA ligase were equal to that for acyl-CoA ligase. Parameter values are listed in table 7.2.

Salicyl-CoA-glycine acyltransferase (SGT) catalyzes the reaction of salicylurate formation from salicyl-CoA (SalCoA) and glycine. The rate law of the enzyme can be described by the following equation:

$$V_{SGT} = \frac{V_f \times SalCoA \times Gly}{K_m^{SalCoA} K_i^{Gly} + K_m^{SalCoA} Gly + K_m^{Gly} SalCoA + SalCoA \times Gly}$$

Here, $V_f$ is the maximal reaction rate; $K_m^{SalCoA}$, $K_m^{Gly}$ are the Michaelis constants for substrates; $K_i^{Gly}$ is the inhibition constant for glycine (Gly). The values of the Michaelis constants and maximal reaction rate values have been estimated from the literature (Forman, Davidson, and Webster 1971). We also assume that the inhibition constant for glycine is equal to its Michaelis constant (see table 7.2 for values of these parameters).

To simulate the influence of salicylate-induced CoA sequestration on the functioning of the Krebs cycle we took SGT and SCL reactions into account.

Uncoupling of oxidative phosphorylation results from salicylate transport via the inner mitochondrial membrane: Salicylate is a weak acid which passes through the membrane in a neutral form only. The neutral form of salicylate is formed by binding of an external proton to the anionic form of salicylate (Skulachev 1989). This means that transport of salicylate into the matrix of mitochondria is accompanied by simultaneous transport of protons in the same direction. It was shown by Haas et al. (1985) that salicylate addition influences the transmembrane potential and delta pH. Two sets of parameter values have been chosen to model the uncoupling effect of salicylate. The first set of parameter values corresponds to the energy state with coupled respiration and oxidative phosphorylation without salicylate addition: $\Delta \psi \cong 0.139$ V, $H_{in} = 5.2$ nM and $H_{out} = 39.8$ nM (Haas et al. 1985). The second set of parameter values describes the functioning of mitochondria in an uncoupled state with added salicylate: $\Delta \psi = 0.135$ V, $H_{in} = 13$ nM; $H_{out} = 39.8$ nM (Haas et al. 1985). Using either the first or the second set of values for the parameters in the system of differential equations (7.1) from chapter 7, and calculating the steady-state values of the intermediate concentrations and fluxes, we can estimate the influence of uncoupling on the functioning of the Krebs cycle.

$\alpha$-ketoglutarate dehydrogenase (KGDH) and succinate dehydrogenase (SDH) can be inhibited by salicylates (Kaplan, Kennedy, and Davis 1954). Since the mechanisms of interaction between salicylates and these enzymes are unknown, we have assumed that both $\alpha$-ketoglutarate dehydrogenase and succinate dehydrogenase are inhibited by salicylates in an uncompetitive manner (Cornish-Bowden 1995). This means that the maximal rates of these enzymes depend on salicylate concentration in accordance with the following equation:

$$V_{\max}^{enzyme}(Sal) = V_{\max}^{enzyme} \Big/ \left(1 + Sal \Big/ K_{i,enzyme}^{Sal}\right) \qquad (9.1)$$

To estimate the values of the inhibition constants of the two enzymes with respect to salicylates, experimental data (Kaplan, Kennedy, and Davis 1954) have been used, and the experiments having been performed as follows: a suspension of mitochondria respiring on $\alpha$-ketoglutarate was incubated in media with and without salicylate, and a time series of

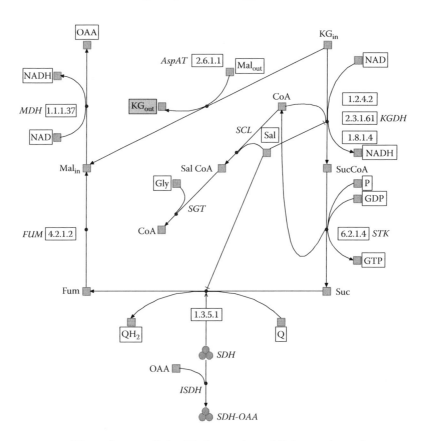

FIGURE 9.2 The scheme of the Krebs cycle oxidising α-ketoglutarate as substrate with the influence of salicylates.

oxygen consumption by these mitochondria was obtained. To quantitatively describe these experiments, a kinetic model of α-ketoglutarate oxidation in the Krebs cycle has been constructed (figure 9.2).

This model was obtained by the following modification of the system of differential equations (7.1) from chapter 7:

1. Elimination of variables (and corresponding differential equations) corresponding to glutamate, aspartate and oxaloacetate concentrations;

2. Elimination of all reactions (and corresponding reaction rates) involved in glutamate transport and degradation (AGC, AspAT and MDH reactions in figure 9.1);

3. Addition of new variables corresponding to the consumed oxygen ($O_2^{consumed}$) and salicyl-CoA (SalCoA) concentration. Time dependences

for these variables are determined by the following differential equations:

$$dO_2^{consumed}/dt = (V_{SDH} + V_{KGDH})/2$$

$$dSalCoA/dt = V_{SCL} - V_{SGT}$$

(9.2)

The new model contains all the mechanisms of salicylate impact on the Krebs cycle (see above and equations (9.1) and (9.2)) and includes oxaloacetate as a parameter. As OAA is not consumed by other reactions (when $\alpha$-ketoglutarate is the only oxidised substrate), the MDH reaction would be at equilibrium. The existence of this reaction would influence OAA concentration. So we have determined the OAA parameter value through the equilibrium constant of the MDH reaction: $OAA = Mal * NAD/(K_{eq}^{MDH} * NADH)$. Time dependences of the oxygen consumption in the experiment with and without salicylate addition were calculated under the following initial conditions: $Mal_{in} = 1.41$ mM; $KG_{in} = 0.08$ µM; SucCoA = 0.02 mM; CoA = 0.98 mM; SalCoA = 0; SDH = 2 µM; SDH-OAA = 48 µM; Suc = 0.12 µM; Fum = 2.4 mM; $O_2^{consumed} = 0$.

Initial values of the intermediate concentrations were calculated as steady-state values in the model of $\alpha$-ketoglutarate oxidation without salicylate addition (Sal = 0; see figure 9.2). To estimate the values of the unknown inhibition constants for the action of salicylate on KGDH and SDH, we fitted the new model to experimentally measured dependences of oxygen consumption. Figure 9.3 demonstrates that experimental data from Kaplan, Kennedy, and Davis (1954; symbols) and theoretical curves corresponding to our model of the Krebs cycle oxidising $\alpha$-ketoglutarate closely coincide. Values of the inhibition constants are listed in table 7.2. The $K_i$ value for succinate dehydrogenase is two orders of magnitude lower than that for $\alpha$-ketoglutarate dehydrogenase, so succinate dehydrogenase should be inhibited more strongly by salicylate.

## Impacts of Different Mechanisms of Salicylate Inhibition on the Total Adverse Effect on the Krebs Cycle

To estimate the contribution of each individual mechanism of Krebs cycle inhibition, we take into account each inhibitory mechanism one by one. We have calculated the steady-state dependence of glutamate influx on external glutamate concentration. The lower the steady-state influx of

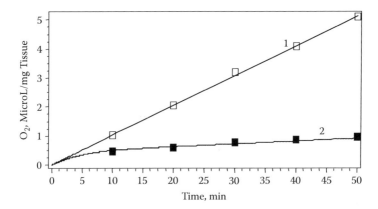

FIGURE 9.3  Inhibition by salicylates of mitochondrial respiration rate on α-ketoglutarate described by the model and experimental points in the following conditions: 1) $KG_{out}$ = 10 mM, $pH_{out}$ = 7.4 (white squares); 2) $KG_{out}$ = 10 mM, $pH_{out}$ = 7.4, Sal = 6.7 mM (black squares).

glutamate, the more significant the individual contribution to the cumulative inhibition of the Krebs cycle has been found to be. External metabolite concentrations have been taken from cytosol data available in the literature (values are listed in table 7.1). Figure 9.4 demonstrates how the steady-state value of the glutamate influx depends on its extra-mito-chondrial concentration without taking into account any mechanism of salicylate influence (curve 1), with incorporation of individual mecha-nisms of salicylate-induced inhibition (curves 2–5), and when all possible inhibitory mechanisms are accounted for (curve 6). Analyzing the results (figure 9.4), we conclude that the CoA-sequestration mechanism of salicy-late (i.e., reactions catalyzed by salicyl-CoA ligase and salicyl-CoA-gly-cine acyltransferase) only changes the glutamate influx slightly (see curve 2). Moreover, uncoupling of oxidative phosphorylation by salicylates (figure 9.4, curve 3) has little effect on the flux. In contrast, all other mech-anisms substantially decrease this flux when they are incorporated into the model individually. Indeed, taking into account either inhibition of α-ketoglutarate dehydrogenase (figure 9.4, curve 4) or succinate dehy-drogenase (figure 9.4, curve 5) decreases the glutamate influx by approxi-mately twenty times. Accounting for all possible inhibitory mechanisms results in a glutamate influx which is near zero (figure 9.4, curve 6). Thus, comparison of individual impacts of different inhibitory mechanisms allows us to conclude that the most substantial contribution of salicylates

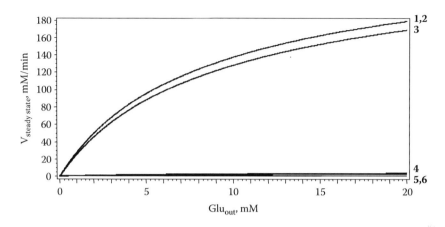

FIGURE 9.4 Influence of different mechanisms of salicylate inhibition on Krebs cycle oxidising glutamate and malate. Dependences of glutamate consumption rates on its extra-mitochondrial concentration were obtained from the models where different mechanisms of salicylate (5 mM) inhibition were taken into account: 1) without salicylate; 2) CoA consumption by salicylate transformation processes; 3) uncoupling effect of salicylates; 4) salicylate inhibition of a-ketoglutarate dehydrogenase; 5) salicylate inhibition of succinate dehydrogenase; 6) all salicylate inhibition mechanisms were included.

to inhibition of the Krebs cycle is due to inhibition of succinate dehydrogenase and $\alpha$-ketoglutarate dehydrogenase.

## Prediction of Possible Ways to Recover Krebs Cycle Functionality

In this section we address the following question: Is it possible to predict changes in the external substrates that would increase the flux through the Krebs cycle and compensate for the salicylate inhibition? We have modified the model of glutamate and malate oxidation to include all salicylate influences (figure 9.1). Figure 9.5 demonstrates that a simultaneous increase in the external malate concentration (from 0.495 mM to 10 mM), a decrease in external $\alpha$-ketoglutarate concentration (from 0.54 mM to 0), and a decrease in internal glycine concentration (from 1 mM to 0.1 nM) result in a significant recovery of steady-state glutamate influx. This result can be explained in the following manner. As shown in the previous section, all mechanisms of salicylate influence (except salicylate uncoupling) that contribute greatly to Krebs cycle inhibition affect the lower part of

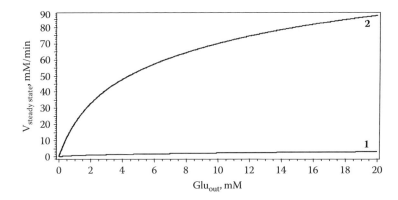

FIGURE 9.5   Effect of changes in concentrations of external substrates on salicylate-inhibited Krebs cycle flux: 1) Sal = 5 mM; $KG_{out}$ = 0.54 mM; $Mal_{out}$ = 0.495 mM; $Asp_{out}$ = 0; Gly = 1 mM; 2) Sal = 5 mM; $KG_{out}$ = 0; $Mal_{out}$ = 10 mM; $Asp_{out}$ = 0; Gly = 0.1 mM.

the Krebs cycle (reactions of KGDH, STK, SDH, FUM in figure 9.1) and do not affect its upper part (reactions AspAT, MDH and KMC). This means that redirection of flux from the lower segment of the Krebs cycle to the $\alpha$-ketoglutarate/malate carrier shunt may result in a substantial increase of flux via the salicylate-inhibited Krebs cycle. The kinetic model allows us to predict changes in concentrations of external metabolites that result in desirable redirection of fluxes and, as a consequence, in substantial recovery of flux in the salicylate-inhibited Krebs cycle (see curves 1 and 2 in figure 9.5). Indeed, decrease in external $\alpha$-ketoglutarate (from 0.54 mM to 0) and increase in external malate (from 0.495 mM to 10 mM) made the $\alpha$-ketoglutarate/malate carrier shunt more competitive in comparison with the $\alpha$-ketoglutarate dehydrogenase for their shared substrate $\alpha$-ketoglutarate. On the other hand, decrease in glycine concentration results in CoA trapping in the complex with salicylate, SalCoA. Substantial decrease of free CoA concentration, resulting from its sequestration with salicylate, leads to the slowing down of $\alpha$ ketoglutarate dehydrogenase operation and, as a consequence, to a weakening of the competition of the enzyme with the $\alpha$-ketoglutarate/malate carrier for $\alpha$-ketoglutarate. Thus, synchronous changes in external concentrations of $\alpha$-ketoglutarate, malate and internal concentration of glycine predicted by the model result in weakening of the inhibitory influence of salicylates on the Krebs cycle.

## MULTIPLE TARGET IDENTIFICATION ANALYSIS FOR ANTI-TUBERCULOSIS DRUG DISCOVERY

Pulmonary tuberculosis is a chronic infection of the lungs, leading in many cases to progressive tissue destruction and death. The causative organism is *Mycobacterium tuberculosis*. The initial stages of infection are thought to involve invasion and multiplication within macrophages, followed by release of large numbers of extracellular bacteria and tissue destruction. Current regimes for the treatment of pulmonary TB involve extended therapy (6–12 months) with multiple antibiotics to prevent selection of resistant strains and to ensure eradication of persisting bacilli from the lungs. Despite this, multiply drug-resistant strains (MDR-TB) are increasingly common. New agents are required which are active against MDR-TB and which are better tolerated than existing drugs, leading to greater patient compliance. The greatest improvements to current therapy and commercial attractiveness would be realised by drugs which are active against persistent organisms, thereby resulting in significantly reduced treatment duration.

The glyoxylate pathway (also called the glyoxylate bypass or shunt) comprises the activities of isocitrate lyase (ICL) and malate synthase (MS). It acts as an alternative route for isocitrate metabolism in the tricarboxylic acid (TCA) cycle, bypassing the steps in which two molecules of carbon are lost as $CO_2$. This enables organisms possessing this pathway to utilise acetyl-CoA as the only input into the TCA cycle and hence permits growth on fatty acids and lipids, which are degraded to acetyl-CoA by beta-oxidation. The glyoxylate pathway is present in bacteria and plants but has not been demonstrated in higher mammals. There is growing evidence that *M. tuberculosis* utilises fatty acids for growth *in vivo* (Graham and Clark-Curtiss 1999,). The ICL gene (icl) has been shown to be massively up-regulated in mycobacteria growing within macrophages and within host organisms (Bentrup et al. 1999). Furthermore, gene disruption of icl in *M. tuberculosis* results in a strain which is unable to cause a persistent, chronic infection in immune-competent mice (McKinney et al. 2000). Wild-type strains multiply within the lungs for the first two weeks of infection until immune activation causes population numbers to plateau i.e., a chronic infection is established. In contrast, the icl strain grows normally for the first two weeks, after which numbers gradually decline for the remainder of the experiment. The same mutant strain was unable to grow in activated macrophages.

It is now known that *M. tuberculosis* has two ICL genes (icl and aceA; Cole et al. 1998). Recent evidence suggests that simultaneously disrupting both of these genes results in greater attenuation of infection than only a single knockout (Graham and Clark-Curtiss 1999). Taken together, the evidence indicates that ICL and perhaps MS are necessary for maintenance of chronic infection by immune-adapted *M. tuberculosis*. In addition, ICL genes have been shown to be up-regulated in a number of bacteria and fungi when growing inside a host cell, suggesting that ICL activity may be part of a wider strategy for intracellular survival by pathogenic microorganisms. Both enzymes are potential targets for chemotherapeutic agents aimed at persisting organisms. Furthermore, the structures of both enzymes have been solved, which makes them amenable to targeted drug discovery strategies (Sharma et al. 2000). Small molecule inhibitors with adequate pharmacokinetics are required to fully validate this hypothesis and there may be broader spectrum applications for such novel inhibitors. Given this information, it would be advantageous to construct a mathematical model of the glyoxylate pathway. The analysis of simultaneous *in silico* inhibitions of both isocitrate lyase (ICL) and malate synthase (MS) could be used to evaluate them as potential target sites for therapeutic drug intervention.

## Construction of a Kinetic Model of the Glyoxylate Shunt in *Mycobacterium tuberculosis*

A mathematical model of the glyoxylate shunt has been constructed with known allosteric regulations, including positive and negative feedbacks (figure 9.6). The resulting mathematical model represents a system of seven differential equations:

$$dMal/dt = v_{MS} + v_{SDH} - v_{MDH}$$

$$dOA/dt = v_{MDH} - v_{CS}$$

$$dICit/dt = v_{CS} - v_{ICL} - v_{IDH}$$

$$dGlOx/dt = v_{ICL} - v_{MS} \tag{9.3}$$

$$dSuc/dt = v_{ICL} - v_{SDH} - v_{Sucefflux}$$

$$dCoA/dt = v_{MS} + v_{CS} - v_{CoArecycling}$$

$$dAcCoA/dt = -v_{MS} - v_{CS} + v_{CoArecycling}$$

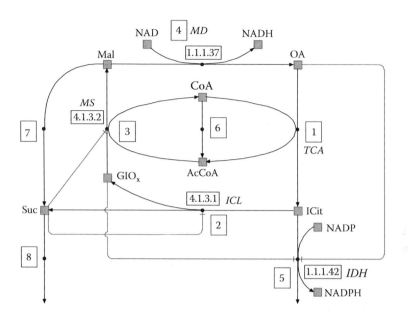

FIGURE 9.6 Glyoxylate shunt model for TB. The consumption of AcCoA in the conversion of oxaloacetate (OA) to isocitrate (process 1), isocitrate lyase (ICL; process 2), malate synthase (MS; process 3), malate dehydrogenase (MDH; process 4), isocitrate dehydrogenase (IDH; process 5), AcCoA production (process 6), conversion of succinate to malate (process 7), succinate utilization (process 8).

eight rate laws and thirty parameters:

1. The consumption of AcCoA in the conversion of oxaloacetate (OA) to isocitrate (ICit; figure 9.6, process 1): The overall reaction corresponds to processes catalyzed by citrate synthase and aconitase. The reaction rate law is in accordance with reversible Michaelis-Menten kinetics:

$$v_{cs} = \frac{V_m^{CS} \times \left(OA \times AcCoA - ICit \times CoA/K_{eq}^{CS}\right)}{K_{m,OA}^{CS} \times \left(1 + ICit/K_{m,ICit}^{CS}\right) + OA}$$

The kinetics for TB are not known, but, after comparison of kinetic constants from different organisms, we conclude that the relevant kinetic parameters from the *Escherichia coli* model could be used to describe these reactions (Drozdov-Tikhomirov, Scurida, and Serganova 1992):

$$V_m^{CS} = 50\,mM\,/\,min$$

$$K_{eq}^{CS} = 10$$

$$K_{m,ICit}^{CS} = 1\,mM$$

$$K_{m,OAt}^{CS} = 0.04\,mM$$

2. Isocitrate lyase (ICL; figure 9.6, process 2): Kinetic information on ICL were taken from Bentrup et al. (1999) in which isoenzymes of ICL from *Mycobacterium tuberculosis* were purified and characterised. The rate equation is:

$$v_{ICL} = \frac{V_m^{ICL} \times \left( ICit - Suc \times GlOx/K_{eq}^{ICL} \right)}{\left( K_{m,ICit}^{ICL} + ICit \right) \times \left( 1 + Suc/K_{i,Suc}^{ICL} \right) \times \left( 1 + Mal/K_{i,Mal}^{ICL} \right)}$$

$K_m$ values for ICit were estimated as $K_{m,ICit}^{ICL} = 0.145\,mM$. Malate (Mal) and succinate (Suc) inhibit ICL by 50 percent at a concentration of about 5 mM. Inhibition is noncompetitive and the estimated inhibition constant is $K_{i,Suc}^{ICL} = 5\,mM$, $K_{i,Mal}^{ICL} = 5\,mM$. The equilibrium constant and maximal rates are $V_m^{ICL} = 30\,mM/min$ and $K_{eq}^{ICL} = 10$.

3. Malate synthase (MS; figure 9.6, process 3): the rate equation is in accordance with reversible Michaelis-Menten kinetics:

$$v_{MS} = \frac{V_m^{MS} \times \left( GlOx \times AcCoA - Mal \times CoA/K_{eq}^{MS} \right)}{K_{m,GlOx}^{MS} \times \left( 1 + Mal/K_{m,Mal}^{MS} \right) + GlOx}$$

with rate constants $K_{m,GlOx}^{MS} = 0.02\,mM, K_{m,Mal}^{MS} = 1\,mM, V_m^{MS} = 50\,mM/min$ and $K_{eq}^{MS} = 10$.

4. Malate dehydrogenase (MDH; figure 9.6, process 4): the kinetics are chosen as reversible Michaelis-Menten:

$$v_{MDH} = \frac{V_m^{MDH} \times \left( Mal \times NAD - OA \times NADH/K_{eq}^{MDH} \right)}{K_{m,Mal}^{MDH} \times \left( 1 + OA/K_{m,OA}^{MDH} \right) + Mal}$$

with the following kinetic parameters:

$$V_m^{MDH} = 50 \, mM/\text{min}$$

$$K_{eq}^{MDH} = 10$$

$$K_{m,Mal}^{MDH} = 0.3 \, mM$$

$$K_{m,OAt}^{MDH} = 0.06 \, mM$$

5. Isocitrate dehydrogenase (IDH; figure 9.6, process 5): the kinetics are irreversible Michaelis-Menten:

$$v_{IDH} = \frac{V_m^{IDH} \times ICit \times NADP}{\left(K_{m,ICit}^{IDH} + ICit\right) \times \left(1 + GlOx/K_{i,GlOx}^{IDH}\right) \times \left(1 + OA/K_{i,OA}^{IDH}\right)}$$

with kinetic constants taken from an experiment in *Mycobacterium phlei* in which IDH was purified and characterised (Dhariwal and Venkitasubramanian 1987). The parameters were estimated as $K_{m,ICit}^{IDH} = 0.07 \, mM$, $V_m^{IDH} = 14 \, mM/\text{min}$. IDH was inhibited by glyoxylate (GlOx) and oxaloacetate (OA). Inhibition was noncompetitive and the estimated inhibition constants were $K_{i,GlOx}^{IDH} = 0.5 \, mM$, $K_{i,OA}^{IDH} = 0.5 \, mM$.

6. AcCoA production (figure 9.6, process 6): The rate law was represented as an irreversible equation according to the mass action law:

$$v_{CoArecycling} = k_{CoArecycling} \times CoA$$

where

$$k_{CoArecycling} = 10 \, \text{min}^{-1}$$

7. Conversion of succinate to malate (figure 9.6, process 7): the overall reaction corresponds to processes catalyzed by succinate dehydrogenase and fumarase using reversible Michaelis-Menten kinetics:

$$v_{SDH} = \frac{V_m^{SDH} \times \left(Suc - Mal/K_{eq}^{SDH}\right)}{K_{m,Suc}^{SDH} \times \left(1 + Mal/K_{m,Mal}^{MDH}\right) + Suc}$$

with the following kinetic parameters:

$$V_m^{SDH} = 20\,mM/min$$

$$K_{eq}^{SDH} = 1$$

$$K_{m,Mal}^{SDH} = 0.1\,mM$$

$$K_{m,Suc}^{SDH} = 0.2\,mM$$

8. Succinate utilization (figure 9.6, process 8): again, we could not find kinetic information and the irreversible equation according to mass action has been used:

$$v_{Sucefflux} = k_{Sucefflux} \times Suc$$

where

$$k_{Sucefflux} = 5\,min^{-1}$$

## Application of the Model to Identify Potential Targets for Therapeutic Drug Intervention

We have analysed the model using bifurcation diagrams. The most interesting results come when examining the behaviours for simultaneous inhibitions of ICL and MS. For these dual inhibitions, the model predicts stable and unstable steady states, multiple steady states, and even has oscillatory regimes (figure 9.7). One conclusion is that the intuitive assumption that the ICL and MS pathway have a linear response is just not true since simultaneous inhibition will result in ICL shunt shutdown. The glyoxylate pathway is inherently not linear from the nonlinear response to the multiple interventions. Effective inhibition is in the range of greater than 50 percent for ICL and over 70 percent for MS; otherwise, the pathway still can be used by the cell for fatty acid utilisation. The pathway could be shut down normally after a 30 percent ICL inhibition, but the functionality could be restored by a simultaneous 50 percent MS inhibition.

The described approach shows that mathematical modelling can be used to generate viable hypotheses about possible intervention strategies. Models of previously studied organisms (*E. coli*) have been used as a starting point for model development in other organisms (in this case TB).

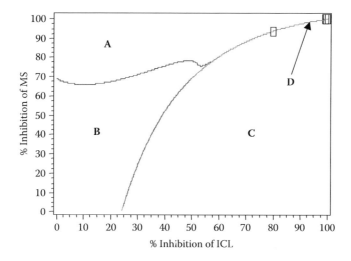

FIGURE 9.7 Multiple sites intervention analysis. Bifurcation diagram has four regions: A, C are nonphysiological regions; B is a physiological region; D is an oscillatory (nonphysiological) region. Effective inhibition is in the range of more than 50 percent ICL and more than 70 percent MS; otherwise, the pathway will be still in a physiological state. The pathway can still exhibit normal functioning after ICL inhibition (~30 percent) by simultaneous MS inhibition (50 percent).

One advantage of starting with an existing model, as was done here, is that it can be used to design a rational experimental program to develop an accurate model for the new target organism. In so doing, and in analysing the resulting experimental data, the model acts as a tool for a fine-grained comparison of the differences in the pathways of the two organisms.

## APPLICATION OF THE KINETIC MODEL OF *ESCHERICHIA COLI* BRANCHED-CHAIN AMINO ACID BIOSYNTHESIS TO OPTIMISE PRODUCTION OF ISOLEUCINE AND VALINE

Valine and isoleucine are branched-chain amino acids that are widely used in biomedicine and biotechnology. Indeed, valine is used in the food industry for flavour additive production. Moreover, this amino acid is one of the sources of cephamycin antibiotic biosynthesis. However, pathways of valine and isoleucine biosynthesis are strongly coupled in *E. coli* because four of the five steps in the valine biosynthesis pathway are catalyzed by enzymes also involved in the isoleucine biosynthesis pathway. This strong coupling between biosynthetic pathways raises the question of whether it is possible to introduce genetic modifications into some chosen strain so as to increase valine production and to decrease production of isoleucine, which possibly contaminates valine production processes in industry. In chapter 7 we developed a kinetic model of the biosynthesis of valine and isoleucine in *E. coli*. In this section we apply this model to predict two strain improvement strategies: the first one enables us to maximise valine yield and to decrease isoleucine production and the second strategy predicts genetic modifications which, on the contrary, should provide an increase in isoleucine production and a decreased valine yield. This means that the aim of the section is merely to demonstrate how to apply kinetic modelling to develop a strategy for improving the production of the useful metabolite.

## Prediction of Possible Genetic Changes That Should Maximise Isoleucine and Valine Production

Modern biotechnology possesses powerful methods to accomplish different genetic modifications of bacterial genomes (mutations, knockouts, amplifications of genes, as well as addition of genes from other organisms to the host genotype). Genetic changes of this type allow us to modify significantly the metabolic system of the cell by means of changes in the activity of certain enzymes. Engineering of new bacterial strains is usually directed at increased synthesis of some products by bacterial cells. However, until recently, the effect of genetic changes performed was sometimes unpredictable. Thus, for example, an incorporated genetic change can lead to a lower improvement than the expected one or even have negative effects on the production rate of the target metabolite. Certain genetic changes seem positive when tested individually but when combined with other improvements they are frequently no longer positive. The difficulties experienced in the prediction of the effects of genetic changes most probably arise from the high complexity of regulation of the corresponding metabolic pathway. In these cases, it seems appropriate to use mathematical models of the corresponding metabolic pathways and to make a computational simulation of the effect of genetic change. We propose the application of kinetic models to the prediction of the effect of genetic changes when new bacterial strains are constructed.

The pathway of branched-chain amino acid biosynthesis is one of these complicated pathways which are subject to rather complex system regulation, including positive and negative feedbacks. The activity of the pathway also depends on the intermediates of linked metabolic pathways such as glycolysis, the Krebs cycle, biosynthesis of certain amino acids, etc. We have tried to use the developed kinetic model of branched-chain amino acid biosynthesis to simulate the effect of possible genetic changes on the production of one of the pathway metabolites. We chose valine and isoleucine as target metabolites and tried to optimise their production in *E. coli* cells.

To simulate the effect of various genetic changes on the specific amino acid production we assume that:

Knockout of a gene means that the corresponding enzyme concentration is zero (or $V_m = 0$).

Mutation in the promoter of a gene means a change (a decrease) in concentration (or $V_m$) of the corresponding enzyme.

Gene amplification means an increase of enzyme concentration (or increase of $V_m$ of the corresponding reaction).

One of the main aims of the model development was to examine two strain improvement strategies: the first one enables us to maximise valine yield and to decrease isoleucine production and the second strategy should predict genetic modifications that, on the contrary, provide increased isoleucine production and decreased valine yield.

The first step was to study the changes in steady-state concentrations of these amino acids in response to changes of pathway enzyme concentrations (parameters of $V_m$ in the model). We assumed that every enzyme could be amplified no more than twenty times. So, to simulate the enzyme amplification we increased the parameters of maximal rates of each enzyme involved in amino acid formation by twenty times (see figure 9.8). Using this approach we could choose the most sensitive stage, which for both amino acids was the stage of NADPH recycling (figure 9.8a, figure 9.8b).

We found that the first step in strain improvement should be amplification of enzymes catalyzing the formation of NADPH involved, for example, in the Krebs cycle. Amplification of NADPH recycling results

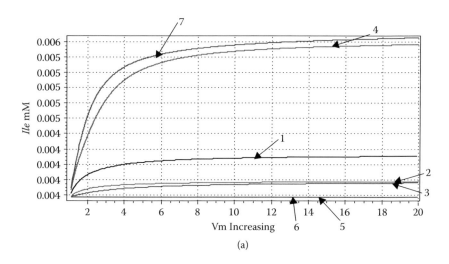

(a)

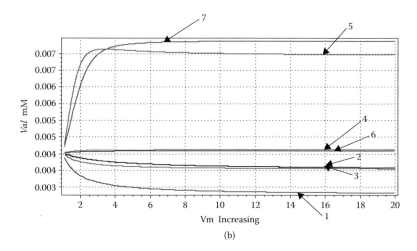

(b)

FIGURE 9.8 Dependence of steady-state concentrations of valine (a) and isoleucine (b) on maximal rates of TDH (line 1), AHAS I (line 2), AHAS III (line 3), IR (line 4), DHAD (line 5), BCAT (line 6) and NADP recycling (line 7).

in a new steady-state level, which is sensitive to changes in concentration of the other enzymes. We studied step-by-step the changes in isoleucine and valine steady-state concentrations resulting from amplifications of the other enzymes and found that amplifications of NADPH recycling enzymes (step 2 in figure 9.9), DHAD (step 3 in figure 9.9), BCAT (step 4 in figure 9.9) and IR (step 5 in figure 9.9) led to maximisation of valine yield and decrease in isoleucine production. On the other hand, the kinetic

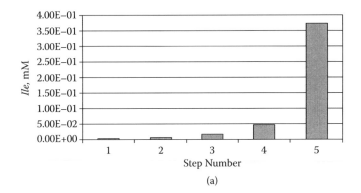

(a)

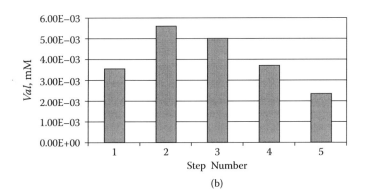

(b)

FIGURE 9.9 Optimization of valine production. Changes in valine (a) and isoleucine (b) steady-state concentrations resulting from different levels of enzyme amplification: 1) the base level; 2) NADPH recycling; 3) DHAD; 4) BCAT; 5) IR.

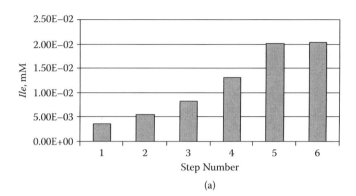

(a)

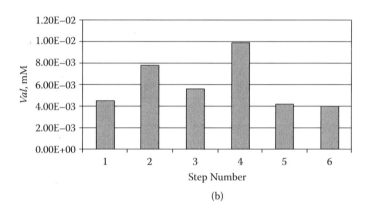

(b)

FIGURE 9.10 Optimization of isoleucine production. Changes in isoleucine (a) and valine (b) steady-state concentrations resulting from different levels of enzyme amplification: 1) the base level; 2) NADPH recycling; 3) TDH; 4) DHAD; 5) IR; 6) AHAS III.

model predicts that amplification of NADPH recycling enzymes (step 2 in figure 9.10), TDH (step 3 in figure 9.10), DHAD (step 4 in figure 9.10), IR (step 5 in figure 9.10) and AHAS III (step 6 in figure 9.10) results in increased isoleucine production and decreased valine yield.

# Conclusion and Discussion

As one can see from this book, kinetic modelling (KM) has been applied to the quantitative description of intracellular interactions and function of organelles. At the same time, the approach has both advantages and disadvantages. Indeed, critics of KM often use two arguments.

Firstly, to identify parameters for a complete biological system, one needs to add reactions one by one, connect them by parameterized equations, and know exact values of all parameters. Furthermore, these connections are subject to change because different reactions can act on the same substrates, under different conditions and localizations. These result in many different equations which all need to be solved, and this may not be practicable.

On the contrary, in this book we have presented a detailed approach showing how to create step-by-step models of enzymes, pathways, organelles, etc., and interconnect them and how to take into account different conditions (pH, temperature). We explain how to make proper use of time and space resolutions, including equilibrium, steady-state and quasi-steady-state assumptions.

The processes of simulation and parameter identification for systems of tens or hundreds of thousands of differential equations are no longer a problem, given our access to high-performance computers with tera- and, in the near future, peta-flop capabilities.

Another criticism is that KM is sensitive to errors in concentrations and parameters. While this is true, we can now calculate global parameter sensitivity, estimate errors in parameters, search for global minima and identify key parameters, so this is no longer a problem. All modelling methods are more or less sensitive to unknowns, but only KM is able to provide a framework in which to explore these unknowns in cases both of unknown interactions or connections and uncertainties in parameter values. KM models can tolerate errors in parameters or concentrations.

This book presents a number of mechanistic models developed using the kinetic modelling approach. These models have a high predictive power and allow analysis and interpretation of multilevel experimental data. However, the applicability of our approach to construction, verification, and study of models has some limitations. Actually, our models are suitable for description of variations in the concentrations of intracellular entities only in the case of systems in which no diffusion gradients exist. Moreover, the application of the law of mass action to describe the kinetics of functioning of enzyme catalytic cycles implies that the number of molecules of a compound involved in the model must be sufficiently high. Otherwise, the law of mass action, which is based on statistical relationships, will be invalid. This means that models of this type are not always applicable to the description of dynamics of biochemical processes occurring inside a single cell. For example, the interaction of a repressor protein with operator within a single cell cannot be described by the mass action law, since the cell contains only a few operators and only a few dozen repressor protein molecules. In order to avoid such ambiguity, we use our models only for description of the dynamics of intracellular processes that occur in all the cells of a cellular assemblage. Therefore, we define the concentration of a compound as the total number of molecules of this compound in all the cells of assemblage divided by the total volume of all these cells. A similar procedure can be applied to modelling at the level of tissue, organ or organism where the number of interacting entities is large enough. Indeed, PK/PD modelling based on similar approximations is being successfully used in pharmaceutical sciences.

In this book, we have not discussed whole organ and organism modelling. We think that whole organism modelling is perfectly feasible and has just started to become a reality. Indeed, the combination of "bottom up" and "top down" approaches looks very promising for whole organism and human modelling. Recently, at an International Forum in Tokyo, the following declaration was made:

> Recent advances in Systems Biology indicate that the time is now ripe to initiate a grand challenge project to create over the next thirty years a comprehensive, molecules-based, multi-scale, computational model of the human ('the virtual human'), capable of simulating and predicting, with a reasonable degree of accuracy, the consequences of most of the perturbations that are relevant to healthcare.

In spite of criticism, we believe that this approach has a future, especially in its application to biotechnology and biomedicine. It is now possible to integrate data from three different scientific disciplines or domains: chemical, biological and medical.

This is why the scientific community is putting a lot of effort into integrated systems approaches to pharmaceutical sciences and the drug research and development process. The KM approach can help in knowledge management of large variable data sets, identification of knowledge gaps, hypothesis generation, mimicking of different therapeutic, environmental, physiological and genetic conditions, and mechanistic understanding of microdosing and combination therapies. Similarly, KM gives insights into drug safety and management of adverse effects, chronotherapy and individual dose prediction, identification of points of intervention (targets) by interrogating mode, and identification of biomarkers for disease progression.

KM can also contribute to our understanding of compound mechanisms of action (MOA) by building compound kinetic models and integrating them with disease kinetic models, understanding compound safety and efficacy, identification of toxicity and efficacy biomarkers, selection of lead compounds by analyzing modelling results, understanding of the pharmacogenomics of patient populations by building mechanism variation models to include population and other differences, and identification of biomarkers for responders and nonresponders.

Finally, KM will enable cheaper and faster methods that are complementary to *in vitro*, *ex vivo* and *in vivo* experiments or animal models. It is synergistic to the European Union's 3Rs approach (replacement, refinement and reduction) to management of animal experiments. Recently, Lee Hood, of the Institute for Systems Biology in Seattle, proposed a new P4 medicine initiative (predictive, personalized, preventive, participatory), and systems biology can help to achieve this goal. We are sure that the KM approach will play a fundamental role in all of these new initiatives by contributing to our understanding of causal mechanisms of disease progression.

In bioengineering, which has now given birth to the new emerging discipline of synthetic biology, the main area of progress is in quantitative studies of biological processes as whole systems. These enable prediction of the consequences of system perturbations; the application of cellular bioengineering to address specific challenges; the prediction of cellular functions; the rational design of pathways, cells, biomarkers and organisms; and the optimization of biocatalysis and fermentation processes for production of vaccines, recombinant proteins, drug precursors, and food production.

The impact of the kinetic modelling approach in the energy field has already been demonstrated and will undoubtedly become more evident through advances in biofuel and biomass production (photosynthesis and growth), application of solar energy for hydrogen generation (green micro-algae, cyanobacteria), and development of microbial fuel cells (*Geobacter*) and enzymatic (glucose oxidase) fuel cells for electricity generation.

Similar impacts will soon be seen in advances in water treatment technology and waste management and utilization through optimization, and improved monitoring and sensing of microbial communities. Already developed areas will also benefit by integration with the KM approach; e.g., food chain modelling, population dynamics, prediction and intervention strategies of climate change, predictions of temperature impact on ecosystems, and the effects of land-use change and deforestation.

# References

## CHAPTER 1

Kitano, H. (2007) A robustness-based approach to systems-oriented drug design. *Nature Reviews Drug Discovery* 6:202–210.

Meyer, E. (1997) The first years of the Protein Data Bank. *Protein Science* 6:1591–97.

Michaelis, L., and M. Menten. (1913) Die kinetik der invertinwirkung. *Biochemische Zeitschrift* 49:333–369.

Philosophical Transactions of the Royal Society of London (2008) Available from http://www.journals.royalsoc.ac.uk/

Reich, J. G., and E. E. Selkov. (1981) *Energy metabolism of the cell. A theoretical treatise.* London: Academic Press.

van Leewenhoek, A. (1979) *Alle de Brieven van Antoni van Leeuwenhoek: The Collected Letters of Antoni van Leeuwenhoek. Edited by L. C. Palm.* Lisse: Swets & Zeitlinger.

Watson, J. D., and F. H. C. Crick. (1953) A structure for deoxyribose nucleic acid. *Nature* 171:737–738.

Westerhoff, H. V. Ed. (2003) *Metabolic engineering in the post-genomic era.* U.K.: Horizon Bioscience, Wymondhan, Norfolk.

Westerhoff, H. V., and B. O. Palsson. (2004) The evolution of molecular biology into systems biology. *Nature Biotechnology* 22:1249–52.

*Wikipedia: The free encyclopedia* (2008)[cited]. Available from http://wikipedia.org/.

WikiWikiWeb: The first wiki (2008)[cited]. Available from http://www.wiki.org/wiki.cgi.

## CHAPTER 2

Almaas, L., et al. (2004) Global organisation of metabolic fluxes in the bacterium *Escherichia coli. Nature* 427:839–43.

Ashburner, M., et al. (2000) Gene ontology: Tool for the unification of biology. The Gene Ontology Consortium. *Nature Genetics* 25(1):25–29.

Biological Pathway Data Exchange format (2008) Available from http://www.biopax.org

Brooksbank, C., et al. (2005) The European Bioinformatics Institute's data resources: Towards systems biology. *Nucleic Acids Research* 33 (Database issue), D46–53.

Burgard, A. P., et al. (2004) Flux coupling analysis of genome scale metabolic network reconstructions. *Genome Research* 14:301–312.

Chvatal, V. (1983) *Linear programming.* W.H. Freeman.& Co, New York.

Covert, M. W., et al. (2004) *Nature* 429:92–96.

EMPProject (2008). Available from http://www.empproject.com

Famili, I., and B. O. Palsson. (2003) The convex basis of the left null space of the stoichiometric matrix leads to the definition of metabolically meaningful pools. *Biophysical Journal* 85:16–26.

Goto, S., et al. (2002) LIGAND: database of chemical compounds and reactions in biological pathways. *Nucleic Acids Research* 30(1):402–404.

Hucka, M., et al. (2003) The Systems Biology Markup Language (SBML): A medium for representation and exchange of biochemical network models. *Bioinformatics* 19(4):524–531.

International Union of Biochemistry and Molecular Biology. IUBMB nomenclature (2008). Available from http://www.chem.qmul.ac.uk/iubmb/.

Kalir, S., and U. Alon. (2004) Using a quantitative blueprint to reprogram the dynamics of the flagella gene network. *Cell* 117:713–720.

KEGG: Kyoto Encyclopedia of Genes and Genomes. Available from http://www.genome.jp/kegg/

Kitano, H., et al. (2005) Using process diagrams for the graphical representation of biological networks. *Nature Biotechology* 23(8):961–966.

Klamt, S., and E. D. Gilles. (2004) Related articles, links minimal cut sets in biochemical reaction networks. *Bioinformatics* 22;20(2):226–234.

Klipp, E., et al. (2007) Systems biology standards—The community speaks. *Nature Biotechnology* 25(4):390–391.

Latterich, M. (2005) Molecular systems biology at the crossroads: To know less about more, or to know more about less? *Proteome Science* 3:8.

Le Novère et al. (2005) Minimum information requested in the annotation of biochemical models (MIRIAM). *Nature biotechnology* 23(12):1509–1515.

Ma, H., et al. (2007) The Edinburgh human metabolic network reconstruction and its functional analysis. *Molecular Systems Biology* 3:135.

Maglott, D., et al. (2007) Entrez gene: Gene-centered information at NCBI. *Nucleic Acids Research* 35:D26–31.

Mazein, A., et al. (2008) Multiple categories for visualisation of biomedical knowledge. *BMC submitted.*

MetaCyc (2008). Available from http://www.metacyc.org/

Nikolaev, E. V., Burgard, A. P., and Maranas, C. D. (2005) Elucidation and structural analysis of conserved pools for genome-scale metabolic reconstructions. *Biophysical Journal.* 88:37–49.

Papin, J., et al. (2004) Comparison of network-based pathway analysis methods. *Cell* 117:689–690.

Schilling, C. H., et al. (2000) Theory for the systemic definition of metabolic pathways and their use in interpreting metabolic function from a pathway-oriented perspective. *Journal of Theoretical Biology* 203:229–248.

Schuster, S., et al. (1999) Detection of elementary flux modes in biochemical networks: A promising tool for pathway analysis and metabolic engineering. *Trends in Biotechnology* 17:53–60.

Schuster, S., et al. (2002) Exploring the pathway structure of metabolism: Decomposition into subnetworks and application to *Mycoplasma pneumoniae*. *Bioinformatics* 18:351–361.

SEED (2008) Available from http://theseed.uchicago.edu/FIG/

Selkov, E., et al. (1996) The metabolic pathway collection from EMP: The enzymes and metabolic pathways database. *Nucleic Acids Research* 24:26–28.

Sorokin, A. A., et al. (2006) The Pathway Editor: A tool for managing complex biological networks. *IBM Journal of Research and Development* 50(6):561–574.

Wagner, A., and D. A. Fell. (2001) The small world inside large metabolic networks. *Proceedings of the Royal Society of London - Series B: Biological Sciences* 268(1478):1803–1810.

Wheeler, D. L., et al. (2006) Database resources of the National Center for Biotechnology Information. *Nucleic Acids Research.* 34(Database issue):D173–180.

Wu, C. H., et al. (2006) The Universal Protein Resource (UniProt): An expanding universe of protein information. *Nucleic Acids Research* 34:D187–191.

## CHAPTER 3

Apache Software Foundation. Apache Derby (2008)[cited]. Available from http://db.apache.org/derby/.

Ariadne Genomics. PathwayStudio (2008)[cited]. Available from http://www.ariadnegenomics.com/products/pscentral/.

Arthorne, J., and C. Laffra. (2004) *Official Eclipse 3.0 FAQs*. New York: Addison-Wesley.

Benson, D., I. Karsch-Mizrachi, D. Lipman, J. Ostell, B. Rapp, and D. Wheeler. (2002) GenBank. *Nucleic Acids Research* 30:17–20.

BioCarta. BioCarta—Charting pathways of life (2008)[cited]. Available from http://www.biocarta.com.

Biocomputation Group, University of Pennsylvania. Bio Sketch Pad (2008)[cited]. Available from http://www.cis.upenn.edu/biocomp/new_html/biosketch.php3.

Biosoft. BioUML framework for systems biology (2008)[cited]. Available from http://www.biouml.org/.

Cassman, M. (2005) Barriers to progress in systems biology. *Nature* 438(7071):1079.

Cook, D., J. Farley, and S. Tapscott. (2001) A basis for a visual language for describing, archiving and analyzing functional models of complex biological systems. *Genome Biology* 2(4):12.1–12.10.

Demir, E., O. Babur, U. Dogrusoz, A. Gursoy, G. Nisanci, R. Cetin-Atalay, and M. Ozturk. (2002) PATIKA: An integrated visual environment for collaborative construction and analysis of cellular pathways. *Bioinformatics* 18(7):996–1003.

Eclipse (2008) Available from http://www.eclipse.org

Edinburgh Pathway Editor (2008). Available from http://www.bioinformatics.ed.ac.uk/epe/

Holford, M., N. Li, P. Nadkarni, and H. Zhao. (2005) VitaPad: Visualization tools for the analysis of pathway data. *Bioinformatics* 21(8):1596–1602.

InNetics. PathwayLab (2008)[cited]. Available from http://innetics.com/pathwaylab_overview.htm.

Institute for Systems Biology. BioTapestry (2008)[cited]. Available from http://labs.systemsbiology.net/bolouri/software/BioTapestry/.

Keck Graduate Institute. Keck computational systems biology (2008)[cited]. Available from http://www.sys-bio.org/.

Kitano, H., A. Funahashi, Y. Matsuoka, and K. Oda. (2005) using process diagrams for the graphical representation of biological networks. *Nature Biotechnology* 23(8):961–66.

Kohn, K. (1999) Molecular interaction map of the mammalian cell cycle control and DNA repair systems. *Molecular Biology of the Cell* 10(8):2703–34.

Kohn, K. (2001) Molecular interaction maps as information organizers and simulation guides. *Chaos* 11(1):84–97.

Ma, H., and A. Zeng. (2003) Reconstruction of metabolic networks from genome data and analysis of their global structure for various organisms. *Bioinformatics* 19(2):270–77.

MySQL AB. MySQL: The world's most popular open source database (2008)[cited]. Available from http://www.mysql.com/.

NetBuilder Development Team. NetBuilder home (2008)[cited]. Available from http://strc.herts.ac.uk/bio/maria/NetBuilder/.

Nomenclature Committee of the International Union of Biochemistry and Molecular Biology. Enzyme Nnomenclature (2008)[cited]. Available from http://www.chem.qmul.ac.uk/iubmb/enzyme/.

Ogata, H., S. Goto, K. Sato, W. Fujibuchi, H. Bono, and M. Kanehisa. (1999) KEGG: Kyoto encyclopedia of genes and genomes. *Nucleic Acids Research* 27(1):29–34.

Protein Lounge. Pathway Builder tool (2008)[cited]. Available from http://www.proteinlounge.com/pathwaybuilder.asp.

Research Collaboratory for Structural Bioinformatics. RCSB Protein Data Bank (2008)[cited]. Available from http://www.rcsb.org.

Sorokin, A., K. Paliy, A. Selkov, O. V. Demin, S. Dronov, P. Ghazal, and I. Goryanin. (2006) The Pathway Editor: A tool for managing complex biological networks. *IBM Journal of Research and Development* 50(6):561–573.

Teranode Corporation. Design automation for life sciences (2008)[cited]. Available from http://www.teranode.com/.

The Systems Biology Institute. CellDesigner: A modeling tool of biochemical networks (2008)[cited]. Available from http://celldesigner.org/.

U.S. National Institutes of Health. PubMed Central (2008)[cited]. Available from http://www.pubmedcentral.nih.gov.

Wu, C., H. Huang, L. Arminski, J. Castro-Alvear, Y. Chen, Z. Hu, R. Ledley, K. Lewis, H. Mewes, B. Orcutt, B. Suzek, A. Tsugita, C. Vinayaka, L. Yeh, J. Zhang, and W. Barker. (2002) The Protein Information Resource: An integrated public resource of functional annotation of proteins. *Nucleic Acids Research* 30(1):35–37.

## CHAPTER 4

Bakker, B. M., et al. Glycolysis in bloodstream form Trypanosoma brucei can be understood in terms of the kinetics of the glycolytic enzymes. (1997) *Journal of Biological Chemistry* 272(6):3207.

Buchholz, A., et al. Metabolomics: quantification of intracellular metabolite dynamics. (2002) *Biomolecular Engineering* 19:5.

Buchholz, A., R. Takors, and C. Wandrey. Quantification of intracellular metabolites in Escherichia coli K12 using liquid chromatographic-electrospray ionization tandem mass spectrometric techniques. (2001) *Analytical Biochemistry* 295:129.

Chassagnole, C., et al. An integrated study of threonine-pathway enzyme kinetics in Escherichia coli. (2001a) *Biochemical Journal* 356:415.

Chassagnole, C., et al. Control of the threonine-synthesis pathway in Escherichia coli: a theoretical and experimental approach. (2001b) *Biochemical Journal* 356:433.

Cornish-Bouden, A. (2001) *Fundamentals of enzyme kinetics*. Portland Press, Cambridge.

Demin, O. V., et al. (2004) Kinetic modelling as a modern technology to explore and modify living cells. In *Modelling in molecular biology*, edited by G. Ciobanu and G. Rozenberg. Springer, Berlin.

Gizzatkulov, N., et al. (2004) DBsolve7: New update version to develop and analyze models of complex biological systems. Paper read at ISMB/ECCB Conference, 31 July–5 August, Glasgow, Scotland.

Goryanin I, Hodgman TC, Selkov E. Mathematical simulation and analysis of cellular metabolism and regulation. (1999) Bioinformatics;15(9):749–58.

Kholodenko, B. N., et al. Quantification of short term signaling by the epidermal growth factor receptor. (1999) *Journal of Biological Chemistry* 274(42):30169.

Lebedeva, G. V., et al. Kinetic model of photosystem II of high plants. (2000) *Journal of Physical Chemistry (Moscow)* 74:1897.

Leung, H. B., and V. L. Schramm. Adenylate degradation in Escherichia coli. The role of AMP nucleosidase and properties of the purified enzyme. (1980) *Journal of Biological Chemistry* 255(22):10867.

Markevich, N. I., et al. Signal processing at the Ras circuit: what shapes Ras activation patterns? (2004) *Systems Biology (Stevenage)* 1(1):104.

Moehren, G., et al. Temperature dependence of the epidermal growth factor receptor signaling network can be accounted for by a kinetic model. (2002) *Biochemistry* 41:306.

Neidhardt, F. C. (1987) Escherichia coli *and* Salmonella typhimurium: *Cellular and molecular biology.* Washington, D.C.: American Society for Microbiology.

Noble, M., et al. The kinetic model of the shikimate pathway as a tool to optimize enzyme assays for high-throughput screening. (2006) *Biotechnology and Bioengineering* 95:560.

Rais, B., et al. Threonine synthesis from aspartate in Escherichia coli cell-free extracts: pathway dynamics. (2001) *Biochemical Journal* 356:425.

Riznichenko, G. Yu., et al. Regulatory levels of photosynthetic processes. (2000) *Biophysics (Moscow)* 45:452.

Schaefer, U., Boos W, Takors R, Weuster-Botz D. Automated sampling device for monitoring intracellular metabolite dynamics. (1999) *Analytical Biochemistry* 270:88.

Selkov, E., et al. The metabolic pathway collection from EMP: the enzymes and metabolic pathways database. (1996) *Nucleic Acids Research* 24:26.

Schomburg, I., A. Chang, and D. Schomburg. BRENDA, enzyme data and metabolic information. (2002) *Nucleic Acids Research* 30:47.

Teusink, B., B. M. Bakker, and H. V. Westerhoff. Control of frequency and amplitudes is shared by all enzymes in three models for yeast glycolytic oscillations. (1996) *Biochimica et Biophysia Acta* 1275(3):204.

Westerhoff, H. V., and K. Van Dam. (1987) *Thermodynamics and control of biological free-energy transduction.* Amsterdam: Elsevier Science Publishers B.V.

## CHAPTER 5

Fed'kina, V. R., and T. B. Bronnikova. (1995) Complex oscillatory regimes in peroxidase-oxidase reaction. *Biophysics* 40:36–47.

Gear, C. W., and L. R. Petzold. (1984) ODE methods for the solution of differential algebraic systems. *SIAM Journal on Numerical Analysis* 21:716–28.

Gizzatkulov, N., A. Klimov, G. Lebedeva, and O. Demin. (2004) DBsolve7: New update version to develop and analyze models of complex biological systems. Paper read at ISMB/ECCB Conference, 31 July–5 August, Glasgow, Scotland.

Goldstein, B. N., and I. I. Goryanin. (1996) Modeling oscillations of two activities--6-phosphofructo-2-kinase/fructoso-2,6-bisphosphatase in rat liver. *Molecular Biology* 30:976-983.

Goryanin, I. (1996) NetSolve: Integrated development environment software for metabolic and enzymatic systems modeling. In *Biothermokinetics of the living cell,* edited by H. V. Westerhoff, J. L. Snoep, F. E. Sluse, J. E. Wijker, and B. N. Holodenko. Biothermokinetics Press: Amsterdam.

Goryanin, I., T. C. Hodgman, and E. Selkov. (1999) DBsolve: Mathematical simulation and analysis of cellular metabolism and regulation. *Bioinformatics* 15(9):749–58.

Goryanin, I., and K. Serdyuk. (1994) Automation of modelling of multienzyme systems using databanks on enzyme and metabolic pathways (EMP). In *Proceedings of the IMACS Symposium on Mathematical Modelling.* Vienna: Austria.

Hindmarsh, A. C. (1983) A systematized collection of ODE solvers. In *Scientific computing,* edited by R. S. Stepleman, M. Carver, R. Peskin, W. F. Ames, and R. Vichnevetsky. Amsterdam: North Holland.

Hucka, M., A. Finney, H. M. Sauro, and H. Bolouri. (2001) Systems Biology Markup Language (SBML) Level 1: Structures and facilities for basic model definitions [cited]. Available from http://www.sbml.org (Date cited: 2008).

ISBSPb - Institute for Systems Biology SPb (2008). Available from http://www. insysbio.ru/

Marquardt, D. W. (1963) An algorithm for least squares re estimation of non linear parameters. *SIAM Journal* 11:431–41.

## CHAPTER 6

Adams, E.; L-Histidinal, A Biosynthetic Precursor Of Histidine J. Biol. Chem.; 217, 325 (1955).

Adams. E.. Synthesis And Properties Of An.-Amino Aldehyde, Histidinal *J. Biol. Chem.* 21'7. 317 (1955).

Alberty, R. A. (1992). Equilibrium calculations on systems of biochemical reactions at specified pH and pMg. *Biophysical Chemistry* 42:117–31.

Ausat, I., G. L. Bras, and J.-R. Garel. (1997) Allosteric activation increases the maximum velocity of *E. coli* phosphofructokinase. *Journal of Molecular Biology* 267:476–80.

Babul, J. (1978) Phosphofructokinases from *Escherichia coli. Journal of Biological Chemistry* 253:4350–55.

Berger, S. A., and P. R. Evans. (1991) Steady-state fluorescence of *Escherichia coli* phosphofructokinase reveals a regulatory role for ATP. *Biochemistry* 30:8477–80.

Blangy, D., H. Buc, and J. Monod. (1968) Kinetic of the allosteric interactions of phosphofructokinase from *Escherichia coli. Journal of Molecular Biology* 31:13–35.

Brandolin, G., J. Doussiere, A. Gulik, T. Gulik-Krzywicki, G. J. M. Lauquin, and P. V. Vignais. (1980) Kinetic, binding and ultrastructural properties of the beef heart adenine nucleotide carrier protein after incorporation into phospholipid vesicles. *Biochimica et Biophysica Acta* 592:592–614.

Burstein, C., M. Cohn, A. Kepes, and J. Monod. (1965) Role of lactose and its metabolic products in the introduction the lactose operon in *Escherichia coli. Biochimica et Biophysica Acta* 95:634–39.

Campos, G., V. Guixe, and J. Babul. (1984) Kinetic mechanism of phosphofructokinase-2 from *Escherichia coli.* A mutant enzyme with defferent mechanism. *Journal of Biological Chemistry* 259:6147–52.

Chassagnole, C., N. Noisommit-Rizzi, J. W. Schmid, K. Mauch, and M. Reuss. (2002) Dynamic modeling of the central carbon metabolism of *Escherichia coli. Biotechnology and Bioengineering* 79:53–73.

Chittur, S. V., T. J. Klem, C. M. Shafer, and V. J. Davisson. Mechanism for acivicin inactivation of triad glutamine amidotransferases (2001) *Biochemistry* 40:876–87.

Cleland, W. W. (1963) The kinetics of enzyme-catalyzed reactions with two or more substrates or products. *Biochimica et Biophysica Acta* 67:104–37.

Cornish-Bowden, A. (2001) *Fundamentals of enzyme kinetics.* London: Portland Press.

Davis, E. J., and W. I. A. Davies-Van Thienen. (1984) Rate control of phosphorylation-coupled respiration by rat liver mitochondria. *Archives of Biochemistry and Biophysics* 233:573–81.

Dean, A. M., and D. E. Koshland Kinetic mechanism of Escherichia coli isocitrate dehydrogenase (1993) *Biochemistry* 32:9302.

Demin, O. V., S. Dronov, I. I. Goryanin, and G. V. Lebedeva. (2004a) Kinetic model of imidazole glycerol phosphate synthetase of *Escherichia coli. Biochemistry (Moscow)* 69(12):1625–38.

Demin, O. V., T. Y. Plyusnina, G. V. Lebedeva, E. A. Zobova, E. A. Metelkin, A. G. Kolupaev, I. I. Goryanin, and F. Tobin. (2005) Kinetic modelling of the *E. coli* metabolism. In *Topics in current genetics,* edited by L. Alberghina and H. V. Westerhoff. Springer, Berlin.

Deville-Bonne, D., R. Laine, and J.-R. Garel. (1991) Substrate antagonism in the kinetic mechanism of *E. coli* phosphofructokinase-1. *FEBS Letters* 290:173–76.

Ewings, K. N., and H. W. Doelle. (1980) Further kinetic characterization of the non-allosteric phosphofructokinase from *Escherichia coli* K-12. *Biochimica et Biophysica Acta* 615:103–12.

Frieden, C., and R. F. Colman. (1967) Glutamate dehydrogenase concentration as a determinant in the effect of purine nucleotides on enzyme activity. *Journal of Biological Chemistry* 242:4045–53.

Froschauer, E. M., M. Kolisek, F. Dieterich, M. Schweigel, and R. J. Schweyen. (2004) Fluorescence measurements of free $[Mg^{2+}]$ by use of mag-fura 2 in *Salmonella enterica. FEMS Microbiology Letters* 237:49–55.

Guixe, V., and J. Babul. (1985) Effect of ATP on phosphofructokinase-2 from *Escherichia coli. Journal of Biological Chemistry* 260:11001–5.

Hoque, M. A., H. Ushiyama, M. Tomita, and K. Shimizu. (2005) Dynamic responses of the intracellular metabolite concentrations of the wild type and *pykA* mutant *Escherichia coli* against pulse addition of glucose or $NH_3$ under those limiting continuous cultures. *Biochemical Engineering Journal* 26:38–49.

Huber, R. E., M. T. Gaunt, and K. L. Hurlburt. (1984) Binding and reactivity at the 'glucose' site of galactosyl-beta-galactosidase (*Escherichia coli*). *Archives of Biochemistry and Biophysics* 234(1):151–60.

Huber, R. E., G. Kurz, and K. Wallenfels. (1976) A quantitation of the factors which affect the hydrolase and transgalactosylase activities of beta-galactosidase (*E. coli*) on lactose. *Biochemistry* 15(9):1994–2001.

Ivanicky, G. R., V. I. Krinsky, and E. E. Sel'kov. (1978) *Mathematical biophysics of the cell.* Moscow: Nauka.

Jobe, A., and S. Bourgeois. (1972) lac Repressor-operator interaction. VI. The natural inducer of the lac operon. *Journal of Molecular Biology* 69:397–408.

Jurgens, C., A. Strom, D. Wegener, S. Hettwer, M. Wilmanns, and R. Sterner. (2000) Directed evolution of a (beta alpha)8-barrel enzyme to catalyze related reactions in two different metabolic pathways. *Proceedings of the National Academy of Sciences USA* 97:9925–30.

Kholodenko, B. N. (1984a) Control of mitochondrial oxidative phosphorylation. *Journal of Theoretical Biology* 107:179–88.

Kholodenko, B. N. (1984b) The mitochondrial carrier of adenylates controls ATP production in the physiological range of respiration rates. *Biofizika (Moscow)* 29:453–58.

Kholodenko, B. N. (1988) Stabilizing regulation in polyenzime systems: Bioenergetic processes modelling. Ph.D. diss., Moscow State University, Moscow.

Klem, T. J., and V. J. Davisson. Imidazole glycerol phosphate synthase: the glutamine amidotransferase in histidine biosynthesis (1993) *Biochemistry* 32:5177–86.

Klingenberg, M. (1980) The ADP-ATP translocation in mitochondria, a membrane potential controlled transport. *Journal of Membrane Biology* 56:97–105.

Klingenberg, M., and H. Rottenberg. (1977) Relation between the gradient of the ATP/ADP ratio and the membrane potential across the mitochondrial membrane. *European Journal of Biochemistry* 73:125–30.

Kornberg, H. L., and H. A. Krebs. Synthesis of cell constituents from C2-units by a modified tricarboxylic acid cycle.(1957) *Nature* 179:988.

Koshland, D. E., G. Nemethy, and D. Filmer. (1966) Comparison of experimental binding data and theoretical models in proteins containing subunits. *Biochemistry* 5:365–84.

Kotlarz, D., and H. Buc. (1982) Phosphofructokinases from *Escherichia coli*. *Methods in Enzymology* 90:60–70.

Kraemer, R., and M. Klingenberg. (1982) Electrophoretic control of reconstituted adenine nucleotide translocation. *Biochemistry* 21:1082–89.

Kurganov, B. I. (1968) Kinetic analysis of dissociative fermentative systems. *Molecular Biology (Moscow)* 2:430–46.

Kurganov, B. I. (1978) *Allosteric enzymes*. Moscow: Nauka.

Loper, J., and E. Adams. (1965) Purification and properties of histidinol dehydrogenase from *Salmonella typhimurium. Journal of Biological Chemistry* 240:788–95.

Lowry, O. H., J. Carter, J. B. Ward, and L. Glaser. The effect of carbon and nitrogen sources on the level of metabolic intermediates in *Escherichia coli* (1971) *Journal of Biological Chemistry* 246:6511.

Metelkin, E., I. Goryanin, and O. Demin. (2006) Mathematical modeling of mitochondrial adenine nucleotide translocase. *Biophysical Journal* 90:423–32.

Metelkin, E., G. Lebedeva, I. Goryanin, and O. Demin. (2008) Kinetic model of beta-galactosidase of *Escherichia coli. Biophysics (Moscow)*.

Miller, S. P., R. Chen, E. J. Karschnia, et al. (2000) Locations of the regulatory sites for isocitrate dehydrogenase kinase/phosphatase. *Journal of Biological Chemistry* 275(2):833.

Mogilevskaya, E. A., G. V. Lebedeva, I. I. Goryanin, and O. V. Demin. (2007) Kinetic model of *Escherichia coli* isocitrate dehydrogenase functioning and regulation. *Biophysics (Moscow)* 52(1):47–56.

Monod, J., J. Wyman, and J. P. Changeux. (1965) On the nature of allosteric transitions: A plausible model. *Journal of Molecular Biology* 12:88–118.

Neidhardt, F. C. (1987) *Cellular & Molecular Biology* 1:3.

Nimmo, H. G. Kinetic mechanism of Escherichia coli isocitrate dehydrogenase and its inhibition by glyoxylate and oxaloacetate (1986) *Biochemical Journal* 234:317.

Pebay-Peyroula, E., C. Dahout-Gonzalez, R. Kahn, V. Trezeguet, G. J. M. Lauquin, and G. Brandolin. (2003) Structure of mitochondrial ADP/ATP carrier in complex with carboxyatractyloside. *Nature* 426:39–44.

Peskov, K., I. Goryanin, and O. Demin. (2008) Kinetic model of phosphofructokinase-1 from *Escherichia coli*. *Journal of Bioinformatics and Computational Biology.* 6:1–25.

Pfaff, E., H. W. Heldt, and M. Klingenberg. (1969) Adenine nucleotide translocation of mitochondria: Kinetics of the adenine nucleotide exchange. *European Journal of Biochemistry* 10:484–93.

Pham, A. S., F. Janiak-Spens and G.D.Reinhart, "Persistent Binding of MgADP to the E187A Mutant of Escherichia coli Phosphofructokinase in the Absence of allosteric Effects," *Biochemistry* 40, 4140–4149 (2001).

Pham A. S. and G.D.Reinhart, "Pre-steady state Quantification of the Allosteric Influence of *Escherichia coli* Phophofuctokinase," *J Biol Chem* 276, 34388–34395 (2001).

Popova, S. V., and E. E. Sel'kov. (1975) Generalization of the model by Monod, Wyman and Changeux fro the case of reversible monosubstrate reaction S↔P. *FEBS Letters* 53:269–73.

Popova, S. V., and E. E. Sel'kov. (1976) A generalization of the Monod-Wyman-Changeux model for the case of multisubstrate reactions. *Molecular Biology (Moscow)* 10:1116–26.

Popova, S. V., and E. E. Sel'kov. (1978) Regulatory reversible enzymic reactions. Theoretical analysis. *Molecular Biology (Moscow)* 12:1139–51.

Reeves, R. E., and A. Sols. (1973) Regulation of *Escherichia coli* phosphofructokinase *in situ*. *Biochemical and Biophysical Research Communications* 50:459–65.

Romani, A. M., and A. Scarpa. (2000) Regulation of cellular magnesium. *Frontiers in Bioscience* 5:D720–34.

Selkov, E., S. Basmanova, T. Gaasterland, I. Goryanin, Y. Gretchkin, N. Maltsev, V. Nenashev, R. Overbeek, E. Panyushkina, L. Pronevitch, and I. Yunis. (1996) The metabolic pathway collection from EMP: The enzymes and metabolic pathways database. *Nucleic Acids Research* 24:26–28.

Shomburg, I., A. Chang, and D. Shomburg. (2002) BRENDA, enzyme data and metabolic information. *Nucleic Acids Research* 30:47–49.

Shuster, R. H., and G. H. Holzhutter. (1995) Use of mathematical models for predicting the metabolic effect of large-scale enzyme activity alterations. Application to enzyme deficiencies of red blood cells. *European Journal of Biochemistry* 229:403–18.

Stueland, C. S., K. R. Eck, Stieglbauer K.T., and D. C. LaPorte. Isocitrate dehydrogenase kinase/phosphatase exhibits an intrinsic adenosine triphosphatase activity. (1987) *Journal of Biological Chemistry* 262:16095.

Stueland, C. S., K. Gorden, and D. C. LaPorte. The isocitrate dehydrogenase phosphorylation cycle. Identification of the primary rate-limiting step. (1988) *Journal of Biological Chemistry* 263:19475.

Sundararaj, S., A. Guo, B. Habibi-Nazhad, M. Rouani, P. Stothard, M. Ellison, and D. S. Wishart. (2004) The CyberCell Database (CCDB): A comprehensive, self-updating, relational database to coordinate and facilitate *in silico* modeling of *Escherichia coli*. *Nucleic Acids Research* 32:293–95.

Taqui Khan MM and A. E. Martell, Metal Chelates of Adenosinediphosphoric and Adenosinemonophosphoric (1962) 84, 3037–3041.

Torres, J. C., V. Guixe, and J. Babul. (1997) A mutant phosphofructokinase produces a futile cycle during gluconeogenesis in *Escherichia coli*. *Biochemical Journal* 327:675–84.

Uhr, M. L., V. W. Thompson, and W. W. Cleland. The kinetics of pig heart triphosphopyridine nucleotide-isocitrate dehydrogenase. I. Initial velocity, substrate and product inhibition, and isotope exchange studies (1974) *Journal of Biological Chemistry* 249:2920.

Umbarger, H. E. (1996) Escherichia coli *and* Salmonella: *Cellular and molecular biology*. Washington, D.C.: ASM Press.

Vignais, P. V., M. R. Block, F. Boulay, G. Brandolin, and G. L. M. Lauguin. (1985) Molecular aspects of structure-function relationships in mitochondrial adenine nucleotide carrier. In *Structure and properties of cell membranes*, edited by V. Bengha. Paris: CRC Press.

Walsh, K., and D. E. Koshland, Jr. Determination of flux through the branch point of two metabolic cycles. The tricarboxylic acid cycle and the glyoxylate shunt. (1984) *Journal of Biological Chemistry* 259:9646.

Waygood, E. B., and B. D. Sanwal. (1974) The control of pyruvate kinases of *Escherichia coli*. *Journal of Biological Chemistry* 249:265–74.

Zalkin, H. The amidotransferases (1993) *Advances in Enzymology & Related Areas of Molecular Biology* 66:203–309.

Zalkin, H., and J. L. Smith. Enzymes utilizing glutamine as an amide donor (1998) *Advances in Enzymology & Related Areas of Molecular Biology* 72:87–144.

## CHAPTER 7

Alberty, R. A. (1961) Fumarase. *The Enzymes* 5(B):531.

Aulabaugh, A., and J. V. Schloss. (1990) Oxalyl hydroxamates as reaction-intermediate analogs for ketol-acid reductoisomerase. *Biochemistry* 29:2824.

Barak, Z., D. M. Chipman, and N. Gollop. (1987) Physiological implications of the specificity of acetohydroxy acid synthase isozymes of enteric bacteria. *Journal of Bacteriology* 169:3750.

Bar-Ilan, A. (2001) Binding and activation of thiamin diphosphate in acetohydroxy acid synthase. *Biochemistry* 40:11946.

Boork, J., and H. Wennerstrom. (1984) The influence of membrane potentials on reaction rates. Control in free-energy-transducing systems. *Biochimica et Biophysica Acta* 767:314.

Cascante, M., and A. Cortes. (1988) Kinetic studies of chicken and turkey liver mitochondrial aspartate aminotransferase. *Biochemical Journal* 250:805.

Cha, S., and R. E. Parks. (1964a) Succinic thiokinase. I. Purification of the enzyme from pig heart. *Journal of Biological Chemistry* 239:1961.

Cha, S., and R. E. Parks. (1964b) Succinic thiokinase. II. Kinetic studies: Initial velocity, product inhibition, and effect of arsenate. *Journal of Biological Chemistry* 239:1968.

Chassagnole, C. (2002) Dynamic modeling of the central carbon metabolism of *Escherichia coli. Biotechnology and Bioengineering* 79:53.

Chunduru, S. K., G. T. Mrachko, and K. C. Calvo. (1998) Mechanism of ketol acid reductoisomerase. Steady-state analysis and metal ion requirement. *Biochemistry* 28:486.

Cleland, W. W. (1963) The kinetics of enzyme-catalyzed reactions with two or more substrates or products. I. Nomenclature and rate equations. *Biochimica et Biophysica Acta* 67:104.

Demin, O. V., et al. (2004) Kinetic model of imidazole glycerol phosphate synthetase of *Escherichia coli. Biokhimiya* 69:1625.

Demin, O. V., et al. Kinetic modelling of the E. coli metabolism. (2005) *Topics in current genetics.* In edited by L. Alberghina and H. V. Westerhoff. Springer Berlin.

Dierks, T., and R. Kramer. (1988) Asymmetric orientation of the reconstituted aspartate/glutamate carrier from mitochondria. *Biochimica et Biophysica Acta* 937:112.

Engel, S., et al. (2000) Determination of the dissociation constant of valine from acetohydroxy acid synthase by equilibrium partition in an aqueous two-phase system. Determination of the dissociation constant of valine from acetohydroxy acid synthase by equilibrium partition in an aqueous two-phase system. *Journal of Chromatography B* 743:225.

Eoyang L, Silverman PM. (1984) Purification and subunit composition of acetohydroxyacid synthase I from Escherichia coli K-12. *Journal of Bacteriology,* 157(1):184–9

Fahien, L. A., and J. K. Teller. (1992) Glutamate-malate metabolism in liver mitochondria. A model constructed on the basis of mitochondrial levels of enzymes, specificity, dissociation constants, and stoichiometry of heteroenzyme complexes. *Journal of Biological Chemistry* 267:10411.

Forman WB, Davidson ED, Webster LT. (1971) Enzymatic conversion of salicylate to salicylurate. *Molecular Pharmacology* 7, 247–259.

Garber, A. J., and R. W. Hanson. (1971) The interrelationships of the various pathways forming gluconeogenic precursors in guinea pig liver mitochondria. *Journal of Biological Chemistry* 246:589.

Greenhut, J., H. Umezawa, and F. B. Rudolph. (1985) Inhibition of fumarase by S-2,3-dicarboxyaziridine. *Journal of Biological Chemistry* 260:6684.

Grivennikova, V. G., et al. (1993) Fumarate reductase activity of bovine heart succinate-ubiquinone reductase. New assay system and overall properties of the reaction. *Biochimica et Biophysica Acta* 1140:282.

Haas, R., et al. (1985) Salicylate-induced loose coupling: protonmotive force measurements. *Biochemical Pharmacology* 34:900.

Hall, T. R., et al. (1993) Branched chain aminotransferase isoenzymes. Purification and characterization of the rat brain isoenzyme. *Journal of Biological Chemistry* 268:3092.

Hamada, M. (1975) A kinetic study of the α-keto acid dehydrogenase complexes from pig heart mitochondria. *Journal of Biochemistry* 77:1047.

Hansford RG, Johnson RN. (1975) The steady state concentrations of coenzyme A-SH and coenzyme A thioester, citrate, and isocitrate during tricarboxylate cycle oxidations in rabbit heart mitochondria. *Journal of Biological Chemistry,* 250(21):8361.

Heyde, E., and S. Ainsworth. (1968) Kinetic studies on the mechanism of the malate dehydrogenase reaction. *Journal of Biological Chemistry* 243:2413.

Hill CM, Duggleby RG (1998) *Escherichia coli* acetohydroxyacid synthase II mutants.*Biochemical Journal* 335:653-661.

Hill CM, Pang SS, Duggleby RG (1997) *Escherichia coli* acetohydroxyacid synthase II. *Biochemical Journal* 327:891–898

Hoek, J. B. (1971) GDH and the oxidoreduction state of nicotinamide nucleotides in rat-liver mitochondria. Ph.D. thesis, Amsterdam.

Holms, W. H. (1986) The central metabolic pathways of *Escherichia coli*: Relationship between flux and control at a branch point, efficiency of conversion to biomass, and excretion of acetate. *Current Topics in Cellular Regulation* 28:69.

Indiveri, C., et al. (1991) Reaction mechanism of the reconstituted oxoglutarate carrier from bovine heart mitochondria. *European Journal of Biochemistry* 198:339.

Inoue, K., et al. (1988) Branched-chain amino acid aminotransferase of *Escherichia coli*: Overproduction and properties. *Journal of Biochemistry* 104:777.

Ivanitzky, G. R., V. I. Krinsky, and E. E. Selkov. (1978) *Mathematical biophysics of the cell.* Moscow: Nauka.

Kaplan, E. H., J. Kennedy, and J. Davis. (1954) Effects of salicylate and other benzoates on oxidative enzymes of the tricarboxylic acid cycle in rat tissue homogenates. *Archives of Biochemistry* 51:47.

Kaufman, S., and S. G. A. Alivisatos. (1955) Purification and properties of the phosphorilating enzyme from spinach. *Journal of Biological Chemistry* 216:141.

Kondrashova, M. N. (1989) Structuro-kinetic organization of the tricarboxylic acid cycle in the active functioning of mitochondria. *Biofizika* 34:450.

Kotlyar, A. B., and A. D. Vinogradov. (1984) Dissociation constants of the succinate dehydrogenase complexes with succinate, fumarate and malonate. *Biokhimiya* 49:511.

Kuramitsu, S. (1985) Aspartate aminotransferase isozymes from rabbit liver. Purification and properties. *Journal of Biochemistry* 97:1337.

La Noue, K. F., et al. (1979) Kinetic properties of aspartate transport in rat heart mitochondrial inner membranes. *Archives of Biochemistry and Biophysics* 195:578.

Lee-Peng, F. C., M. A. Hermodson, and G. B. Kohlhaw. (1979) Transaminase B from *Escherichia coli*: Quaternary structure, amino-terminal sequence, substrate specificity, and absence of a separate valine-α-ketoglutarate activity. *Journal of Bacteriology* 139(2):339.

Limberg, G., W. Klaffke, and J. Thiem. (1995) Conversion of aldonic acids to their corresponding 2-keto-3-deoxy-analogs by the non-carbohydrate enzyme dihydroxy acid dehydratase (DHAD). *Bioorganic & Medicinal Chemistry* 3:487.

Massey, V. (1960) The composition of the α-ketoglutarate dehydrogenase complex. *Biochimica et Biophysica Acta* 38:447.

McCormack, J. G., and R. M. Denton. (1979) The effects of calcium ions and adenine nucleotides on the activity of pig heart 2-oxoglutarate dehydrogenase complex. *Biochemical Journal* 180:533.

Mogilevskaya, E., O. Demin, and I. Goryanin. (2006) Kinetic model of mitochondrial Krebs cycle: Unraveling the mechanism of salicylate hepatotoxic effects. *Journal of Biological Physics* 32(3–4):245.

Myers, J. W. (1961) Dihydroxy acid dehydrase: An enzyme involved in the biosynthesis of isoleucine and valine. *Journal of Biological Chemistry* 236:1414.

Panov, A. V., and Scaduto, R. C., Jr. (1995) Influence of calcium on NADH and succinate oxidation by rat heart submitochondrial particles. *Archives of Biochemistry and Biophysics* 316:815.

Parlo, R. A., and P. S. Coleman. (1984) Enhanced rate of citrate export from cholesterol-rich hepatoma mitochondria. *Journal of Biological Chemistry* 259(16):9997.

Reynolds, J. A., E. A. Johnson, and C. Tanford. (1985) Incorporation of membrane potential into theoretical analysis of electrogenic ion pumps. *Proceedings of the National Academy of Sciences USA* 82:6869.

Ricks CA, Cook RM. (1981) Regulation of volatile fatty acid uptake by mitochondrial acyl CoA synthetases of bovine liver. J. Dairy Sci. 64, 2324–2335.

Siess, E. A., R. I. Kientsch-Engel, and O. H. Wieland. (1984) Concentration of free oxaloacetate in the mitochondrial compartment of isolated liver cells. *Biochemical Journal* 218:171.

Smith, C. M., J. Bryla, and J. R. Williamson. (1974) Regulation of mitochondrial α-ketoglutarate metabolism by product inhibition of α-ketoglutarate dehydrogenase. *Journal of Biological Chemistry* 249:1497.

Umbarger, H. E. (1996) Escherichia coli *and* Salmonella: *Cellular and molecular biology.* Washington, D.C.: ASM Press.

Vessey D.A., Hu J., Kelly M. (1996) Interaction of salicylate and ibuprofen with the carboxylic acid: CoA ligases from bovine liver mitochondria. *Journal of Biochemical Toxicology.* 11, 73–78.

Vinogradov, A. D. (1986) Succinate-ubiquinone reductase of the respiratory chain. *Biokhimiya* 51:1944.

Vyazmensky, M., et al. (1996) Isolation and characterization of subunits of acetohydroxy acid synthase isozyme III and reconstitution of the holoenzyme. *Biochemistry* 35:10339.

Wessel, P. M., et al. (2000) Evidence for two distinct effector-binding sites in threonine deaminase by site-directed mutagenesis, kinetic, and binding experiments. *Biochemistry* 39:15136.

Williamson, D. H., P. Lund, and H. A. Krebs. (1967) The redox state of free nicotinamide-adenine dinucleotide in the cytoplasm and mitochondria of rat liver. *Biochemical Journal.* 103:514.

Wilson, D. F., D. Nelson, and M. Erecinska. (1982) Binding of the intramitochondrial ADP and its relationship to adenine nucleotide translocation. *FEBS Letters* 143:228.

## CHAPTER 8

Alberty, R. A. Standard Gibbs free energy, enthalpy, and entropy changes as a function of pH and pMg for several reactions involving adenosine phosphates (1969) *Journal of Biological Chemistry* 244:3290.

Barbour, R. L., and S. H. Chan. (1981) Characterization of the kinetics and mechanism of the mitochondrial ADP-atp carrier. *Journal of Biological Chemistry* 256:1940.

Bohnensack, R. (1982) The role of the adenine nucleotide translocator in oxidative phosphorylation. A theoretical investigation on the basis of a comprehensive rate law of the translocator. *Journal of Bioenergetics and Biomembranes* 14:45.

Boork, J., and H. Wennestrom. (1984) The influence of membrane potentials on reaction rates. Control in free-energy-transducing systems. *Biochimica et Biophysica Acta* 767:314.

Boveris, A., and B. Chance. (1973) The mitochondrial generation of hydrogen peroxide. General properties and effect of hyperbaric oxygen. *Biochemical Journal* 134:707.

Bowyer, J. R., and B. L. Trumpower. (1981) Rapid reduction of cytochrome c1 in the presence of antimycin and its implication for the mechanism of electron transfer in the cytochrome b-c1 segment of the mitochondrial respiratory chain. *Journal of Biological Chemistry* 256:2245.

Brandolin, G., Y. Dupont, and P. V. Vignais. (1982) Exploration of the nucleotide binding sites of the isolated ADP/ATP carrier protein from beef heart mitochondria. 2. Probing of the nucleotide sites by formycin triphosphate, a fluorescent transportable analogue of ATP. *Biochemistry* 21:6348.

Brown, G. C., and M. D. Brand. (1986) Changes in permeability to protons and other cations at high proton motive force in rat liver mitochondria. *Biochemical Journal* 234:75.

Brzezinski, P., and P. Adelroth. (1998) Pathways of proton transfer in cytochrome c oxidase . *Journal of Bioenergetics and Biomembranes* 30:99.

Demin, O. V., et al. (2001) Kinetic modeling of energy metabolism and superoxide generation in hepatocyte mitochondria. *Molecular Biology (Moscow)* 35(6):1095.

Demin, O. V., B. N. Kholodenko, and V. P. Skulachev. (1998) A model of O2.-generation in the complex III of the electron transport chain. *Molecular and Cellular Biochemistry* 184:21.

Demin, O. V., H. V. Westerhoff, and B. N. Kholodenko. (1998) Mathematical modelling of superoxide generation with the bc1 complex of mitochondria. *Biochemistry (Moscow)* 63(6):634.

Drachev, L. A., et al. (1989) An investigation of the electrochemical cycle of bacteriorhodopsin analogs with the modified ring. *Biochimica et Biophysica Acta* 973:189.

Dupont, Y., G. Brandolin, and P. V. Vignais. (1982) Exploration of the nucleotide binding sites of the isolated ADP/ATP carrier protein from beef heart mitochondria. 1. Probing of the nucleotide sites by Naphthoyl-ATP, a fluorescent nontransportable analogue of ATP. *Biochemistry* 21:6343.

Garlid, K. D. (1980) On the mechanism of regulation of the mitochondrial K+/H+ exchanger. *Journal of Biological Chemistry* 225:11273.

Green, D. E., and D. C. Wharton. (1963) Stoichiometry Of The Fixed Oxidation-reduction Components Of The Electron Transfer Chain Of Beef Heart Mitochondria. *Biochemische. Zeitschrift.* 336:335.

Grivennikova, V. G., and A. D. Vinogradov. (1982) Kinetics of ubiquinone reduction by the resolved succinate: ubiquinone reductase. *Biochimica et Biophysica Acta* 682:491.

Hansford, R. G., B. A. Hogue, and V. Mildaziene. (1997) Dependence of $H_2O_2$ formation by rat heart mitochondria on substrate availability and donor age. *Journal of Bioenergetics and Biomembranes* 29:89.

Harris, E. J., and J. A. Bangham. (1972) Titration of mitochondrial buffer by accumulated anions. *Journal of Membrane Biology* 9:141.

Jones, D. (1986) Intracellular diffusion gradients of $O_2$ and ATP. *American Journal of Physiology* 250:C663.

Kholodenko, B. N. (1988) Stabilizing regulation in multienzyme systems: Modelling of bioenergetics processes. Ph.D. thesis, Moscow (in Russian), Moscow State University.

Kingsley, P. B., and G. W. Feigenson. (1981) 1H-NMR study of the location and motion of ubiquinones in perdeuterated phosphatidylcholine bilayers. *Biochimica et Biophysica Acta* 635:602.

Korshunov, S. S., et al. (1998). Fatty acids as natural uncouplers preventing generation of O2.- and H2O2 by mitochondria in the resting state *FEBS Letters* 435:215.

Korshunov, S. S., V. P. Skulachev, and A. A. Starkov. (1997) High protonic potential actuates a mechanism of production of reactive oxygen species in mitochondria. *FEBS Letters* 416:15.

Korzeniewski, B. (1996) Simulation of state 4 --> state 3 transition in isolated mitochondria . *Biophysical Chemistry* 57:143.

Korzeniewski, B. (1998) Is it possible to predict any properties of oxidative phosphorylation in a theoretical way? *Molecular and Cellular Biochemistry* 184:345.

Korzeniewski, B., and W. Froncisz. (1991) An extended dynamic model of oxidative phosphorylation. *Biochimica et Biophysica Acta* 1060:210.

Kotlyar, A. B., and A. D. Vinogradov. (1984) Interaction of the membrane-bound succinate dehydrogenase with substrate and competitive inhibitors. *Biochimica et Biophysica Acta* 784:24.

Kramer, R., and M. Klingenberg. (1982) Electrophoretic control of reconstituted adenine nucleotide translocation. *Biochemistry* 21:1082.

Kunz, W., R. Bohnensack, G. Bohme, U. Kuster, G. Letko, and P. Schonfeld. (1981) Relations between extramitochondrial and intramitochondrial adenine nucleotide systems. *Archives of Biochemistry Biophysics* 209:219.

La Noue, K. F., and A. C. Schoolwerth. (1984) Effect of pH on glutamate efflux from rat kidney mitochondria. *Bioenergetics.* Amsterdam: Elsevier Science.

Lawson, J. W. R., and R. L. Veech. (1979) Effects of pH and free Mg2+ on the Keq of the creatine kinase reaction and other phosphate hydrolyses and phosphate transfer reactions. *Journal of Biological Chemistry* 254:6528.

Letko, G., et al. (1980) Investigation of the dependence of the intramitochondrial [ATP]/[ADP] ratio on the respiration rate. *Biochimica et Biophysica Acta* 593:196.

Ligeti, G., et al. (1985) Kinetics of Pi-Pi exchange in rat liver mitochondria. Rapid filtration experiments in the millisecond time range. *Biochemistry* 24:4423.

Liu, S.-S. (1997) Generating, partitioning, targeting and functioning of superoxide in mitochondria. *Bioscience Reports* 17:259.

Liu, S.-S., and J. P. Huang. (1996) Co-existence of" reactive oxygen cycle" with cycle and H+ cycle in respiratory chain. In *Proceedings of the International Symposium on Natural Antioxidants: Molecular Mechanisms and Health Effects,* edited by D. Moores. Champaign, Ill.: AOCS Press.

Markin, V. S., and U. A. Chizmadgev. (1974) *Induced ion transport.* Moscow: Nauka (in Russian).

Martin, W. H., A. D. Beavis, and K. D. Garlid. (1984) Identification of an 82,000-dalton protein responsible for K+/H+ antiport in rat liver mitochondria. *Journal of Biological Chemistry* 259:2062.

Michel, H. (1998) The mechanism of proton pumping by cytochrome c oxidasex. *Proceedings of the National Academy of Sciences USA* 95:12819.

Nakashima, R. A., R. S. Dordick, and K. D. Garlid. (1982) On the relative roles of $Ca_2+$ and $Mg_2+$ in regulating the endogenous K+/H+ exchanger of rat liver mitochondria. *Journal of Biological Chemistry* 257:12540.

Nicholls, D. G. (1974) The influence of respiration and ATP hydrolysis on the proton-electrochemical gradient across the inner membrane of rat-liver mitochondria as determined by ion distribution. *European Journal of Biochemistry* 50:305.

Paula, S., et al. (1999) Proton and electron transfer during the reduction of molecular oxygen by fully reduced cytochrome c oxidase: a flow-flash investigation using optical multichannel detection. *Biochemistry* 38:3025.

Reich, J. G., and K. Rohde. (1983) On the relationship between Z delta pH and delta psi as components of the protonmotive potential in Mitchell's chemiosmotic system. *Biomedica Biochimica Acta* 42:37.

Reynolds, I. A., E. A. Johnson, and C. Tanford. (1985) Incorporation of membrane potential into theoretical analysis of electrogenic ion pumps. *Proceedings of the National Academy of Sciences USA* 82:6869.

Rich, P. R. (1984) Electron and proton transfers through quinones and cytochrome bc complexes. *BBA* 768:53.

Rosing, J., and E. C. Slater. (1972) The value of G degrees for the hydrolysis of ATP. *Biochimica et Biophysica Acta* 267:275.

Schwerzman, K., et al. (1986) Mitochondria and aging. *Journal of Cell Biology* 102:97.

Semenov, A. Y. (1993) Electrogenic steps during electron transfer via the cytochrome bc1 complex of Rhodobacter sphaeroides chromatophores. *FEBS Letters* 321:1.

Skulachev, V. P. (1988) *Membrane bioenergetics*. Berlin: Springer.

Skulachev, V. P. (1997) Role of uncoupled and non-coupled oxidations in maintenance of safely low levels of oxygen and its one-electron reductants. *Quarterly Reviews of Biophysics* 29:169.

Skulachev, V. P., and I. A. Kozlov. (1997) *Proton adenozinethreephosphatase: molecular biological generators of electric currency.* Moscow: Nauka (in Russian).

Srere, P. A. (1981) Macromolecular interactions: tracing the roots. *Trends in Biochemical Sciences* 6:4.

Trumpower, B. L. (1981) Function of the iron-sulfur protein of the cytochrome b-c1 segment in electron-transfer and energy-conserving reactions of the mitochondrial respiratory chain. *BBA* 639:129.

Tsukihara, T., et al. (1995) Structures of metal sites of oxidized bovine heart cytochrome c oxidase at 2.8 A. *Science* 269:1069.

Tsukihara, T., et al. (1996) The whole structure of the 13-subunit oxidized cytochrome c oxidase at 2.8 A. *Science* 272:1136.

Van Dam, K., et al. (1980) Relationship between chemiosomotic flows and thermodynamic forces in oxidative phosphorylation. *Biochimica et Biophysica Acta* 591:240.

Westerhoff, H. V., and K. Van Dam. (1987) *Thermodynamics and control of biology free energy transduction.* Amsterdam: Elsevier.

Wrigglesworth, J. M., et al. (1988) Activation by reduction of the resting form of cytochrome c oxidase: tests of different models and evidence for the involvement of CuB. *Biochimica et Biophysica Acta* 936:452.

Yoshikawa, S., et al. (1998) Redox-coupled crystal structural changes in bovine heart cytochrome c oxidase. *Science* 280:1723.

## CHAPTER 9

Bentrup, K. H. Z., et al. (1999) Characterization of activity and expression of isocitrate lyase in *Mycobacterium avium* and Mycobacterium tuberculosis. *Journal of Bacteriology* 181:7161.

Bjorkman, D. (1998) Nonsteroidal anti-inflammatory drug-associated toxicity of the liver, lower gastrointestinal tract, and esophagus. *American Journal of Medicine* 105(5A):17S.

Cole, S. T., et al. (1998) Deciphering the biology of *Mycobacterium tuberculosis* from the complete genome sequence. *Nature* 393:537.

Cornish-Bowden, A. (1995) *Fundamentals of enzyme kinetics.* London: Portland Press.

Demin, O. V., et al. (2004) Kinetic Modelling as a Modern Technology to Explore and Modify Living Cells. Modelling in molecular biology. In Natural computing series, edited by G. Ciobanu and G. Rozenberg. Springer, Berlin.

Demin, O. V., et al. (2005) Kinetic modelling of the E. coli metabolism. *Topics in current genetics,* edited by L. Alberghina and H. V. Westerhoff. Springer, Berlin.

Dhariwal, K. R., and T. A. Venkitasubramanian. (1987) NADP-specific isocitrate dehydrogenase of *Mycobacterium phlei* ATCC 354: Purification and characterization. *Journal of General Microbiology* 133:2457.

Drozdov-Tikhomirov, L. N., G. I. Scurida, and V. V. Serganova. (1992). Flux Stoichiometric Models of Cell Metabolism. Reports of International Conference Modeling and Computer Methods in Molecular Biology and Genetics. Nova Science Publishing. p.329–334.

Forman, W. B., E. D. Davidson, and L. T. Webster. (1971) Enzymatic conversion of salicylate to salicylurate. *Molecular Pharmacology* 7:247.

Fromenty, B., and D. Pessayre. (1995) Inhibition of mitochondrial beta-oxidation as a mechanism of hepatotoxicity. *Pharmacological Therapeutics* 67:101.

Goryanin, I. I., O. V. Demin, and F. Tobin. (2003) Applications of Whole Cell and Large Pathway Mathematical Models in the Pharmaceutical Industry. (2003) *Metabolic engineering in the post genomic era,* edited by B. Kholodenko and H. Westerhoff. Horizon Bioscience, UK, Oxford, p.103–129.

Graham, J. E., and J. E. Clark-Curtiss. (1999) Identification of *Mycobacterium tuberculosis* RNAs synthesized in response to phagocytosis by human macrophages by selective capture of transcribed sequences (SCOTS). *Proceedings of the National Academy of Sciences USA* 96:11554.

Haas, R., et al. (1985) Salicylate-induced loose coupling: Protonmotive force measurements. *Biochemical Pharmacology* 34:900.

Kaplan, E. H., J. Kennedy, and J. Davis. (1954) Effects of salicylate and other benzoates on oxidative enzymes of the tricarboxylic acid cycle in rat tissue homogenates. *Archives of Biochemistry* 51:47.

McKinney, J. D., et al. (2000) Persistence of *Mycobacterium tuberculosis* in macrophages and mice requires the glyoxylate shunt enzyme isocitrate lyase. *Nature* 406:735.

Miyahara, J. T., and R. Karler. (1965) Effect of salicylate on oxidative phosphorylation and respiration of mitochondrial fragments. *Biochemical Journal* 97:194.

Mogilevskaya, E., O. Demin, and I. Goryanin. (2006) Kinetic model of mitochondrial Krebs cycle: Unraveling the mechanism of salicylate hepatotoxic effects. *Journal of Biological Physics* 32(3–4):245.

Ricks, C. A., and R. M. Cook. (1981) Regulation of volatile fatty acid uptake by mitochondrial acyl CoA synthetases of bovine liver. *Journal of Dairy Science* 64:2324.

Sharma, V., et al. (2000) Structure of isocitrate lyase, a persistence factor of *Mycobacterium tuberculosis*. *Nature Structural & Molecular Biology* 7:663.

Skulachev, V. P. (1989) *Energetics of biological membranes*. Nauka, Moscow.

Vessey, D. A., J. Hu, and M. Kelly. (1996) Interaction of salicylate and ibuprofen with the carboxylic acid: CoA ligases from bovine liver mitochondria. *Journal of Biochemical Toxicology* 11:73.

# Index